Fundamentals of Machine Design

Edited by
Waldemar Karaszewski

Fundamentals of Machine Design

Special topic volume on Fundamentals of Machine Design

Edited by

Waldemar Karaszewski

TTP

Trans Tech Publications Ltd
Kreuzstrasse 10
CH-8635 Durnten-Zurich
Switzerland
http://www.ttp.net

Volume 490 of
Key Engineering Materials
ISSN 1013-9826
Full text available online at *http://www.scientific.net*

***Distributed** worldwide by*

Trans Tech Publications Ltd.
Kreuzstrasse 10
CH-8635 Durnten-Zurich
Switzerland

Fax: +41 (44) 922 10 33
e-mail: sales@ttp.net

and in the Americas by

Trans Tech Publications Inc.
PO Box 699, May Street
Enfield, NH 03748
USA

Phone: +1 (603) 632-7377
Fax: +1 (603) 632-5611
e-mail: sales-usa@ttp.net

printed in Germany

Preface

This book is a collection of the accepted papers concerning Fundamentals of Machine Design. All accepted papers were subjected to strict peer-reviewing by 2 expert referees.

The resultant dissemination of the latest research results, and the exchanges of views concerning the future research directions to be taken in this field makes the work of immense value to all those having an interest in the topics covered. The present book represents a cooperative effort to seek out the best strategies for effecting improvements in the quality and the reliability of machines and machine parts and to widen their application.

Our sincere appreciation and thanks go to these authors for their contributions to this book. We are grateful to all referees for their meticulous work and their constructive comments.

Prof. Michał Wasilczuk

Dr. Waldemar Karaszewski

Table of Contents

Key Engineering Materials Vol. 490 (2012) pp 1-7
© (2012) Trans Tech Publications, Switzerland
doi:10.4028/www.scientific.net/KEM.490.1

A General Approach to the Equalisation of Partial Contact Ratios in 3K Transmissions

Andrzej Golenko

Wrocław University of Technology,

Wyb. Wyspiańskiego 27, Wrocław, Poland

andrzej.golenko@pwr.wroc.pl

Keywords: 3K Paradox, efficiency, contact ratio, equalisation

Abstract. The equalisation of the approach and recess contact ratios is one of the methods aiming at higher efficiency in any meshing toothed pair. Hori, Kurita and Hayashi implemented this method to the design of a 3K planetary gear with a common planet gear (the Mechanical Paradox) by a purposeful modification of the addenda of each of the 3 meshing toothed pairs. The author verified this method in one of the previous publications and proposes here a more general design approach aiming at better equalisation still at the stage of selecting the numbers of teeth in the transmission. The results confirm the purposefulness of this approach.

Introduction

3K transmissions have been known for robust design, high transmission ratios available and unfortunately, unsatisfactory efficiency. Their origins could be attributed to an English clockmaker James Ferguson, who designed and constructed an orrery (Fig. 1) with this type of gearing at the end of the 18th century. As the author himself wrote in [1]: 'Make four wheels, one wheel as thick as the other three. Cut teeth in all wheels. Set the thick wheels in line. Set the thick wheel so its teeth take into the three ones. Turn the thick wheel: One of the three wheels rotates in one way, the second in the other way and the third does not rotate at all'. An interesting account of the history of this transmission together with a theological discussion prompted by this invention among those skilled in the art of chronology at that time was presented by prof. Gawrysiak in [2]. Its scheme is presented in Fig. 2a.

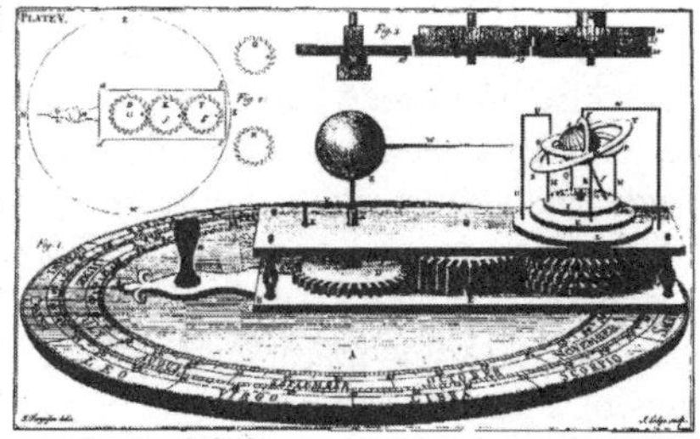

Fig. 1. Ferguson's orrery, 1773 [1]

Fig. 2b shows a traditional combined planetary gear (Wolfrom) and Fig 2c, its modified version with a common planet gear. The latter one (also referred to as the Mechanical Paradox) features heavily diversified approach and recess contact ratios in the two most important toothed pairs of the common planet gear with the two annulus gears, which is detrimental to the efficiency of meshing.

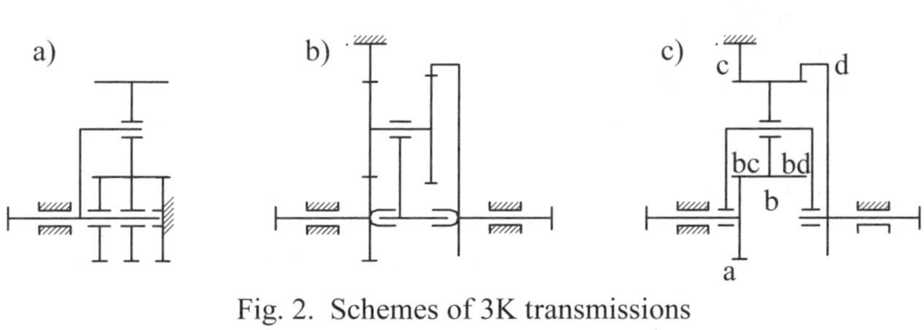

Fig. 2. Schemes of 3K transmissions

The very idea of enhancing efficiency in 3K planetary gear transmissions by equalising the approach and recess contact ratios was proposed by Horia and Kurita in [3]. By modifying addenda of the four meshing gears it was possible to improve this disadvantageous situation and to raise the overall efficiency of the transmission. For the clarity of the following analysis, the most important points of this approach have been repeated here in a brief summary.

A starting point for the modification was a conventional 3K transmission with the common planet gear in which the number of teeth, module etc. had been already established (Fig. 3). The left part of this combined drawing represents the toothed pair *b* – *d*; and the right side, the toothed pair *b* – *c*. The pressure lines are traced on both sides to show the respective high and low points of meshing (*A, E*). The conventional transmission is shown in dashed lines. A strong shift in the position of the operating pitch points, either towards the addendum (pair *b-c*; the right side), or towards the dedendum (pair *b-d*; the left side) is clearly visible. The addendum diameter of gear *b* is a maximum value that offers the minimum required value of the bottom clearance in mesh with the three other gears. As the centre distance must remain the same, any change in the profile shift coefficients that improves situation in one toothed pair makes it worse in the other one.

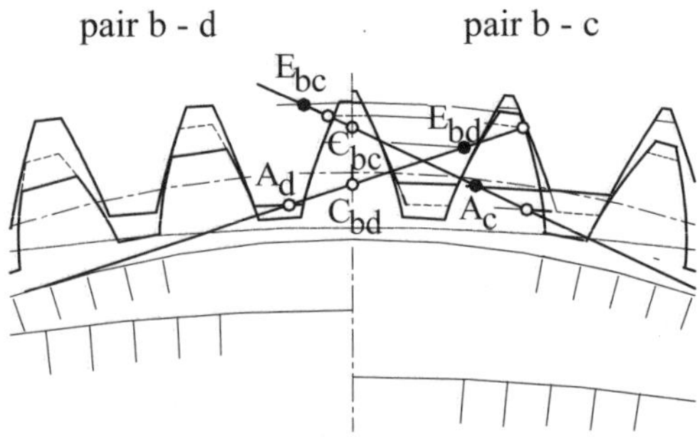

Fig. 3. Meshing representation in a conventional (dashed lines) and modified (solid lines) 3K transmission with a common planet gear (after [3]).

A modification proposed by Hori and Kurita (solid lines) consists in the reduction of these disproportions in meshing. In the modified gear the tooth addendum in gear *bc* (a section of *b* in mesh with *c*) was increased and in the gear *c*, decreased. In toothed pair *bd-d* the addendum of gear *bd* was decreased and in gear *d*, it was left unchanged. The idea was implemented and tested by the authors. The results showed an approximately 10% increase of efficiency for a 351 transmission ratio gear.

As the idea proved to be successful and worth further studies, Hori and Hayashi [4] proposed a systematic design approach aiming at the equalisation of the approach and recess contact ratios. The procedure proposed is separate for each of the meshing toothed pairs. The idea of the method is to sequentially change the addendum coefficient in gear *b* seeking the maximum addendum diameter in gear *bc* while preserving the necessary minimum value of the top land width. The search shall be carried out within an area defined by six constraints, the most important of them being the absence of undercutting and interference. Based upon the proposed procedure Hori and Hayashi established optimum meshing parameters for a 351 (z_a = 12; z_b = 47; z_c = -105; z_d = -108; m = 1.25) transmission ratio gear with the contact ratio equal to 1.2. These were: x_b = 0.36; x_a = 0.308; x_c = -1.627; x_d = 0 (the ISO designation convention for internal gearing applies here). These values correspond to a pointed tooth limit of the planet gear (minimum top land width). The analysis made by Hori and Hayashi was based upon an assumption that as long as the pointed tooth limit diameter is larger than (or the same as) the diameter based on the condition of equal approach/recess contact ratios, it is also possible to generate such gearing, which is not always true. Tooth proportions obtained using this method are shown in Fig. 4 (void points denote the equal approach/recess contact ratio). These tooth proportions violate both the generation and meshing constraints even if they were explicitly stated in the method. To avoid detailed explanations, see that toothed pair *a-bc* does not preserve the fundamental law of toothed gearing and needed is an ex-post modification of the teeth geometry in all toothed pairs.

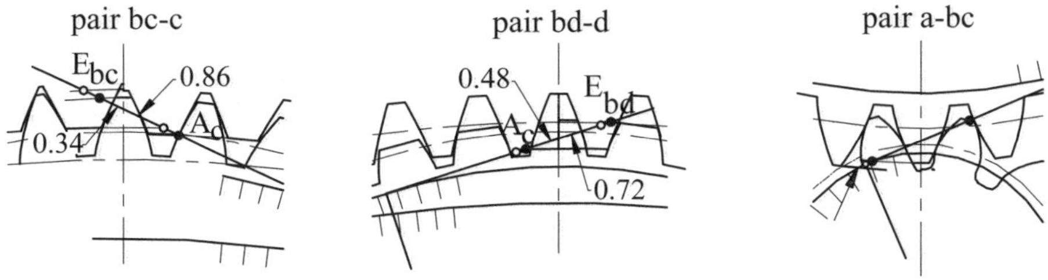

Fig. 4. Equalisation in agreement with the Hori and Hayashi's method

The author verified in [5] some aspects of this method and modified it including the control of all generation and meshing constraints at all steps of the iterative procedure proposed by Hori and Hayashi. This new approach accounted for the tool type and its regrinding status.

The quality of equalisation is controlled with an objective function taking into account all meshing gears. As the efficiency of toothed pair *a–bc* is of small importance to the overall efficiency of the gear train, a simplified formula, including only toothed pairs *bc–c* and *bd–d*, has been used by the author in all analytical calculations:

$$FC = 2 - \varepsilon_{\alpha 1}^{bc} - \varepsilon_{\alpha 2}^{bc} - \varepsilon_{\alpha 1}^{bd} - \varepsilon_{\alpha 2}^{bd} + \left(\varepsilon_{\alpha 1}^{bc}\right)^2 + \left(\varepsilon_{\alpha 2}^{bc}\right)^2 + \left(\varepsilon_{\alpha 1}^{bd}\right)^2 + \left(\varepsilon_{\alpha 2}^{bd}\right)^2$$

where the right side terms of the above equation represents partial (approach and recess) contact ratios for toothed pairs *bc-c* and ***bd-d*.** With the ideal equalisation this simplified function takes a value of 1.04. In the above analysed case this function is equal to 1.24 (after an elementary modification, i.e. without tool optimisation). Dark points in Fig. 4 denote the limits of equalisation with tool optimisation ($FC = 1.20$; see numerical values of partial contact ratios indicated by leader lines). When departing from a Hori and Hayashi's recommendation of leaving gear d without any modification, it was possible to further reduce the value of *FC* (1.17 [6]). The results are shown graphically in Fig. 5 ($x_d = 0.2$).

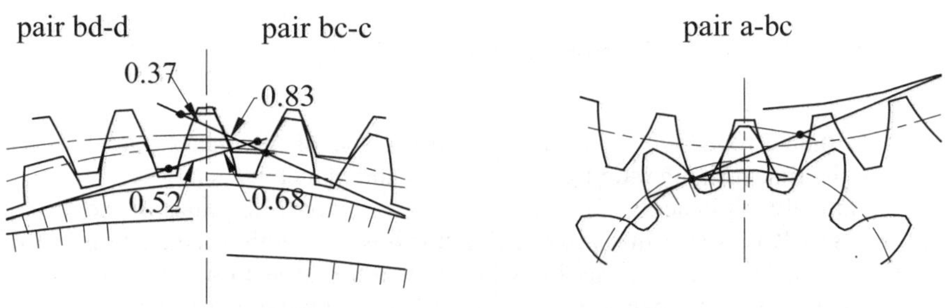

Fig. 5. Equalisation pattern obtained in [5]

All in all, the proposed method is a shift in the traditional method of designing 3K transmissions with a common planet gear, i.e. in a close relationship with the cutting tool parameters, to a method in which these parameters are included in the constraints only. This approach has been known as the Direct Gear Design [7].

The method proposed by Hori and Hayashi and its modification proposed by the author were focused on the distribution of the addendum coefficients for each of the involved gears. The principal objective of the present paper is to discuss the equalisation problem from a further perspective, still at a moment when the number of teeth for a given transmission ratio has not been selected.

Conceptual rationale

The full equalisation of the approach and recess contact ration in 3K Paradox transmissions is difficult not only due to generation and meshing constraints. For a given transmission ratio the two operating pitch points are spaced by a small distance Δ. When improving equalisation in one toothed pair, the same proportions in the second meshing pair are compromised. As the limiting points of the two meshing zones are established by the intersection of the operating pressure lines with the addendum circles of the two gears, it was presumed that meshing with a reduced value of the operating pressure angles in the two external/internal toothed pairs would yield better equalisation. The second factor that may contribute to a further reduction in these disproportions is the distance between the two operating pitch points. This distance is not constant for a given transmission ratio. To avoid complex subscripts, the following simplified notation was employed: $R_1 = R_{a/bc}$ (the operating pitch radius of gear b in mesh with a). Consequently, $R_{21} = R_{bc/a}$; $R_{23} = R_{bc/c}$; $R_3 = R_{c/bc}$; $R_4 = R_{bd/d}$; $R_5 = R_{d/bd}$.

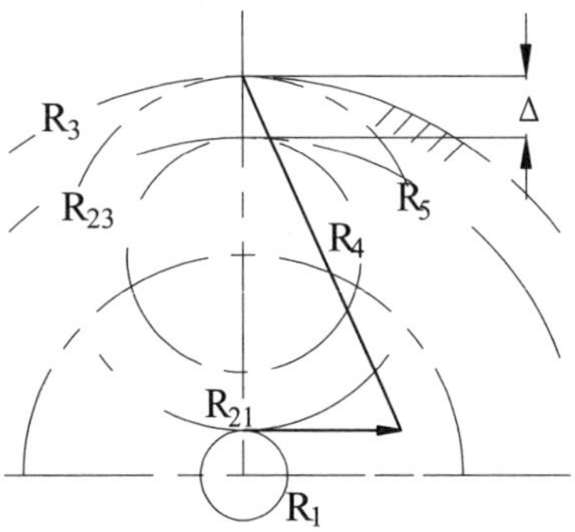

Fig. 6. Operating pitch radiuses in the transmission

The overall transmission ratio can be expressed as:

$$u_{15} = \frac{v_1}{R_{12}}\frac{R_5}{v_5} \quad \text{but} \quad \frac{V_1}{R_{21}+R_{23}} = \frac{V_5}{\Delta}.$$

As $R_5 = R_{12} + R_{21} + R_{23} - \Delta$; after transformations:

$$\Delta = \frac{\frac{R_{21}+R_{23}}{u_{15}R_{12}}\left(R_{12}+R_{21}+R_{23}\right)}{1+\frac{R_{21}+R_{23}}{u_{15}R_{12}}};$$

if $R_{21} \approx R_{23} = R_2$ and $R_2/R_{12} = i_{12}$ then finally: $\Delta = \dfrac{2R_2\left(1+2i_{12}\right)}{u_{15} - 2i_{12}}$.

As $u_{15} >> i_{12}$ then a reduction in Δ is possible if the size of the planet gear R_2 is reduced. This will also reduce the transmission ratio between the sun and planet gears (i_{12}).

This is qualitatively explained in Fig. 7, which shows numerical values of the distance Δ in the meshing of the Hori-Kurita transmission (a), in a transmission with a reduced operational pressure angles (b) [6] and finally, in a transmission with the number of teeth of the planetary gear reduced by one (c) traced for the same centre distance as in 'b'. A favourable trend, especially in the last case, is well pronounced.

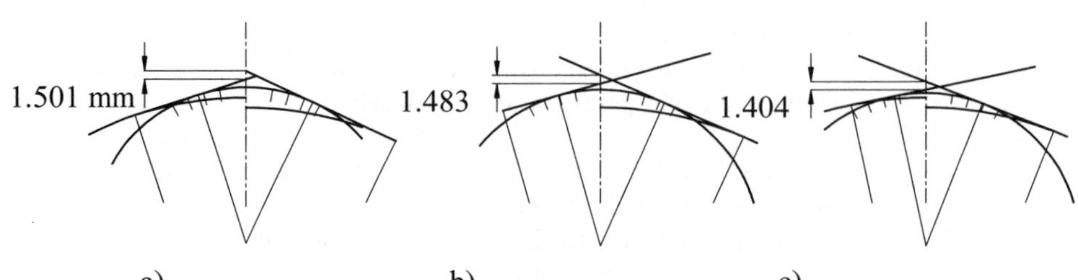

a) b) c)

Fig. 7. Positions of the operating pitch point vs. gearing specifications

If, however, the approach depicted in Fig. 7b was limited by a threat of undercutting in the sun gear, a reduced size of the planet gear *bc* would undoubtedly result with an insufficient contact ratio in the meshing of the sun and planet gears. The purpose of this paper is to evaluate quantitatively the approach proposed in Fig. 7c using the same analytical evaluation tools as those used in [6].

Analysis

The analysis was carried out for a transmission with the number of teeth in the common planet gear reduced by one (i.e. 46). The calculation algorithm included the following steps. All generation procedures are done with the use of pinion shaped cutters, so controlled are all possible interference forms in the meshing of a tool with the generated gear (undercutting, tip chamfering by the non-involute part of the cutter fillet (small cutters), chamfering during the radial feed of a tool etc.). Steps that are in agreement with the Hori-Hayashi method are denoted by HH. Some specific conditions will be specified in the following list.

1. Selection of the parameters of a pinion shaped cutter (number of teeth, regrinding status, addendum height coefficient). Based upon previous studies [6], the following cutters were selected: gear *a*, a new cutter (24, 0.104, 0.3); gear *b*, a heavily reground cutter (30, -0.28, 0.3); gear *c*, a new cutter (30, 0.105, 0.3); gear *d*, a slightly reground cutter (40, 0, 0.25).
2. Assumption of the addendum modification for gear *d* and *b*. In agreement with the proposed approach, values of x_d are positive. Limiting values for x_b and x_d shall be such that the operating pressure angle in this pair is still positive. Once x_b and x_d are assumed it is possible to calculate the operating centre distance a_w, x_c and x_a.
3. Calculations of addendum radiuses in *bc* and *c* based upon a condition of equal approach and recess contact ratios (assumed $\varepsilon_\alpha = 1.2$; HH).
4. Calculations of the maximum diameter of gear *bc* at which the width of the top land is still greater than 0.4 m (HH). If a radius from 3 is greater than that from 4, contrary to HH (a change in the value of x_b), the radius is reduced to a value calculated in 4.
5. The generation of gear *b*. The most important in this step is the radius of the first involute point at the base of the gear *b* (actually in *bd*) for the calculation of the elongated addendum in gear *d* (the addendum diameter) and the dedendum diameter (the bottom clearance). An additional checkout if the elongated addendum in *bc* will not be too long for the dedendum in the tool.
6. The generation of gear *c*. Important here is the last involute point at the base of gear *c* (interference with the addendum in *bc*) and its dedendum diameter (the bottom clearance with *bc*). Is the bottom land width sufficient large to accommodate the cutter?
7. Meshing control in toothed pair *bc-c*: interference (shall be absent) and the bottom clearance (0.25m min). If interference is present or the bottom clearance is too small, then the addendum diameter of gear *bc* is reduced from a value calculated in point 3 until the two conditions are met.
8. Calculation of the approach and recess contact ratios in toothed pair *bc-c*. If the recess contact ratio is less than 0.6 than the difference is compensated by the elongation in the addendum of gear *c* such that the resultant value is equal to 1.2 (HH).
9. The generation of gear *d* and the control of meshing with gear *bd* (interference, bottom clearance). Especially important here is the first involute point in the tool and the absence of chamfering of the elongated addendum in gear *d* by the first non-involute point in the cutter and tooth thickness at the addendum circle.
10. Calculation of the approach contact ratio, compensation (if needed) by the elongation of the addendum in gear *bd* (HH).
11. Calculations of the objective function (excluding possible impacts from meshing in *a-bc*).
12. Generation of gear *a* (addendum modification, operating pressure angle, etc.).

13. Meshing control in toothed pair *a-bc* (interference and bottom clearance at the base of gear *a*). If any of the two conditions is not met, the addendum diameter in gear *bc* is reduced and calculation repeated from step 4. Calculations of the approach contact ratio in the toothed pair *a-bc*.
14. Calculation of the maximum addendum diameter in gear *a* based on the required bottom clearance with *bc*, and absence of interference with the first involute point in *bc*.
15. Calculations of the top land width in *a*. If not sufficient (0.25m), the addendum in *a* is reduced from a value calculated in point 14.
16. Calculation of the recess contact ratio in toothed pair *a-bc*. At the moment, a deficiency in this value cannot be compensated. If this value is less than a predetermined minimum value (1.1), the analysed combination of data is not acceptable and calculations are repeated for a new value of x_b and x_d.

The results of calculations are plotted in Fig. 8. A small deficiency in the contact ratio in the toothed pair *a-bc* is shown with dashed lines. Minimum values of the objective function *FC* are very close to this constraint in each of the depicted graphs.

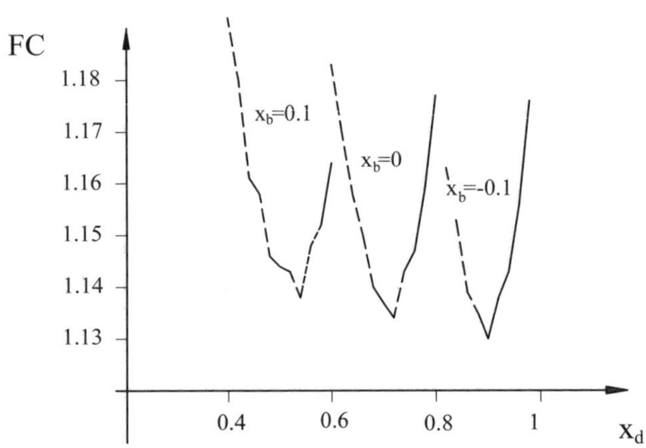

Fig. 8. The objective function *FC* as a function of the addendum modification in gear *b* and *d*.

Fig. 9 shows a meshing arrangement for the minimum value of *FC* with no modification in the planet gear (x_b = 0; x_d =0.72). The value of the objective function for this layout is equal to 1.134 with a fractional deficiency in the contact ratio in the pair *a-bc* (1.09). With a slightly negative modification in *b*, the target value (1.1) is easily obtained.

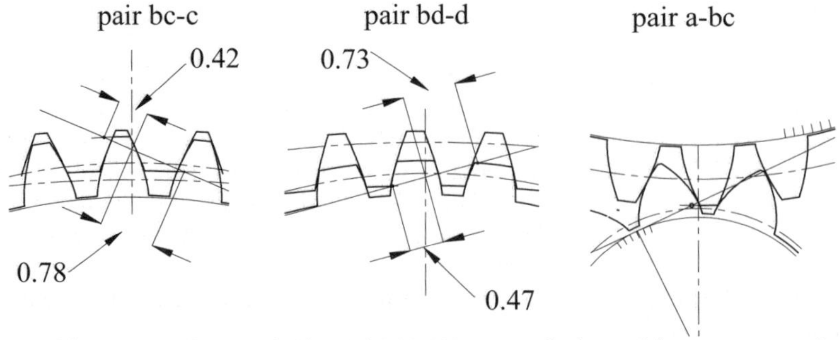

Fig. 9. Meshing layout in a 12/46/-105/-108 3K transmission with a common planet gear

Conclusions

1. When selecting the number of teeth for a 3K transmission with a common planet gear for equalized approach and recess contact ratios, the following procedure is to be recommended:
    a. Select the number of teeth in the sun gears *a* (12).
    b. Select the numbers of teeth for gears *c* and *d* such that the transmission ratio is close to the target value (assembly constraints satisfied).
    c. Of a few possible numbers of teeth for the planet gear select that closest a minimum acceptable.
    d. Based upon standard angular modification formulas, find the total addendum modification to toothed pair *bd-d* for which the operating pressure angle in this pair is still positive.
    e. Follow the procedure given in the analysis chapter to find the best equalisation within the selected range of values for x_d and x_b.
2. The toothed pair *a-bc* features a heavily misbalanced approach and recess contact ratios. The impact of this situation on the overall efficiency of the transmission is however close to nil.
3. Contrary to the HH method, with the approach proposed, the equalisation process is not limited by a threat of undercutting in the sun gear but by a deficiency in the contact ratio between the sun and planet gears.

References

[1] J. Ferguson, Select Mechanical Exercises, 1773

[2] M. Gawrysiak, A fascinating history of planetary gearing (in Polish), 24th Conference on Fundamentals of Machine Design, Białowieża, 2009

[3] K. Hori, Y. Kurita, Improvements in 3K Mechanical „Paradox" Planetary Gear Efficiency, Power Transmission and Gearing Conference, (ASME), vol. 88, 1989

[4] K. Hori, I. Hayashi, Improvement of Efficiency of 3K-Type Mechanical Paradox Planetary Drive by Equalising Approach and Recess Contact Ratios. DE - vol. 48, Power Transmission and Gearing Conference (ASME), 1996

[5] A. Golenko, Verification of an Approach and Recess Contact Ratio Equalisation Method (Hori and Hayashi), XLVI International Conference of Machine Design Departments, Sedmihorky, Czech Republic, 2005, pp. 83 to 86,

[6] A. Golenko, 3K Mechanical Paradox transmissions: the shaping of the meshing zone for better efficiency, Archive of Civil and Mechanical Engineering, vol. IX, no 2, 2009, pp 39 to 46

[7] A.L. Kapelevich, T.M. McNamara: Introduction to Direct Gear Design®, the Global Power Train Congress 2005, September 2005

Key Engineering Materials Vol. 490 (2012) pp 8-15
© (2012) Trans Tech Publications, Switzerland
doi:10.4028/www.scientific.net/KEM.490.8

A New Technique of Spine Disease Treatment Supported by ADRSK1 Device

Szymon Grymek[1, a] and Mateusz Pawelec[2,b]

[1]Gdańsk University of Technology, Department of Medical Engineering, Narutowicza 11/12, 80-233 Gdańsk, Poland

[2]Swissmed Health Center, Wileńska 44, 80-215 Gdańsk, Poland

[a]sgrymek@pg.gda.pl, [b]pawelec.mateusz@gmail.com

Keywords: scoliosis, rehabilitation, physiotherapy, rehabilitation device, ADRSK1

Abstract. Scoliosis is one of the most frequently occurring defects of human body. Its effective rehabilitation requires a suitable methodology of corrective exercises and devises which support the implementation of these exercises. The subject of scoliosis and its rehabilitation is brought up in this paper. The most popular rehabilitation devices as well as the new ADRSK1 apparatus from Gdańsk University of Technology are presented. The structure, the principle of operation and advantages compared to existing devices of the ADRSK1 are described.

Introduction

Scoliosis is one of the most frequently occurring defects of human body. It affects more than 2% of the population [1, 2] and is considered to be disease associated with the progress of civilization.
That determines the action to understand the aetiology of the disease and the development of methods of prevention and rehabilitation. Early detection of changes in the structure of the spine allows stopping the development of scoliosis and avoiding a dangerous and costly surgery. Unfortunately, the existing design solutions of most devices for scoliosis rehabilitation exercises rarely produce tangible effects. Therefore, there is a need for the continuous improvement of existing facilities and searching for new structural solutions.

In 2000 in Gdańsk, Andrzej Stolarz, an experienced physiotherapist and a specialist in the field of combating dysfunctions of the spine, created an apparatus named Skol-as [3]. Together with a special plan of exercises it enabled prevention of the spine's pathomechanics progress in children and young people under the age of 20. In 2010, the creator of the Skol-as device undertook the cooperation with the Department of Medical Engineering, Gdańsk University of Technology. Its objective was to adapt the structure of the Skol-as apparatus to the introduction of new sets of rehabilitation exercises, elimination of defects in the apparatus revealed in the course of nearly ten years of operation, the preparation of the so-called portable version (which enables rehabilitation at home) and, ultimately, its automation. As a result of collaboration, an innovative device ADRSK1 supporting the process of rehabilitation of scoliosis was designed and manufactured in 2011.

Scoliosis

Correct structure of the spine depends on three mechanisms:
- passive stabilisation mechanism – the vertebra, the posterior ligament complex and the intervertebral discs;
- active stabilisation mechanism – the local muscles, the global muscles and the abdominal muscles;
- nervous control system – the central nervous system.

Interactions of these mechanisms are very complex. Muscles are attached to each of the vertebrae, adjusting and stabilising their relative position. It is possible due to muscle-tension, which in turn is controlled by the central nervous system.

Scoliosis (known as lateral curvature) is a disease involving three-dimentional deviation of the spine from its correct state [4]. Scoliosis consists of (refer with Fig. 1) [3, 5]:

- changes in sagittal plane – changes manifest themselves through the shallowing or deepening of the natural curvature of the spine leading respectively to lordosis or kyphosis;
- curvature in the coronal plane – a lateral curvature of the spine, resulting from the nonparallel placement of the surfaces of adjacent vertebrae. Disturbances in the two described planes change the shape and functionality of the vertebrae. This dysfunction and axial loads overlapping it cause progression of the pathology of mechanics of the spine;
- curvature in the transverse plane – vertebrae's cores rotation. It is related to the excessive load of appendages and the vertebral appendices and arches and thus deepens the bending forward of the spine. The consequence of this process is an imbalance of passive stabilisation mechanism characterized by the extension of a column of the vertebrae's cores, and asymmetric expansion and displacement of intervertebral disc in the direction of curvature. A further consequence of the curvature in the transverse plane is a rib hump or a lumbar hump.

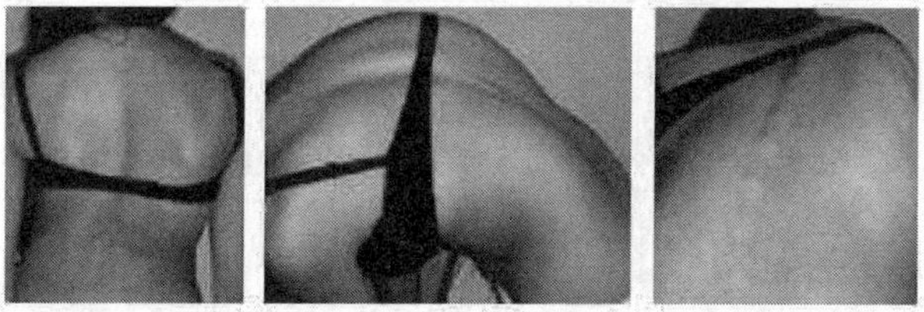

Fig.1. Spine's dysfunction in the coronal plane, the sagittal plane and the transverse plane

In addition to these there are also secondary effects, making the scoliosis systemic disease. If left untreated, it leads to impaired cardiovascular, respiratory and inefficiency of the internal organs [2].

Spine Disease Treatment Supporting Devices

Nowadays there are two leading products on the world market, which improve the health status of patients with scoliosis: orthopaedic corsets [6, 7] and Sastre's apparatus using the FED method [8 – 10]. There are also a number of devices, which are not yet widely used, e.g. apparatus DELFIN developed by Healthy Back Academy in Poland [11].

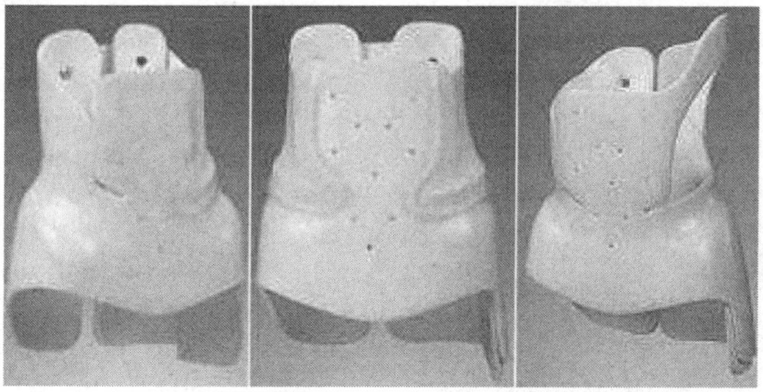

Fig.2. Orthopaedic corsets – The Boston System
thoraco-lumbo-sacral orthosises [6]

Orthopaedic Corsets. There are many different types of orthopaedic corsets. This diversity stems from the existence of many theories about the progression of scoliosis. Most commonly used ones are the Boston type corsets (see Fig. 2) and the Blounta – Milwaukee type corsets. The general operating principle of corsets consists in passive correction of posture, which is due to relieve of vertebrae on the concave side of the spine. Thanks to that muscles and tendons on the convex side, which previously were overly stretched, return to their original state. After prolonged forcing the correct orientation of the spine, muscles and tendons take over the function of corset and prevent the progression of scoliosis. However, it does not cause regression. Corsets are intended for people whose skeleton is flexible enough and susceptible to correction. They do not stimulate central nervous system, at all. Construction of the corset (its size) and the materials from which it is made (Young's modulus) are selected individually for each patient. Using of orthopaedic corset is cumbersome – it significantly reduces patient’s motion skills, as it should be worn at least 20 hours per day [6, 7].

The FED System Device. Currently, it is the only device effectiveness of which in the rehabilitation of spine diseases is documented by a series of X-rays photographs. The initiator of the construction of this device was Prof. Santos Sastre Fernandez from Spain. Nowadays, it is already the third generation of this device (see Fig. 3). It is intended, just like corsets and other here mentioned devices, for people whose skeleton is still flexible enough.

Fig. 3. The FED system device [8]

The FED system methodology assumes the impact in three planes, where scoliosis exists, and takes into account the stimulation of the three mechanisms important for proper functionality and structure of the spine. An arm driven by the pneumatic actuator forces hypercorrection of the spine and directly affects on the passive stabilisation mechanism. During the arm’s withdrawing patient is required to maintain the corrected position using a force of his own muscles. It stimulates the active stabilisation mechanism and the nervous control system.

However, it seems that the impact on the nervous control system requires refinement. When the pneumatic arm is withdrawn and the patient seeks to maintain the corrected position, the force generated by the muscle may be insufficient to maintain the desired orientation of the spine. This automatically changes the hypercorrected position of the spine without the patient's awareness. The problem can be solved by introducing the control-measuring system with a display, which advises the patient, in which position he should enforce his muscles.

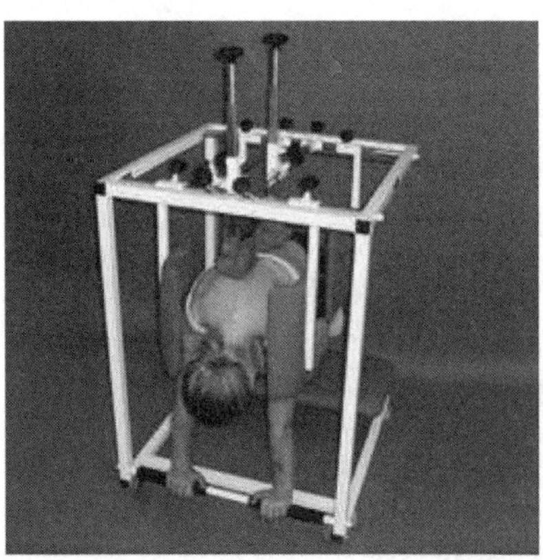
Fig. 4. Derotator and scoliosis corrector DELFIN [10]

Derotator and Scoliosis Corrector DELFIN. It is a device designed and used by specialists from the Healthy Back Academy (Poland). The device realises an idea of scoliosis correction in the three planes using the shoulder girdle stabilisers and stabilisers of the pelvis (refer with Fig. 4). While patient is kneeling down he performs an exercise known as “cat's back” and encounters the resistance of the moving heads. This allows one to adjust the orientation of the spine in the sagittal plane and the horizontal plane. Exercises performed with the use of the DELFIN device stimulate only the passive stabilisation mechanism and they have no influence on the active stabilisation mechanism and the nervous control system.

Structure of the ADRSK1 Device

A device for supporting spine disease treatment was designed in the Department of Medical Engineering, Gdańsk University of Technology in 2011 (see Fig. 5) and was named ADRSK1. The effectiveness of the therapy conducted with the use of this device derives from the fact that the structure of its individual components allows for the implementation of planned exercises as well as for the change of the patient's position in the device space forced by a therapist.

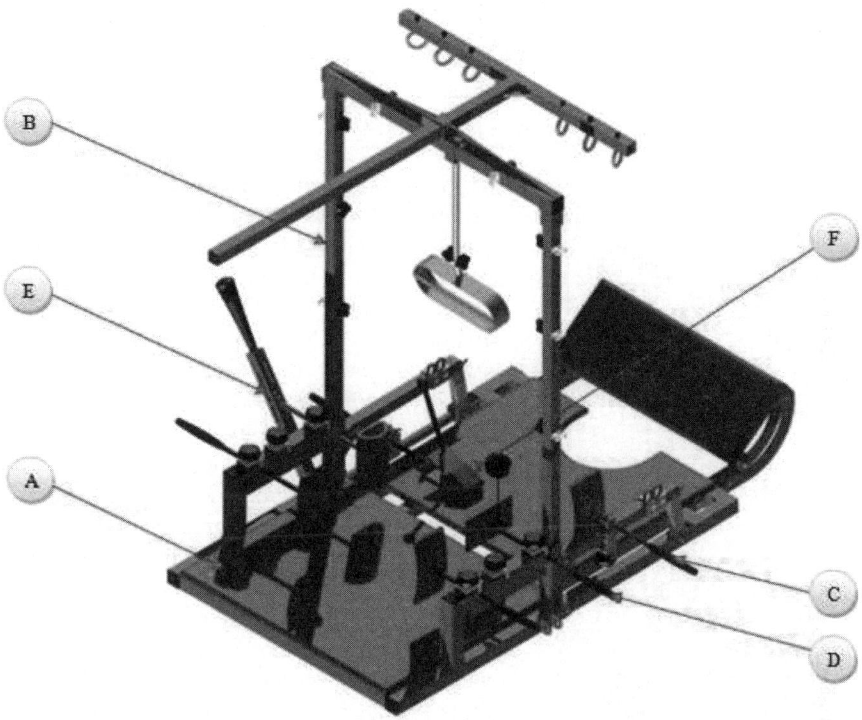

Fig.5. Structure of the ADRSK1 device

The ADRSK1 consists of: the base (A), the gate (B), the arms: stabilising (C), measuring (D) and reclining (E), and the derotating lever (F).

Within *the base* one can highlight: a fixed part, a part having the possibility, thanks to using establishing bolts, to change the tilt angle relative to the ground and a connecting profile. The connecting profile allows to install the arms in the device. A position of the arms in the lateral axis of the device is determined thanks to knobs. The fixed part and the moving part are covered by upholstery. Selection of material from which it is made – ecoskin – helps to maitain asepsis and its structure allows the mounting of pressure pads – elements of the control-measuring system. Also within *the base* additional sockets enabling the assembly of *the gate* are provided.

The gate is an element that has not direct contact with the body of the patient. It allows to join a number of key elements for the process of rehabilitation: springs to change angle of pectoral support, a longitudinal jib with an elastic band placed under the patient's head and mounting elements for the pressure gauges used to stimulate the nervous system through external replacement loopback.

Elements having direct contact with the patient are: *the stabilising arms*, *the measuring arm* and *the reclining arm*. All the arms have unified fixing mandrels. This increases the possibility of adapting the device to the specificity of disease, and thus the scope of its applicability. Spatulate part of each arm varies depending on its function.

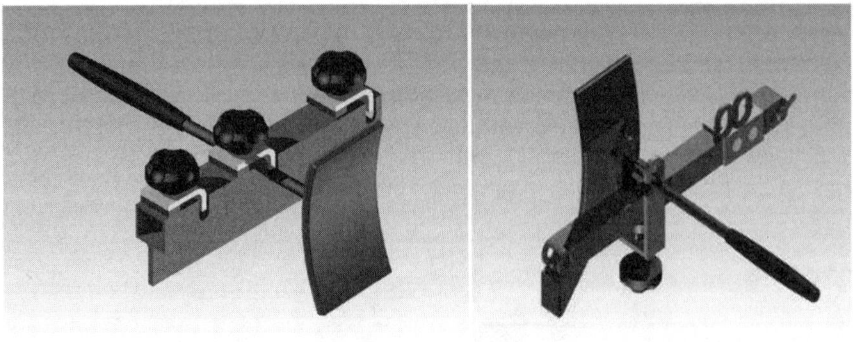

Fig.6. Stabilising arms: non-sliding and sliding

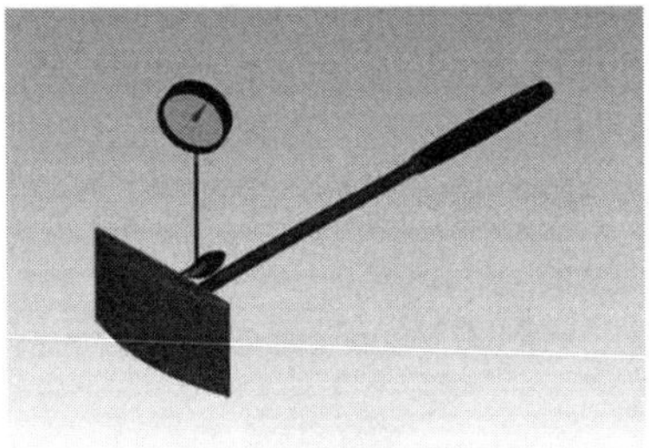

Fig.7. Measuring arm

Fig.8. Reclining arm

Fig.9. Derotating lever

The stabilising arms (refer with Fig. 6) are applied for positioning the patient in device space. They are also points of support during execution of the reclination process. In order to increase the contact surface between the patient's body and the stabilising arms their scapular parts are specially profiled. The scapular part of *the measuring arm* (refer with Fig. 7) is covered by upholstery with a sewn-in latex pressure pad. The pressure pad has two tubes where pump and pressure gauge are attached. By means of a specially designed holder, the pressure gauge is visible for both: patient and therapist. *The reclining arm* (see Fig. 8) applies the impact force on the curvature of the spine in its thoracic or lumbar part. The force operates in a plane perpendicular to the axis of the spine (the transverse plane); movement along the axis of the spine is not required. The shape and dimensions of the scapular part of this arm are very important. They must ensure, for each patient, the impact in the area of two adjacent vertebrae segments of thoracic or lumbar spine. As a result of the analysis it was found that the optimum diameter of the convex part of the scapular part should be 50 mm.

The derotating lever (see Fig. 9) enables rehabilitation in a transverse plane. Its design allows to have an impact on the surface of the patient's body with a force three times greater than the force applied. Additionally, one can very precisely determine its position in the space of the ADRSK1 device depending on the physical condition of the patient.

The ADRSK1 device has been designed with a view to the rehabilitation in low position (recumbent posture). However, thanks to the high degree of adaptiveness of individual components of the device it is also possible to adapt it to support the therapeutic process in high position (seating posture).

Principle of Operation of the ADRSK1 Device

Structure of the ADRSK1 device allows rehabilitation of the spine pathologies according to commonly recognized methods for the prevention of the spine's pathomechanics progress. This means that it supports in three planes the active and passive rehabilitation simultaneously stimulating the nervous system.

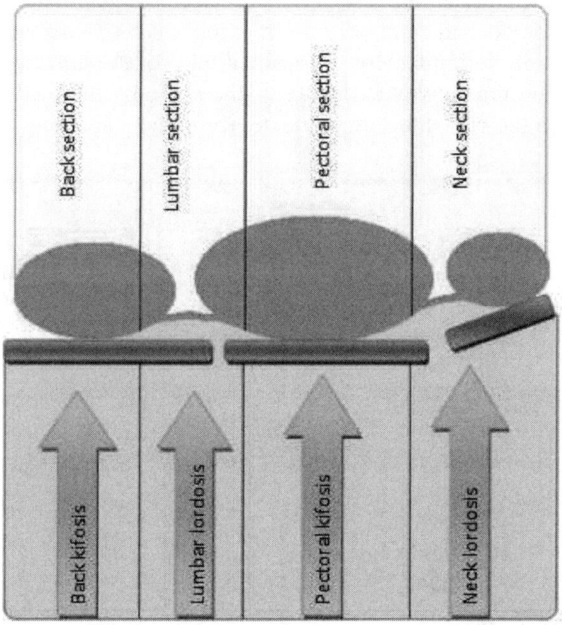

Fig.10. Spine correctly oriented in the sagittal plane

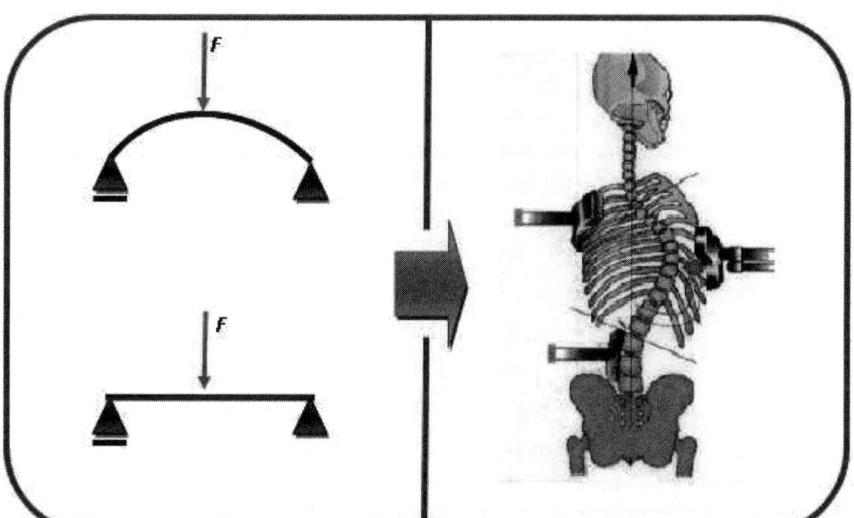

Fig.11. Idea of reclination process

Operation in the sagittal plane. The patient is placed within the device's space in recumbent position. Thanks to the movement of asymmetrically placed stabilising arms it is possible to adjust the position of the pelvis. It forces the correct orientation of the spine in its thoraco-lumbar part. The simultaneous use of bolster allows the mapping of the natural curves for the entire length of the spine (refer with Fig. 10).

Operation in the coronal plane. The stabilising arm, correcting the position of the pelvis, is also the first support for the spine in its lumbar part. The second stabilising arm is placed below the glenohumeral joint as the second support (the moving one). Thanks to such location of the stabilising arms it is possible to implement the reclination process (refer with Fig. 11). During this process the impact force on the convex curvature of the spine by the reclining arm is applied. The reclining arm is located on the opposite side of the device to the stabilising arms.

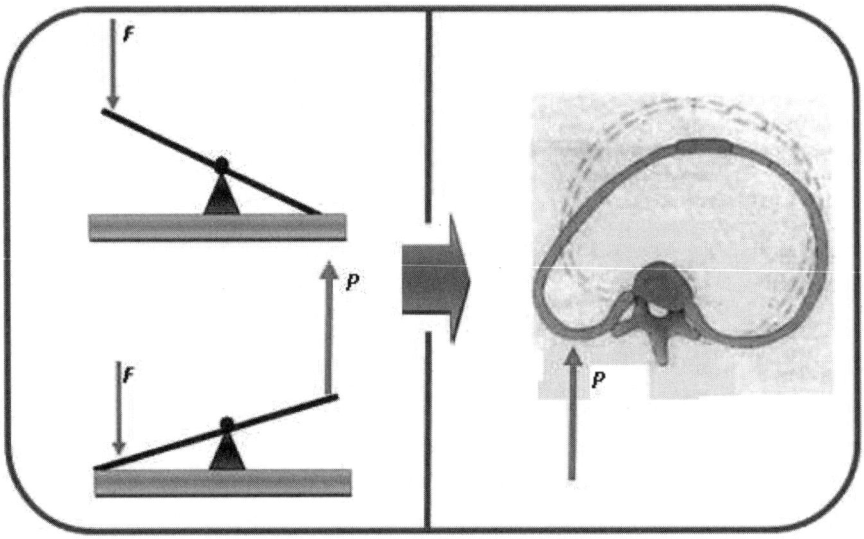

Fig.12. Idea of derotation process

Operation in the transverse plane. The derotating lever is situated in the moving part of the device at the level of pectoral part of the spine. This location was chosen due to the fact that the vertebrae's cores in the region are experiencing the greatest rotation in horizontal plane. This rotation results in a formation of rib hump. In order to implement derotation process one should, when the patient is in recumbent position, insert the working tip of the derotation lever between a spinous appendix and a transverse one. Applying a small force to the other end of the lever results in derotation of the vertebrae (refer with Fig. 12).

Summary

The ADRSK1 device for supporting spine condition treatment was designed in the Department of Medical Engineering, Gdańsk University of Technology. Its structure utilises the theoretical foundation of scoliosis's rehabilitation – the simultaneous stimulation of passive stabilising mechanism, active stabilising mechanism and nervous control system.

The FED system device is the only one comparable to the ADRSK1 with regard to its therapeutic potential. It is more and more recognised among the European physiotherapists. But in authors' opinion the FED system device, in comparison to the ADRSK1, seems to have three drawbacks:

- the most serious one is that the effectiveness of stimulation of nervous control system is in doubt;

- in practice it is only suitable for use only in the rehabilitation facilities; compactness and ergonomic structure of the ADRSK1 make it easy to transport and enable its usage at home, which is especially important in the case of the treatment of children from villages and smaller towns;
- it is a costly device – a rehabilitation clinic from Zgorzelec (Poland) bought it for about 35 000 €, while the cost of the ADRSK1 *prototype* was less than 2 000 €.

Now the ADRSK1 device is, inter alia, utilised at the Academy of Phisical Education and Sport in Gdańsk. Although the device has been exploited there for a few months, reviews of the local physiotherapists, as well as their patients, are very positive. It confirms the soundness of structural solutions adopted in the device.

A weak point of the ADRSK1 device, comparing to the FED system device, is the lack of mechanical drive. It is not critical for rehabilitation, but requires that all power enforcements and their control are applied by a physiotherapist. Currently, the ADRSK1 is under intensive development in order to create fully mechatronic device with the pneumatic driving system and the computer control system. After the implementation of the mechatronic solution a long-term research programme on the effectiveness of rehabilitation supported by the ADRSK1 device is planned.

There is a hope that creation of an efficient, cheap and mobile device for scoliosis rehabilitation, which the ADRSK1 is, will decrease the morbidity rate of that threatening systemic disease.

References

[1] J. Wilczyński, Lateral spinal curvature – identification and correction, Wydawnictwo Wszechnica Świętokrzyska, 2000 (in Polish).

[2] H.-R. Weiss, Spinal deformities rehabilitation – state of the art review, Scoliosis 2010, 5:28 (24 December 2010).

[3] A. Stolarz, Private notes (not published).

[4] C. Richardson, Kinesitherapy in the stabilization of lumbar-pelvic complex, Wydawnictwo Elsevier Urban&Partner, Wrocław 2009 (in Polish).

[5] J. Nowotny, Exercises based on surrogate feedback on postural re-education, Postępy Rehabilitacji, Tom II, 1988 (in Polish).

[6] Information on http://www.skolioza.com

[7] G. Ingiot, Scoliosises – recognition, causes, Biuletyn nr 2004/4 – pismo Okręgowej Izby Lekarskiej, Rzeszów 2007 (in Polish).

[8] Information on http://www.sastre-roca.com

[9] S. Fernandez, Method of treatment of scoliosis, kyphosis and lordosis, Wydawnictwo Markmed, Rehabilitacja, 2008 (in Polish).

[10] S. Sastre, J. P. Lapuente, C. Barrios, Benefits of F.E.D. Treatment in Scheurmann's Disease, I.R.Spinal Deformities, IOS Press, 3, Oxford, 2002.

[11] Information on http://www.scoliosisstop.zdroweplecy.org

Key Engineering Materials Vol. 490 (2012) pp 16-24

doi:10.4028/www.scientific.net/KEM.490.16

An Analysis of Weld Damage in Punching Press

Jan Ryś[1,a] , Stanisław Łaczek[1,b]
[1]Politechnika Krakowska, 31-864 Kraków, al. Jana Pawła II 37, Poland
[a] Szymon@mech.pk.edu.pl, [b] Laczek@mech.pk.edu.pl

Keywords: Machine Design, Welding, Punching press, FEM, Dynamic Analysis

Abstract. This article deals with a description of weld damage in a punching press. Static and modal analyses were performed in ANSYS v.11. Dynamic stresses in welds were evaluated using the Laplace's transformation for a simplified model with a single mass on a spring.

Introduction - failure description

The break-down of machines can be very expensive especially in automatically working system, but on other hand it gives opportunity to find requirements and main criteria for appropriate calculations. For these reasons individual components and its joints should have adequate strength, rigidity and effectively resist vibration induced by external and internal forces actions.

Possible causes of weld damage in the punching press are analyzed in this paper. The press has the ability of punching holes in plates under maximum force of 220 T. Plates are made of S355J0 steel, which has tensile strength R_m= 470 – 620 MPa. Hole diameter is from 25 to 32 mm, plate thickness is from 15 to 20 mm. While the pres was at work, a damage of a weld occurred (on the left side). The weld joined a table (board) with the press body (support). A view of the press with damaged weld is shown in Fig.1a.

There was a crack of 12 mm depth across the weld, practically on the lower part of the weld. The crack had smooth fracture, which is typical for fatigue structure. As a result of the weld damage there was a small skewing of the board against the horizontal direction (Fig.1b).

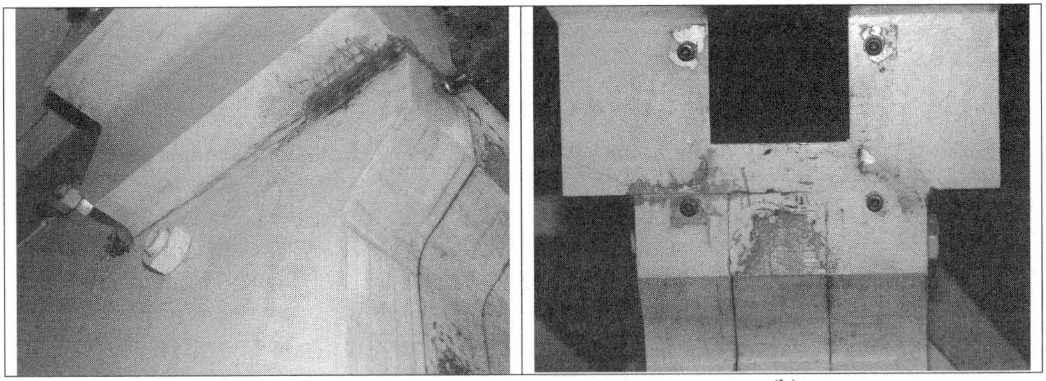

(a) (b)

Fig.1. Fracture damage of welding joint

To determine stress level in welds joining the board (table) to base, following parameters were assumed in calculations:

thickness of the weld a = 15 mm,

length of the weld L= 275 mm,

maximum force acting on the table from actuator during punching holes P=200 T.

Dimensions of press frame were reconstructed from basic documentation and measurements of some elements of the press (Fig.2).

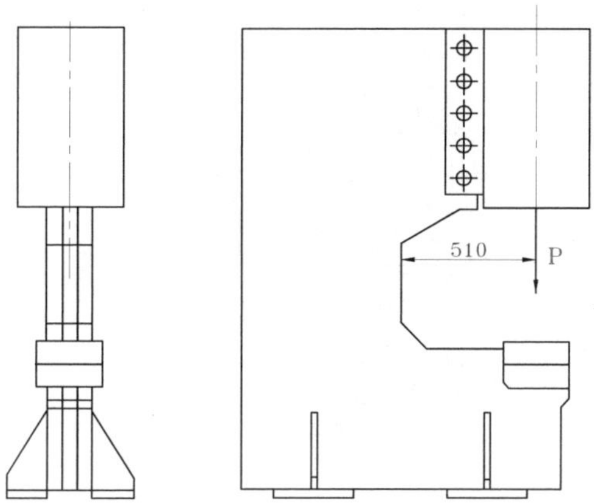

Fig.2. General view of the press body

Equivalent stresses in welds in a case of a gap under the board

In order to evaluate the influence of a gap between the board and the press body (created in the process of welding during assembly) a numerical simulation was performed. One-sided gap of 0.3 mm was assumed (Fig.3). The finite-element mesh of the press support with welds is presented in Fig.4.

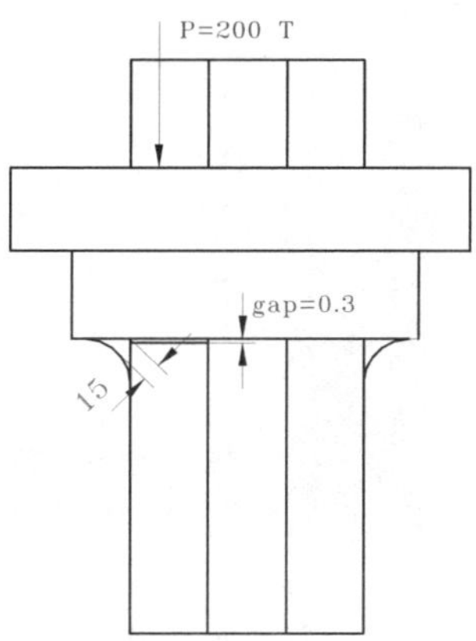

Fig.3. One side gap between the board and the press body (support)

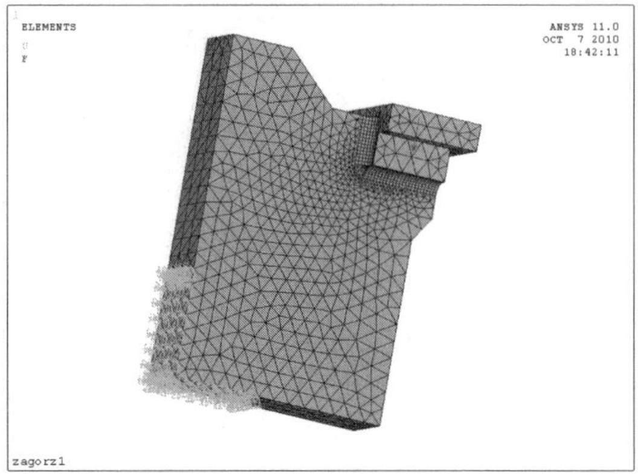

Fig.4. The finite-element mesh of the board and the support

A small value of the gap was assumed in calculations (for instance 0.3 mm) in order to show that even a small gap can generate high stresses in the weld, which may result in creating a crack. An equivalent stress in the weld in case of such imperfection is shown in Fig.5.

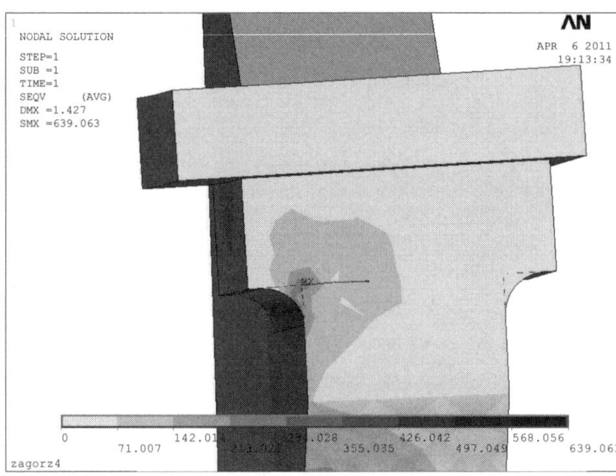

Fig.5. Equivalent stress distribution in the weld and the support

In the outer part of the weld equivalent stress equals 280 MPa, which is less than the stress value near the gap. Maximum stress of 639 MPa is in a vicinity of the gap, causing local weld material strengthening. Micro cracks can appear after multiple loading and unloading of the press board. They evolve in a crack and encompass full cross section of the weld. After unloading (decreasing punching force P to zero) tensile stresses can appear in the weld. It can depend on pressure pad system mounting the plates during punching. Upon carrying out analysis it can be seen how important accuracy of performance and welding board to press body and dimensions of the weld are. Equivalent stresses under punching force P= 200 T in the press without the gap are shown in Fig.6. Displacement (deformation) distribution of the press body is shown in Fig.7a and Fig.7b, according to 'z' and 'y' direction.

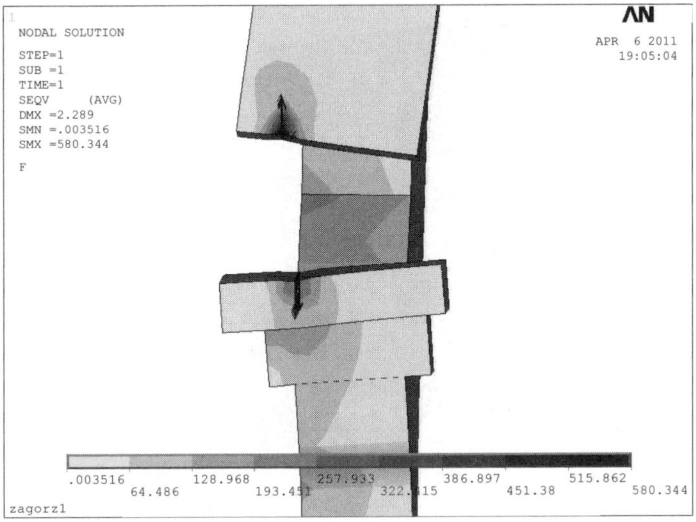

Fig.6. Static stress distribution in the press without a gap

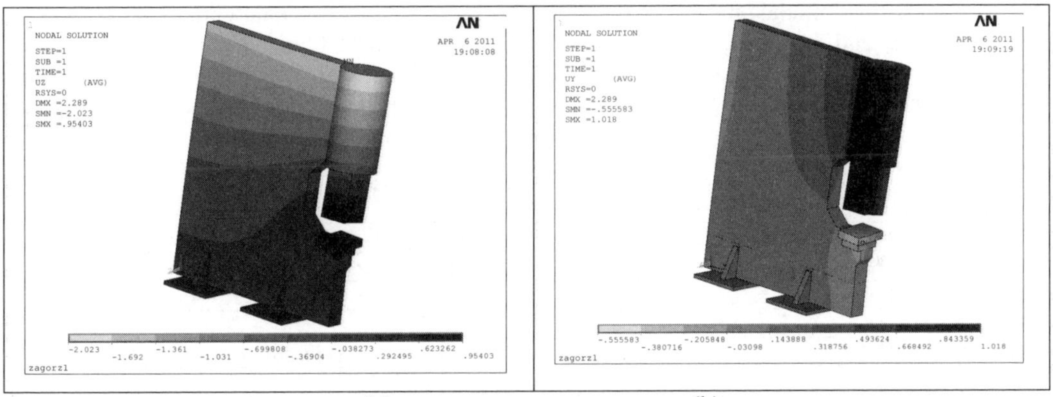

(a) (b)

Fig.7. Static deformation of the punching press in z- and y- direction under P force

Vibrations and equivalent stresses in welds after force has fallen to zero

The second cause of the press failure is the dynamic unloading after punching. Eigenvalue problem should be solved in order to determine free form vibrations. The lower frequencies are concerned with transversal vibrations of the press body in z- direction (as it is seen in Table1 and Fig.8). Higher frequencies are related to vertical vibrations (in y- direction) of the press body. If pressure pad system releases holding down the plates after punching then elastic energy cumulated in the press body causes harmonic vibrations with the amplitude of 0.555 mm (calculated from static analysis). Values of displacements related to modal form are presented in Fig.9 (f=935,4 Hz) and displacement plot versus time is shown in Fig.10. Vibrations quickly disappear; nevertheless (in the first moment when the force decreases) tension stresses will occur in the weld, which is shown in Fig.11. Amplitude of the board center was assumed as 0.42 mm.

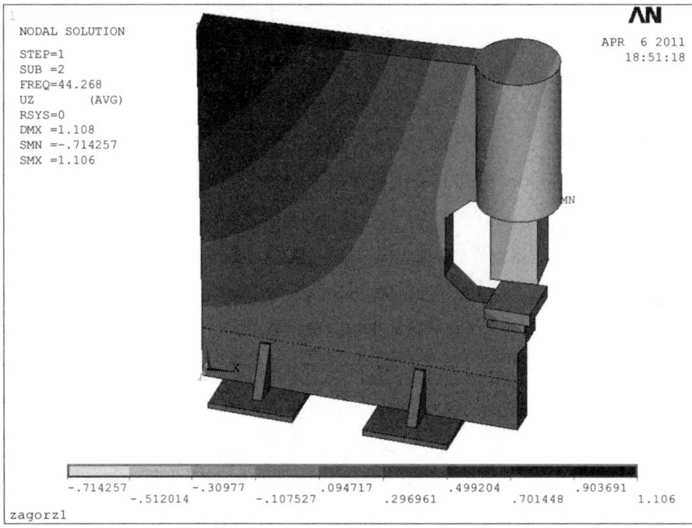

Fig.8. Modal analysis of the press (the second mode under f=44.2 Hz)

It is characteristic of modal analysis of the press that lower frequencies deal with bending form in Z direction (for f_1 and f_2). Medium frequencies are concerned with blended modal form (in Z-X plane for f_3, f_4 and Z-X-Y plane for f_{19} frequency). An interesting frequency (for the dynamic analysis of the press board) is f_{20}=935.4 Hz mainly in X-Y plane. List of frequencies obtained in modal analysis is presented in Table 1. Table 2 also contains components of dimensionless displacements from modal analysis.

Table 1. Free form vibrations of the press

f_1 = 18.688 Hz (vibrations in Z direction)	f_2 = 44.268 Hz (vibrations In Z direction)
f_3 = 105.12 Hz (vibrations in Z –X plane)	f_4 = 118.16 Hz (vibrations in Z –X plane)
f_5 = 184.28 Hz	f_6 = 202.44 Hz
f_7 = 250.55 Hz	f_8 = 257.22 Hz
f_9 = 397.06 Hz	f_{10} = 426.29 Hz (vibrations in Z –Y plane)
f_{11} = 453.77 Hz	f_{12} = 506.17 Hz
f_{13} = 561.68 Hz	f_{14} = 634.51 Hz
f_{15} = 686.47 Hz	f_{16} = 747.11 Hz
f_{17} = 762.40 Hz	f_{18} = 795.67 Hz
f_{19} = 808.61 Hz (vibrations in X-Y-Z dir.)	f_{20} = 935.43 Hz (vibrations inY-X plane)

Table 2. Components of dimensionless displacements from modal analysis

Number of the form	Freq. Hz	Usum	Uz	Uy	Ux
f_1	18.68 Hz	0 0.767	-0.011 0.767	-0.0946 0.0946	-0.0289 0.0289
f_2	44.26	0 1.108	-0.714 1.106	-0.0446 0.0446	-0.213 0.213
f_{10}	426.4	0 1.323	-0.981 1.323	-0.269 0.270	-0.146 0.146
f_{19}	808.6	0 1.89	-1.033 1.752	-0.684 0.706	-0.457 0.514
f_{20}	935.4	0 1.282	-0.1299 0.0677	-0.742 0.861	-0.801 0.949
Usum – means total normalized displacement from 0 do max Ux, Uy, Uz – components of normalized displacements from min to max					

The press board has a mass and directional stiffness in X, Y, Z directions which are connected to resonance frequencies and modal forms. Lower frequencies in Y direction can be suppressed by hydraulic system, mounting the plate and can be reduced by internal damping of the material of the press arm. Higher frequencies in Y direction with big relative amplitude induce inertial force, which cause alternative stresses in welds mounting the board.
One of the dangerous frequencies is f=935.4 Hz, which generates a significant inertial force of the board in upward Y direction after decreasing P force to zero. This case is presented in Fig.9.

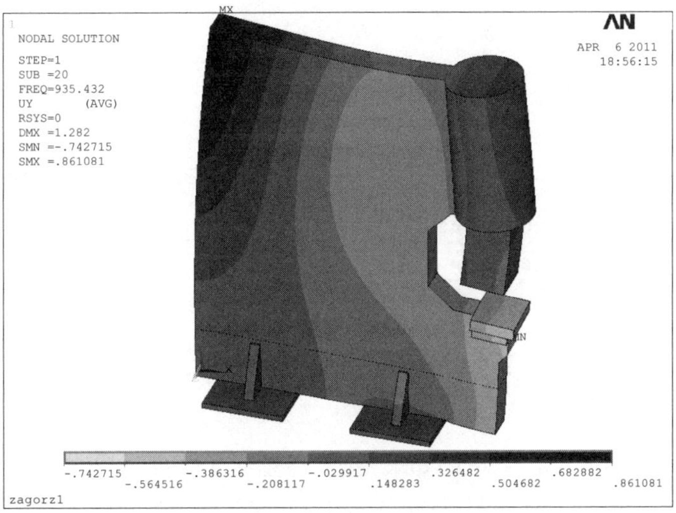

Fig.9. Dimensionless u_y displacement distribution at 935.4 Hz

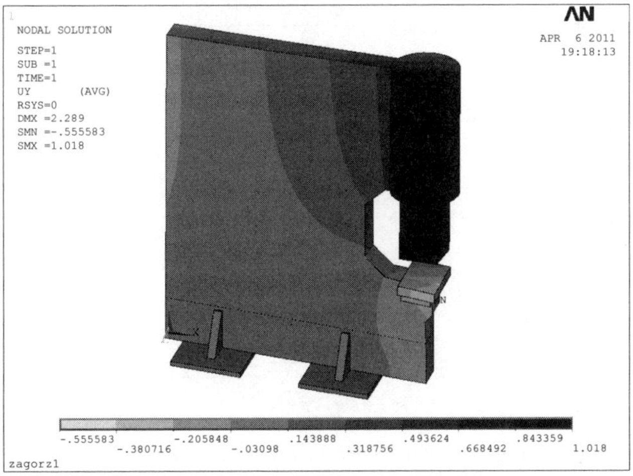

Fig.10. Nodal static solution for displacement u_y

4. Stress evaluation in welds

The values of actual stresses in weld joint connecting the frame and the board of the press are the results of relative vibration between frame and the board of press. The most dangerous are the high frequencies related to the free frequencies of the connections.

In a modeling of such phenomenon we assumed that the mass of the board 'm' has elastic foundations with rigidity 'k' with 'b' damping property. Dynamic equation for single mass on a spring with damping under short rectangular impulse can be described as follows:

$$m\ddot{y} + b\dot{y} + ky = P_o[\Phi(t) - \Phi(t-\tau)] \ , \tag{1}$$

or in such form:

$$\ddot{y} + 2\xi\omega_o\dot{y} + \omega_o^2 y = p_o[\Phi(t) - \Phi(t-\tau)], \tag{2}$$

where

$$p_o = \frac{P_o}{m}, \qquad \omega_o^2 = \frac{k}{m}, \qquad 2\xi\omega_o = \frac{b}{m} \ , \quad \Phi(t) \text{ – Heaviside function= 1(t).}$$

For the Laplace's transformation we obtain equation [2], [3]

$$Y(s) = \frac{p_o\left(1 - e^{-\tau s}\right)}{s(s^2 + 2\xi\omega_o s + \omega_o)} = f(s)\,. \tag{3}$$

For τ – time of force impulse, ζ<1, $\omega_1 = \omega_o\sqrt{1-\xi^2}$, the displacement X(t) has a form:

$$\frac{y(t)\cdot\omega_o^2}{p_o} = \frac{x(t)}{u_{\max}} = X(t)\,. \tag{4}$$

$$X(t) = \Phi(t-\tau)\cdot e^{-\xi\cdot\omega\cdot(t-\tau)}\cdot\cos[\omega_1\cdot(t-\tau)] - \Phi(t)\cdot e^{-\xi\cdot\omega\cdot t}\cdot\cos[\omega_1\cdot t] + \Phi(t) - \Phi(t-\tau) \tag{5}$$

The resolution X(t) is shown on Fig.11. Following parameters were put down for this figure:

$$f = 935\,[Hz], \qquad \omega = 2\cdot\pi\cdot f\,[1/s], \qquad \tau = 0.5\,[s], \qquad \xi = 0.005$$

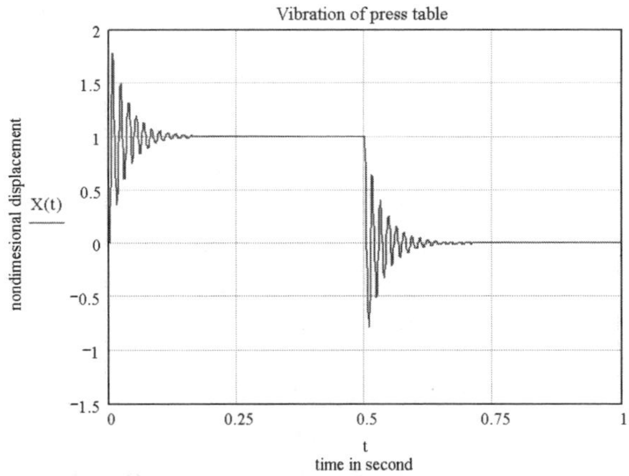

Fig.11. Nondimensional displacement of the board center in a function of time

In next calculations we omit damping because ζ factor is unknown.

Maximum punch force is P=200 T= 2000000 N and maximum static displacement in vertical direction of the press support under force P (numerically calculated in ANSYS system) is u_y=-0.555 mm. Stiffness of the press support

$$k=P/u_y= 2000000/0.555=3603603 \text{ N/mm} = 3.604\cdot 10^9 \text{ N/m} \tag{6}$$

Mass of the board m = 97.9 kg was calculated in ANSYS system.

When modeling the press as a single mass on a spring, resonance circular frequency of the board equals

$$\omega = \sqrt{\frac{k}{m}} = \sqrt{\frac{3.6042 \cdot 10^9}{97.9}} = 6067.4 \quad \text{rd/s} \tag{7}$$

Approximate frequency of the press support resonance is

$$f = \frac{\omega}{2\pi} = 965.6 \text{ Hz}$$

The nearest frequency of free form vibration in X-Y plane (numerically calculated) equals f_{20}= 935 Hz (Fig.9), which is comparable with an approximate frequency f=965.6 Hz.
We can describe displacement of the table (without damping), starting from the moment of decreasing the punching force to zero:

$$y(t) = u_y \cdot \cos(\omega t) \tag{8}$$

An average displacement of the table u_y=0.42 mm can be accepted for dynamic calculations as a vibration amplitude of the board. Dynamic inertial force acting on the support after punching hole [1]:

$$P = m \cdot y'' = -m \cdot u_y \cdot \omega^2 \cdot \cos(\omega t) \tag{9}$$

Maximum inertial force in upward Y- direction is equal to

$$P_{max} = m \cdot u_y \cdot \omega^2 = 97.9 \cdot 0.00042 \cdot 34.51 \cdot 10^6 = 14.19 \cdot 10^5 \text{ N} \tag{10}$$

Cross section of the welds is

$$F_s = (l_1 + l_2) \cdot 2 \cdot a = (155+120) \cdot 2 \cdot 15 = 8250 \text{ mm}^2, \tag{11}$$

where: l_1, l_2 – corresponding length of horizontal and vertical welds

a – thickness of the welds.

Amplitude of the shear stresses in weld is

$$\tau_a = \frac{P_{max}}{F_s} = \frac{14.19 \cdot 10^5}{8250} = 171.99 \text{ MPa} . \tag{12}$$

This formula for shear stresses in weld is valid while the board is moving up under inertial loading (reaction forces of the support are not acting in this moment).

In fact, maximum dynamic stresses may be lower because of damping forces in actuators of mounting system of the plates (in which holes are perforated). If the mounting system is working correctly with a delay of 1 to 2 seconds, pressing the plate with force Q=10 T, then displacement u_y and τ_a can be diminished. Even if nominal stresses were 50 % lower than calculated, they would still be dangerous and could damage the weld.

It has been shown that due to vibrations in welds tensile stresses can occur. As a result of vibration there is full cycle of fatigue in welds, which can be measured strain gauge method.
Vibration amplitude as well as the table acceleration can be measured using piezoelectric sensors. Presented method of phenomenon of modeling fatigue stresses in welds is the base of conclusion.

Conclusions

Damage of the press was due to:

a. The welds connecting press table with the support were damaged because of fatigue,
b. Initiation of weld crack began from inner edge of weld, caused by a gap between the board and the support during welding,
c. After punching holes in plates vertical vibrations of the table arise, which causes fatigue stresses in welds,
d. The analyzed punching press welds do not have appropriate strength for assuming structure of welds and performance technology.

References

1. *Norton R. L.,* Machine Design, Impact loading, Pearson Prentice Hall, New Jersey, 2006.
2. *Hagel R., Zakrzewski J.,* Dynamic Measuring (Miernictwo dynamiczne), WNT, Warsaw, 1984.
3. *Deutsch G.*, Practice of the Laplace Transformation (Praktyka przekształcenia Laplace'a), WNT, Warsaw, 1964
4. ANSYS-ver.11- Help Manual (Anshelp.chm), 2008.

Key Engineering Materials Vol. 490 (2012) pp 25-32
© (2012) Trans Tech Publications, Switzerland
doi:10.4028/www.scientific.net/KEM.490.25

Measurement of normal force in magnetorheological and ferrofluid lubricated bearings

J.Salwiński[1, a], W.Horak[1,b]

[1] AGH University of Science and Technology, Faculty of Mechanical Engineering and Robotics, Cracow, Poland

[a] jsalwin@agh.edu.pl, [b]horak@agh.edu.pl

Keywords: magnetorheological fluid, ferrofluid, bearing, normal force.

Abstract. Preliminary analysis of magnetorheological fluid usability in fluid lubricated bearings has been described in the present study. Results of the study aimed at rheological properties of chosen fluids, which possess magnetic properties (both ferrofluids and magnetorheological fluids) with respect to their application in slide bearings have been presented Preliminary analysis of potential advantages related with the magnetic fluid bearing construction was carried out. Results of measurements of normal force developed within magnetorheological fluid and ferrofluid in result of magnetic field action at various shear rate values have been presented.

Introduction

Magnetic fluids are suspensions of ferromagnetic microscopic particles in carrier liquids (usually water or oil). Depending on the particle size, two types of magnetic fluids: ferrofluids (FF) [1], and magnetorheological fluids (MRF) [2] can be distinguished. Unique properties of magnetic fluids result from ability of reversible, almost immediate change of their flow properties under the influence of magnetic field. Particles of ferrofluids contain single magnetic domains, what makes their behaviour considerably different than behaviour of magnetorheological fluids. Moreover, ferrofluids have high sedimentation stability. Because of different behaviour of both substances, they can be used in different technical applications. In case of mechanical devices, ferrofluids are used primarily in seal constructions [3], whereas magnetorheological fluids are applied in vibration dumpers, brakes and clutches [4].

Plain bearings lubricated with magnetic fluids.

Hydrodynamic pressure generated in result of mutual movement of bearing surfaces having suitably shaped geometry is a source of the plain bearing capacity. Capacity of such bearing is a function of geometry, relative velocity of bearing surfaces and lubricant rheological properties (especially viscosity). Bearing geometry and journal rotational speed are parameters forced by the bearing construction and working conditions, but application of magnetic fluid as a lubricant allows effective influence on the bearing characteristics by changing rheological properties of the lubricant. It has been proved [5] that magnetic fluid-lubricated plain bearings may characterize with better capacity and vibration damping then oil-lubricated bearings. It has also been proved that adding magnetic particles into the oil improves its lubrication properties. Kuzhir [6] pays attention to the ferrofluid self-sealing effect and levitation force acting on a rotating shaft, as advantageous feature resulting from application ferrofluids as a lubricant in plain bearings. Moreover, additional capacity resulting from reduction of side lubricant leakages can be obtained. In other study [7], comparative analysis of fast-speed ball bearing used in hard discs driving units and a ferrofluid-lubricated bearing has been executed. Better dumping capacity of ferrofluid-lubricated bearings was proved.

Thrust magnetic fluid-lubricated bearings

Generation of hydrostatic film is one of commonly used solutions for lubrication of the thrust bearing. Flow resistance of the lubricant between bearing sliding surfaces is a source of bearing capacity. The flow can be forced by external pressure source (pump), or in short time intervals, it can result from rapid approach of the bearing mating surfaces. The second solution occurs also in case of the plain bearing lubrication.

In case of pump-forced flow lubrication, the lubricant is pumped under pressure between the bearing operating surfaces, forming a lubricant film. Assuring constant thickness of the bearing oil gap during changed bearing load, or during changed lubricant viscosity resulting from temperature change is one of important problems related with operation of hydrostatic slide bearings. Keeping constant thickness of the lubricant film, i.e. assuring high bearing stiffness, is possible thanks to application of suitable flow resistors [8] and flow regulators in the oil supplying system. Control of conventional hydrostatic bearing operation via changes of the lubricant flow rate is related with necessity of application of complex and exposed to failure systems.

Results of the magnetic fluid-lubricated thrust bearing studies were presented in [9]. The results indicate that application of magnetorheological fluid as a lubricant allows reaching high stiffness independently on the bearing oil gap, what in turn allows reduction of the bearing sliding surfaces tolerance. Feedback system allowing effective control of the bearing operation has also been discussed.

Results of magnetic fluid-lubricated bearing tested without external supplying pump have been presented in the study [10]. Bearing capacity was achieved thanks to the bearing self-sealing effect. This effect is related with keeping the magnetic fluid in position by magnetic field. Suitable bearing surface shape reduction of lubricant leakage via the bearing oil gap.

Magnetic fluid film baring capacity change under influence of magnetic field is a complex issue. In the study [11] the attention was paid to three possible sources of the magnetic fluid lubricated bearing uplift pressure, in particular with magnetorheological fluid: flow resistance resulting from the fluid viscosity generated chain-like structures of magnetic particles within the fluid and the fluid inertia. Moreover, because of the possibility of magnetic fluid apparent viscosity change, the attention should be paid to development of normal stresses within magnetic fluids [12,13], as well as development of the pressure related with influence of magnetic field gradient onto the liquid [9].

Development of normal force in magnetorheological fluids can be described on the microscope level with a model shown in Fig.1.

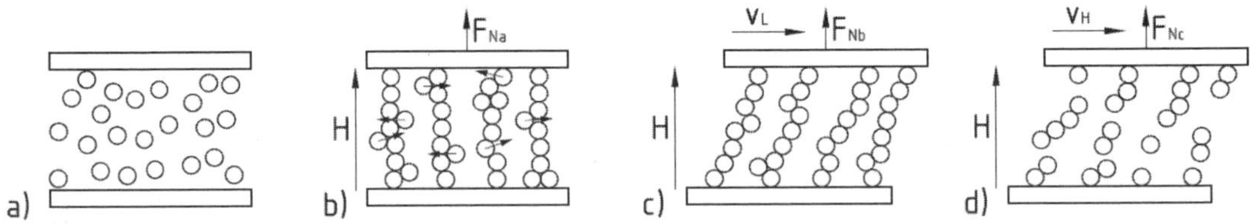

Fig. 1. Model of changes taking place within the structure of the magnetorheological fluid in result of magnetic field and shear rates action, where: H- magnetic field intensity, F_N- normal force (F_{Na}≠F_{Nbc}≠F_{Nc}), v_L- low velocity, v_H- high velocity,

Magnetic particles, which are not under the influence of magnetic field move within the liquid in chaotic manner- Fig.1.a. In result of the magnetic field action, particles of magnetorheological fluid are arranged into chains like structures directed along the magnetic field lines. The developed chains generate forces having direction and turn aligned with direction of the field action- Fig.1.b. If the fluid is exposed to shear, the particles forming chains are deformed- Fig.1.c., and at high shear rates make the chains disrupted- Fig.1.d. This affects the value of the exerted normal force. Phenomenon of the fluid structure destruction and recovery is cyclic and depends both on magnetic field intensity and velocity of the fluid deformation.

Research procedure and results

The present study was aimed at normal force, which is generated within chosen magnetic fluids in result of magnetic field action. The examinations were conducted with use of rheometer MCR 301 manufactured by Anthon Paar equipped with a magnetic field testing cell, allowing generation of magnetic field, which is perpendicular to the tested sample surface. Magnetic field intensity was controlled via electromagnet current changes. Construction of the magneto-rheological device used for examinations is schematically shown in Fig.2.a.

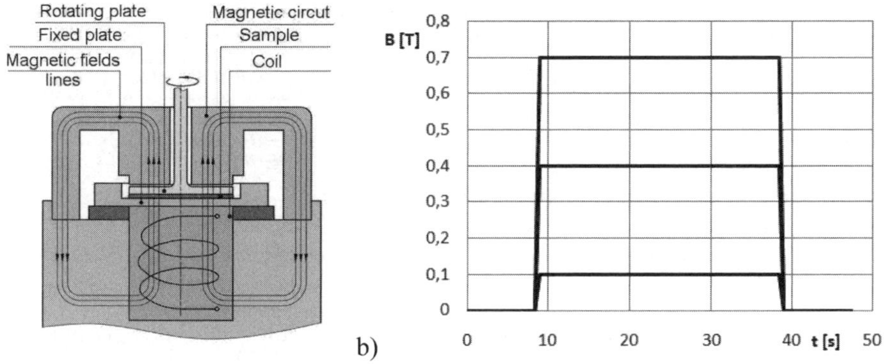

a) b)

Fig. 2. a) Scheme of rheometer measuring geometry, b) changes of magnetic flux density B in function of time t, for step measurement.

Two different magnetic fluids have been tested. Magnetorheological fluid MRF122-ED manufactured by LORD, and ferrofluid FLA040-04 by Ferrolabs. The measurements were made in conditions of thermal stabilization in temperature of 20°C.

All measurements were conducted with measuring plate of diameter 20 mm and constant gap 0,5 mm. In order to assure uniform distribution of the sample on measuring plate surface, each measurement was preceded with 1 minute pre-shear, with shear rate (SR) of 50 s^{-1}, and then 1 minute period of the sample stabilization. All measurements were carried out for three different values of shear rate (10, 100, 1000 s^{-1}) and during stopped measuring plate rotation (SR=0 s^{-1}).

In the first stage, characteristics of viscosity and normal force changes in function of magnetic flux density were determined in order to characterize influence of magnetic field onto the tested fluids.

Results of viscosity and normal force changes obtained for tested ferrofluid, determined for various values of shear rate, are shown in Fig. 3.

Exposing of the tested ferrofluid to magnetic field action resulted in the increase of tested fluid viscosity, whereas viscosity change in result of magnetic flux density action considerably depended on shear rate, at which the measurement was made.

Within tested magnetic induction range, for tested ferrofluids, increase of the normal force up to the value of 2 N was observed, whereas the force value was not dependent on shear rate, at which the measurements were made.

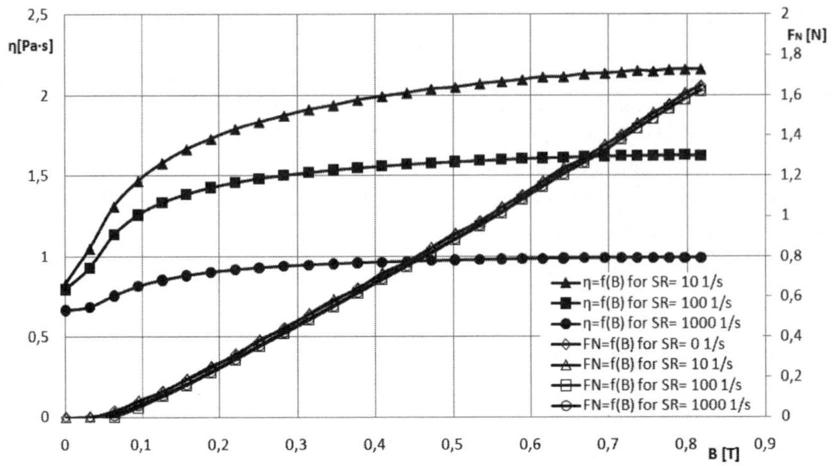

Fig. 3. Changes of viscosity η and normal force F_N of tested ferrofluid in function of magnetic flux density values.

Velocity and normal force changes of the tested magnetorheological fluid determined for various shear rate values are shown in Fig. 4.

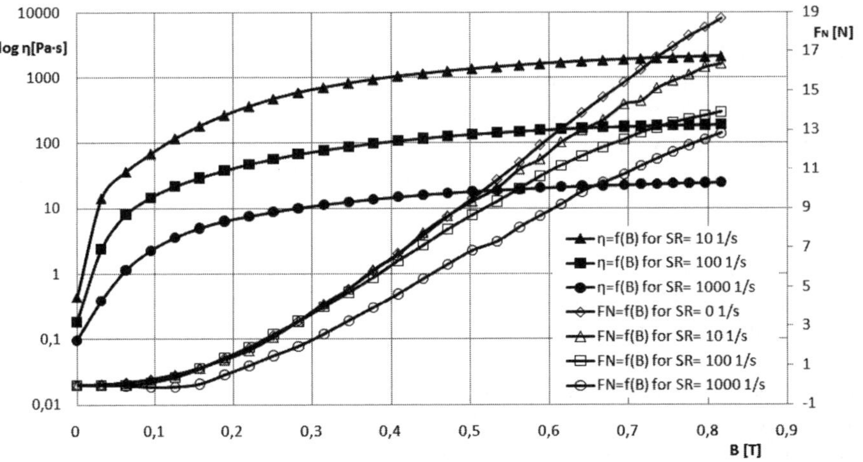

Fig. 4. Viscosity curves (logarithmic scale) of the magnetorheological fluid determined for shear rate 10, 100 and 1000 s^{-1} and diagrams of normal force changes in function of magnetic flux densities for different shear rate values.

The tested fluid revealed properties, which are characteristic for non-Newtonian fluids. Influence of the shear rate onto obtained measurement results is significant. At low shear rate (10 s^{-1}) viscosity of tested fluid reach values of over 2100 Pa·s, whereas the same liquid exposed to shear rate of 1000 s^{-1}, at the same value of magnetic induction reaches viscosity of about 25 Pa·s.

Range of measured normal force, which is acted onto the plate surface also depends on shear rate. Increase of shear rate results in reduction of obtained normal force value.

In the second part of experiments, measurements of magnetic field step action were made. Magnetic flux density changes as function of time are shown in Fig. 2.b. Measurements were made at three steps of magnetic flux density 0-0.1-0; 0-0.4-0 and 0-0.7-0 T. Value of normal force, exerted by the examined fluid onto the measurement plate was measured.

Measured results of normal force obtained for magnetorheological fluid, determined by induction stream 0-0.1-0 T change, determined for various shear rate values are shown in Fig.5.

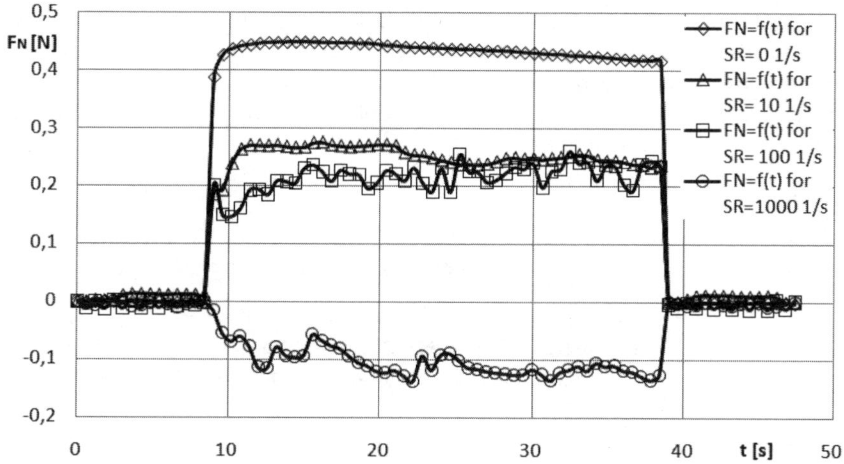

Fig. 5. Measured normal force value - 0.1 T step changes of magnetic field.

The highest normal force value was obtained when rotation of measurement plate was stopped. Decrease of measured normal force was connected with increased shear rates, until drop of normal force value below value measured for a sample, which was not exposed to action of magnetic field (SR=1000 s^{-1}). Normal force value decrease is probably related with destruction of developed chains of magnetic particles, which due to relatively low values of the magnetic field intensity are relatively "weak". Non-uniform courses of measured normal force values accompanying increasing shear rate is characterized with greater scatter of the measured results, what can be interpreted as a result of destruction and rebuilding of developed chains.

Results of measured normal force values obtained for examined magnetorheological fluid, determined for induction change 0-0.4-0 T are shown in Fig.6.

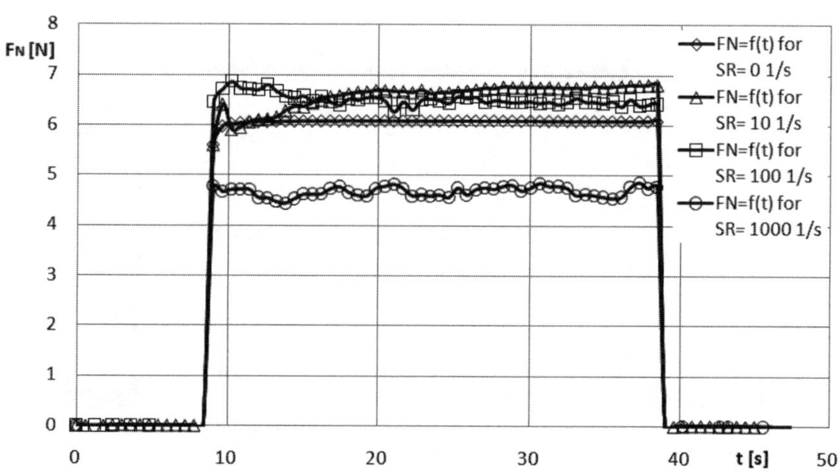

Fig. 6. Measured normal force value for 0.4 T step changes of magnetic field.

In case of a sample exposed to magnetic flux density 0.4 T, the highest normal force value was registered for measurements, in which measurement plate rotation rate was moderate. The highest normal force value was obtained for shear rate of 10 s^{-1}, then 100 s^{-1}, whereas the lowest value occurred at the shear rate of 1000 s^{-1}.

Increase of registered value of the normal force, within low range of shear rates can be explained by the fact that relative movement of magnetorheological fluid boundary surfaces, within range of limited shear rates, conductive to development of chains of magnetic particles. It can also be proved by the fact that faster stabilization of the normal force value in initial phase of magnetic field action was observed for measurement carried out with shear rate of 100 s^{-1}.

Values of normal forces observed for magnetorheological fluid versus changing magnetic flux density range 0-0.7-0 T, are shown in Fig.7

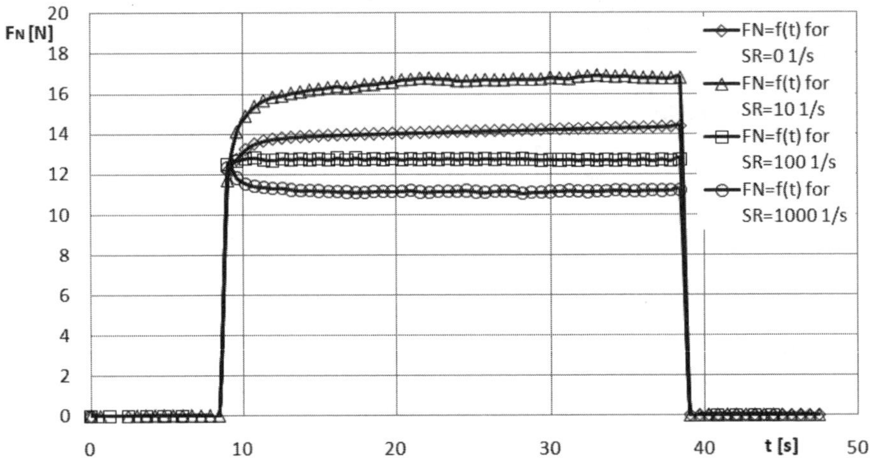

Fig. 7. Diagram showing measured normal force value for 0.7 T step changes of magnetic field.

The highest normal force value was registered at the shear rate of 10 s^{-1}. As was observed in preceded measurements, the lowest values of the force were obtained at shear rate of 1000 s^{-1}.

Cumulated results of normal force measurements obtained for ferrofluid, determined for changing magnetic flux density within ranges 0-0.1-0 T, 0-0.4-0 T, 0-0.7-0 T are shown in Fig.8.

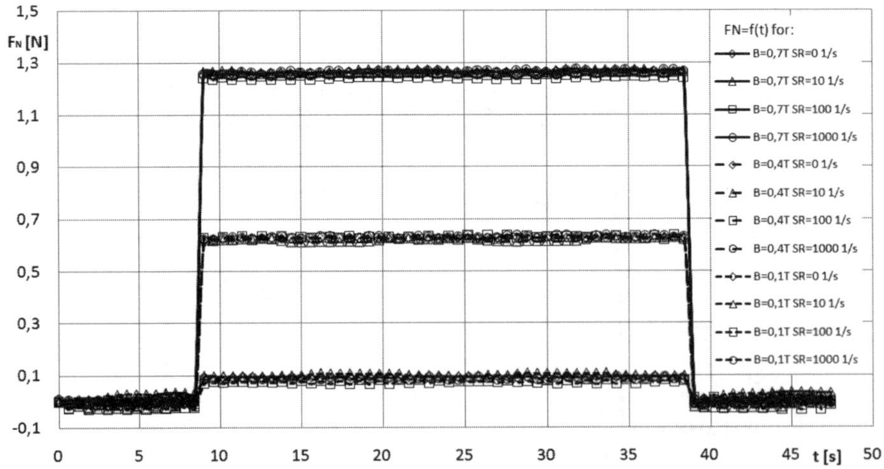

Fig. 8. Values of measured normal force for magnetic field step - value of 0.1T, 0.4T and 0.7 T determined for ferrofluid for various shear rates (0,10,100,1000 s^{-1}).

Values of measured normal force obtained for ferrofluid samples are lower than the corresponding values obtained in the same conditions for magnetorheological fluid. The highest value of the normal force is not higher than 1.3 N. Moreover, no considerable influence of shear rate onto values of measured normal force was observed, what indicates for different nature of the normal force development, which is probably not related with formation of chains of magnetic particles, as was observed in case of magnetorheological fluids.

Summary

Application of magnetic fluids, both ferrofluids and magnetorheological fluids, in slide bearing construction derive number of potential advantages, like: possibility of the bearing operation control via change of lubricant rheological properties influenced by external magnetic field. There is a possibility of construction of easy-implemented feedback loop systems allowing obtaining high stiffness and high accuracies of the bearing operation.

However, number of issues should be taken into account:
- Necessity of selection of bearing and shaft materials having appropriate magnetic properties.
- Need of special pump application in hydrostatic bearings.
- High cost of magnetic fluids.
- Limited durability of magnetic fluids [14,15].
- Sedimentation of magnetic particles (only magnetorheological fluid).
- Possibility of abrasive effect of magnetic particles.
- Possibility of deposition of magnetic particles on the bearing surfaces influencing bearings characteristics.
- Environmental influence of the magnetic field

Bearings lubricated with magnetic fluids have self-sealing abilities what irrespectively on the improvement of bearing capacity resulting from limitation of the lubricant leakage allows simplification of the bearing system construction (elimination of additional sealing elements).

Magnetic fluids allow obtaining better slide bearing capacity thanks to levitation phenomenon, as well as development of additional normal stresses within the fluid (magnetic field gradient action).

Maximum values of additional pressure related with generated normal force for used measurement geometry (parallel plate) amount for 54kPa in case of examined magnetorheological fluid and 4kPa for ferrofluid, respectively. It should be underlined that it is only one of possible sliding bearing capacity sources.

References

[1] S. Odenbach, Ferrofluids-magnetically controlled suspensions, Colloids and Surface, 2003, 217.

[2] L. Vekas, Ferrofluids and Magnetorheological Fluids, Advances in Science and Technology Vol. 54 (2008), 127.

[3] K. Raj, B. Moskowitz, R. Casciari, Advances in ferrofluid technology, Journal of Magnetism and Magnetic Materials,1995, 149.

[4] A. G. Olabi, S. Grunwald, Design and application of magneto-rheological fluid, Materials and Design. 2007, 28.

[5] A. Miszczak, Analysis of hydrodynamic lubrication of journal bearings, Gdynia, Foundation for the Development of the Gdynia Maritime University, 2006.

[6] P. Kuzhir, Free boundary of lubricant film in ferrofluid journal bearings, Tribology International, 4, 2008, 41.

[7] M. Miwa, H. Harita, R. Kaneko, H. Ishizaki, Frequency characteristics of stiffness and damping effect of a ferrofluid bearing, Wear, 2003, 254.

[8] F. T. Barwell, Bearing systems, Oxforrd University Press, 1979.

[9] J. M. Guldbakke, J. Hesselbach, Developmnent of bearings and a damper based on magnetically controllable fluids, Journal of Physics, Volume 18, Issue 38, 2959, 2006.

[10] K. Nagayaa, S.Takedaa, A. Satoa, A. Ikaia, H. Sekiguchia, N.Saitoa. Thrust bearing using a magnetic fluid lubricant under magnetic fields, Tribology International 1, 1993, vol. 26.

[11] A. Farjoud, R. Cavey, M. Ahmadian, M. Craft. Magneto-rheological fluid behavior in squeeze mode, Smart Materials and Structures, 2009, 9.

[12] M. T. López-López, P. Kuzhir, J. D. G. Durán, G. Bossis, Normal stresses in a shear flow of magnetorheological suspensions: viscoelastic versus Maxwell stresses, Journal of Rheology, 2011, 54.

[13] H. M. Laun, C. Gabriel, G. Schmidt, Primary and secondary normal stress differences of a magnetorheological fluid (MRF) up to magnetic flux densities of 1 T, Journal of Non-Newtonian Fluid Mechanic, 2008, 148.

[14] J. C. Ulicny, C. A. Hayden, P. M. Hanley, D. F. Eckel, Magnetorheological fluid durability test-Organics analysis. Materials Science and Engineering, 2007, 464.

[15] J. C. Ulicny, M. P. Balogh, N. M. Potter,R. A. Waldo, Magnetorheological fluid durability test-Iron analysis. Materials Science and Engineering, 2007, 4443.

Key Engineering Materials Vol. 490 (2012) pp 33-44
© (2012) Trans Tech Publications, Switzerland
doi:10.4028/www.scientific.net/KEM.490.33

Compensation of pressure peaks in PWK-type hydraulic pumps

Leszek Osiecki[1, a], Piotr Patrosz[1,b], Tomasz Zawistowski[2,c],
Bettina Landvogt[3,d], Janusz Piechna[4,e], Bartek Żyliński[5,f]

[1] Gdansk University of Technology, ul. Narutowicza 11/12, 80-233 Gdansk, Poland

[2] Space Research Center of the Polish Academy of Sciences ul. Bartycka 18A, 00-716 Warszawa, Poland

[3] Fraunhofer Institute for Algorithms and Scientific Computing SCAI, Schloss Birlinghoven 53754 Sankt Augustin, Germany

[4] Warsaw University of Technology, Pl. Politechniki 1, 00-661 Warszawa, Poland
[5]Bandak Engineering, Postboks 1505 Ålesund, Norway

[a] losiecki@pg.gda.pl, [b] piotr.patrosz@pg.gda.pl, [c] tomasz.zawistowski@cbk.waw.pl,
[d] bettina.landvogt@scai.fraunhofer.de, [e] jpie@mail.pw.edu.pl, [f] bartek.zylinski@bandakgroup.com

Keywords: Hydraulic pumps, axial pumps, pumps research, FEA analysis, CFD analysis

Abstract. Hydraulic axial pumps equipped with cam-driven commutation unit (PWK pumps) proved their high efficiency up to 55 MPa and ability to work self-sucking, even at high speed. Displacement of PWK pump may easily be changed by moving its control cam. Full discharge from hydrostatic forces makes such control possible by direct action of a low-energy actuator like proportional electromagnet or stepping motor. That eliminates heavy and costly hydraulic servomechanism, necessary in other variable displacement machines. Such a control mechanism was positively evaluated in prototype tests in Department of Hydraulics and Pneumatics of Gdańsk University of Technology. The new commutation unit generates however harmful pressure peaks. The paper presents the compensation method for pressure peaks by use of the elastic compensation chamber. It shows also results of multidisciplinary FEA and CFD analysis necessary to properly design the compensation chamber.

Introduction

Axial pumps with cam-driven commutation unit (PWK pumps) were developed as a result of a research project conducted in the Department of Hydraulics and Pneumatics of the Gdansk University of Technology. The new design was granted both Polish and European patents.

Fig. 1: Constant displacement pumps PWK-78 (upper) and PWK-27.

Currently produced constant displacement PWK pumps (Fig. 1) showed good performance like: working pressure up to 55 MPa, overall efficiency - 94%, torque efficiency exceeding 98% and high power density.

The most important, strategic goal however is the creation of the variable displacement version of the new pump. According to theoretical analysis [1], thanks to the full discharge of it's commutation mechanism from hydrostatic forces, the variable pump can be directly controlled by low-energy actuator like stepping motor or proportional electromagnet. It means that costly and complicated hydraulic servomechanism (applied in all present variable displacement pumps) is no longer necessary. It allows to reduce pump's cost and dimensions and to simplify the whole circuit, thanks to elimination of auxiliary pump required to supply the servomechanism.

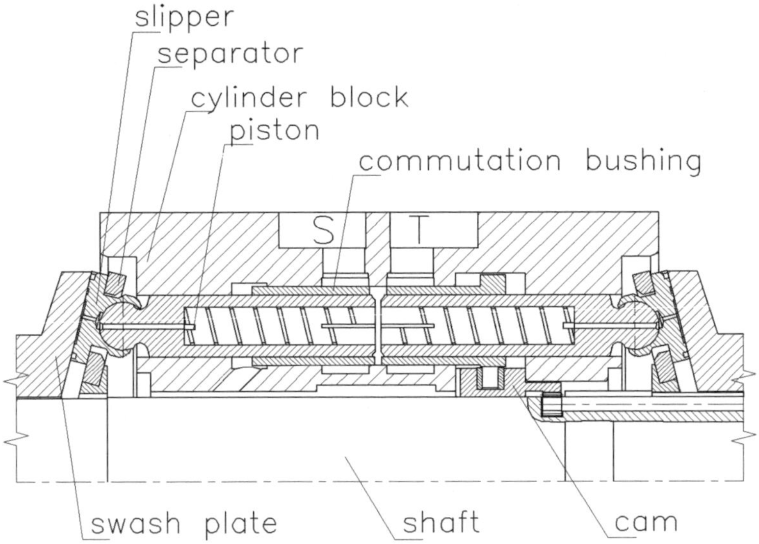

Fig. 2: Main elements of PWK pump.

Displacement control of the PWK pump requires changing the angular position of the control cam in relation to the pump's shaft (Fig. 2). The main difficulty is to assure precise control despite fast rotation of both elements. To make it possible a special planetary gearbox was created (Fig. 3). Such a system allows precise control of the pump's delivery within the $-Q_{max} \div +Q_{max}$ range. The mechanism is directly driven by the stepping motor.

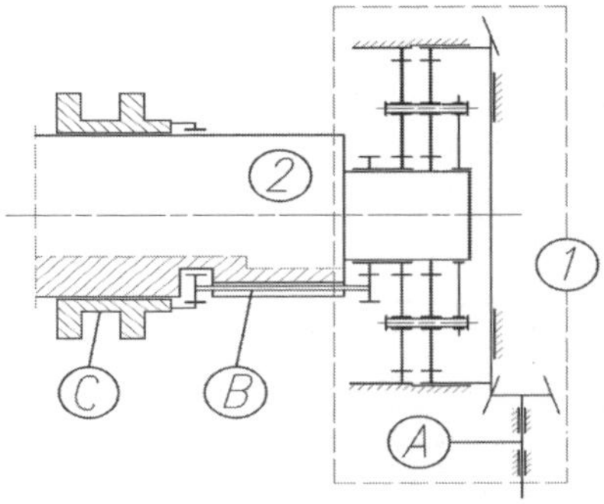

Fig. 3. Displacement control mechanism driven by stepping motor. 1. gearbox 2. shaft A. step motor's shaft B. control gear C. cam

Pressure peaks

Tests of PWK variable pump equipped with the new mechanism confirmed the possibility of such a control, but some harmful dynamic phenomena (pressure peaks in cylinder chamber) were observed (Fig. 6-9). Those peaks are caused by temporary disconnection of the cylinder chamber from both inlet and outlet channels.

Amplitude of pressure peaks can be affected by many different factors (Fig. 4). The most important of them are displacement adjustment, rotational speed of the pump's shaft and leakage. In tests no significant pressure peaks were observed within the cylinder chamber when pump capacity was maximum (Fig.5). Additionally peaks value is increasing with the decrease of pumps displacement (Fig.6-7). It occurs because with the decreasing displacement, the velocity of the pistons at the moment of chamber disconnection is increasing. This causes greater compression of the fluid and rapid pressure buildup. In the fig.6-7 it is visible, that the value of the pressure peaks is also dependent on the direction of the control cam rotation during displacement adjustments. The reduction of pump's capacity can be realized by turning the cam; turning it clockwise will cause higher pressure peaks than counterclockwise.. Also the position of the peaks will be affected, because, depending on the cam turning direction, they will appear at the beginning or at the end of the pumping phase.

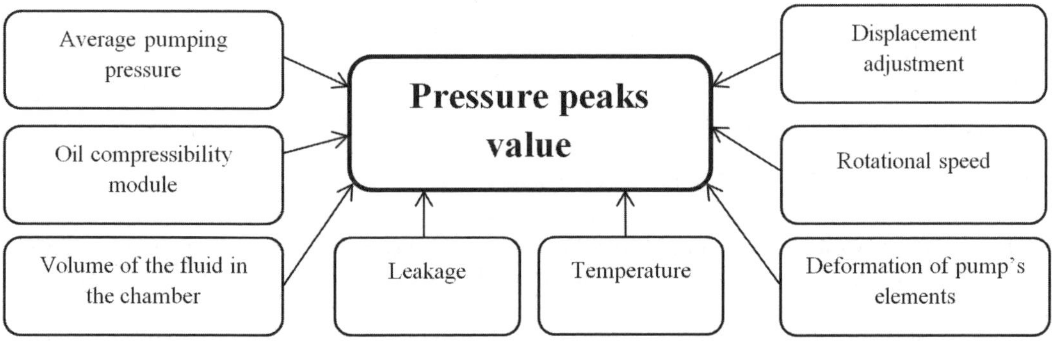

Fig. 4. Factors influencing pressure peaks value.

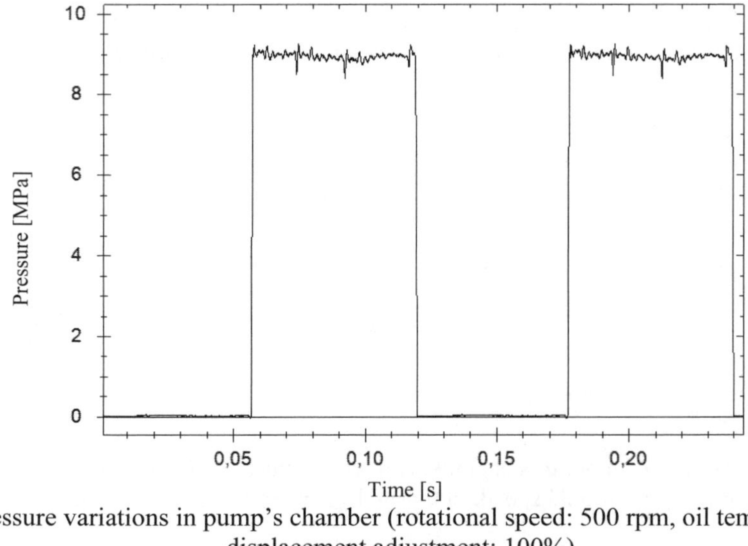

Fig. 5. Pressure variations in pump's chamber (rotational speed: 500 rpm, oil temperature: 33°C, displacement adjustment: 100%).

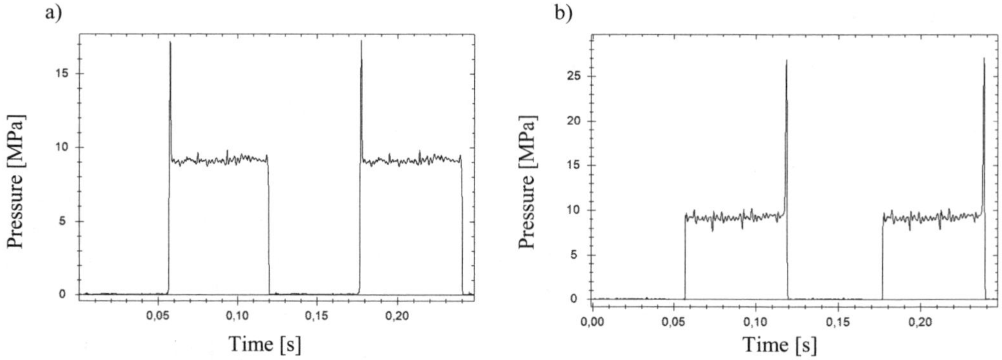

Fig. 6. Pressure variations in pump's chamber (rotational speed: 500 rpm, oil temperature: 33°C, displacement adjustment: 60%): a)counterclockwise cam turn, b)clockwise cam turn.

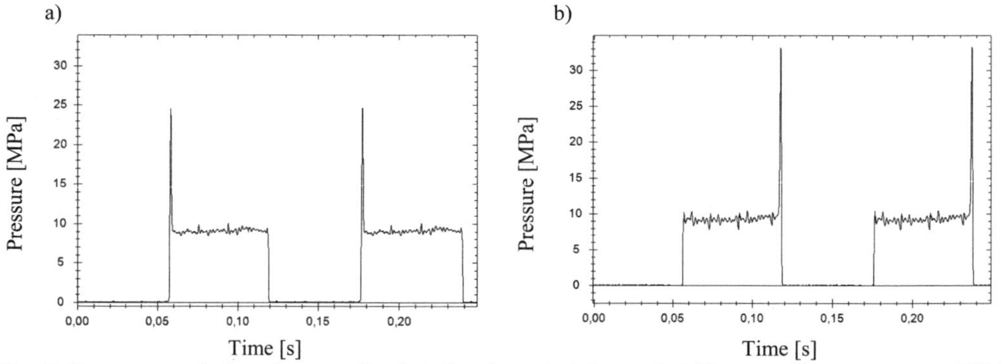

Fig. 7. Pressure variations in pump's chamber (rotational speed: 500 rpm, oil temperature: 33°C, displacement adjustment: 20%): a)counterclockwise cam turn, b)clockwise cam turn.

Growth of the rotational speed causes rise of pressure peak value. This phenomenon is inseparably connected with the leakage of oil from the disconnected chamber. The leakage decreases the amount of the fluid to be compressed, however the increase of pump's shaft rotational speed accelerates the movement of pistons. This shortens the process of the compression and reduces the influence of the leakage, which causes the increase of pressure peak value (Fig. 8).

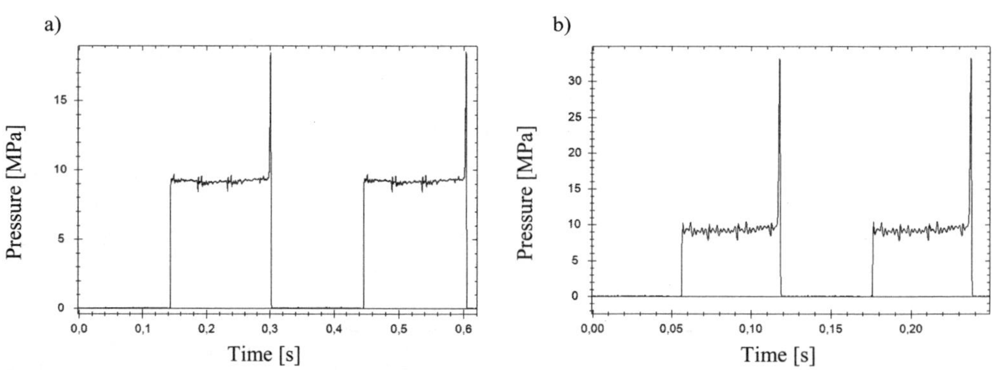

Fig. 8. Pressure variations in pump's chamber (oil temperature: 33°C, displacement adjustment: 20%): a)- rotational speed: 200 rpm, b)- rotational speed: 500 rpm.

There are several locations in the working pump where leakage can occur. The most important of them are:

- ring gap between commutation bushing and piston,
- hydrostatic bearing under the piston foot,
- gap between the window in the bushing and suction and pumping channels.

The sum of all leaks always decreases the amount of fluid to be compressed and eventually the pressure peak value. The leakage through the gaps is the function of gap dimensions and the pressure. However the pressure affects the gap height. It was calculated that some gaps, like the ring gap between commutation bushing and piston, can double their height with the pressure increase. This intensifies the leakage, which affects the pressure value, leading to a complicated feedback, very hard to calculate using non-numerical methods.

The leakage is significantly affected by oil temperature. The rising temperature decreases oil's viscosity, which increases the amount of leaking fluid and decreases pressure peaks during chamber disconnection (Fig. 9).

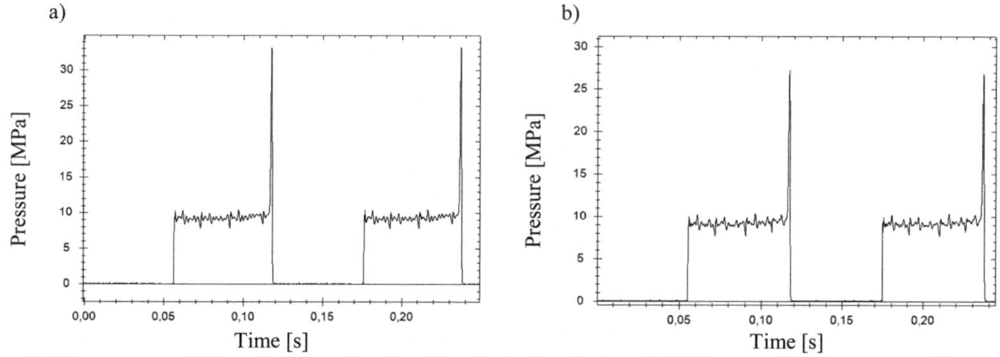

Fig. 9. Pressure variations in pump's chamber (rotational speed: 500 rpm, displacement adjustment: 20%): a) oil temperature: 33°C, b) oil temperature: 58°C.

The pressure peaks value can be estimated using the following equation [2]:

$$p = p_o + K\frac{\Delta V}{V} \quad (1)$$

p – pressure peak value,
p_o – average pumping pressure,
K – compressibility modulus,
ΔV – real fluid compression,
V – initial volume of the disconnected chamber.

The equation (1) indicates that the compressibility modulus is an important factor affecting pressure peak value. The modulus is dependent on the temperature, viscosity, pressure and the content of the air dissolved in oil [3] and can vary from 1000 up to 2500 MPa. The other important factor, not mentioned before is the initial volume of the fluid in the disconnected chamber. It is in the position of denominator in the equation so the greater it will be, the lower pressure peak value will be expected.

To include the leakage in the equation (1), it is necessary to subtract the volume of the fluid, which leaked from the disconnected chamber, from the theoretical fluid compression. The theoretical fluid compression is the geometrical change of volume during chamber disconnection. The equation (2) represents the real value of fluid compression after the subtraction of leakage.

$$\Delta V = \Delta V_t - (Q_{HB} + Q_{SP} + Q_W) \cdot t_d \quad (2)$$

ΔV_t – theoretical fluid compression
Q_{HB} – leakage flow in hydrostatic bearing
Q_{SP} – leakage flow in commutation bushing -piston gap
Q_W – leakage flow in window-channels gap
t_d – chamber disconnection time

Pressure peak compensation

Peak amplitude exceeds 20 MPa and must be compensated to avoid pump's damage and high noise emission. To solve the problem an elastic compensation chamber was used. For the moment of disconnection from pump's main channels the cylinder chamber is momentarily connected to elastic chamber which compensates pressure peak (Fig. 10). Thanks to alternating connection one compensation chamber may reduce peaks in all cylinder chambers of the pump.

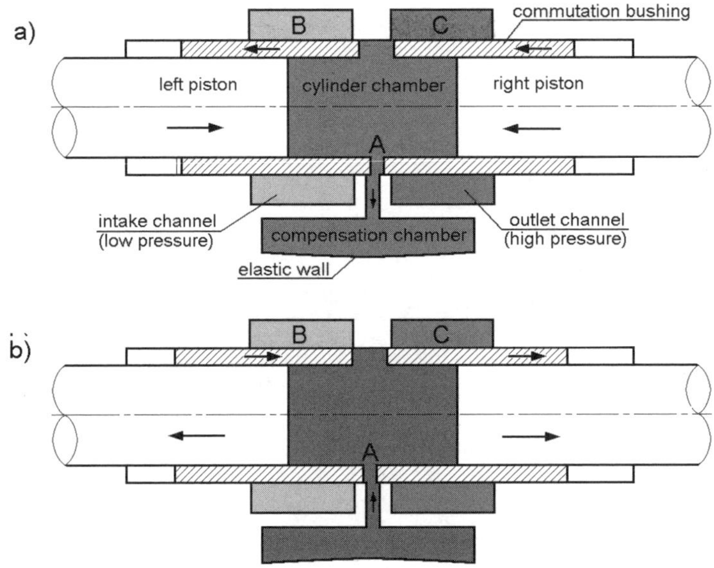

Fig. 10. Pressure peak compensation(a) pumping phase b) suction phase).

Figure 10 shows two cases of pressure compensation. In case a) cylinder chamber volume is decreasing while commutation bushing disconnects it from high pressure channel. Compensation chamber, connected at the same moment to the cylinder chamber, absorbs the surplus fluid and reduces the amplitude of pressure peak. Case b) shows opposite situation: fluid from compensation chamber flows into cylinder chamber, disconnected from low pressure channel at the moment of volume growth.

Preliminary tests show promising results (Fig. 11). The use of the elastic compensation chamber resulted in the decrease of pressure peaks by nearly 50%, not affecting pump's efficiency. The further research will optimize the shape, elasticity and volume of the compensation chamber lowering the pressure peaks.

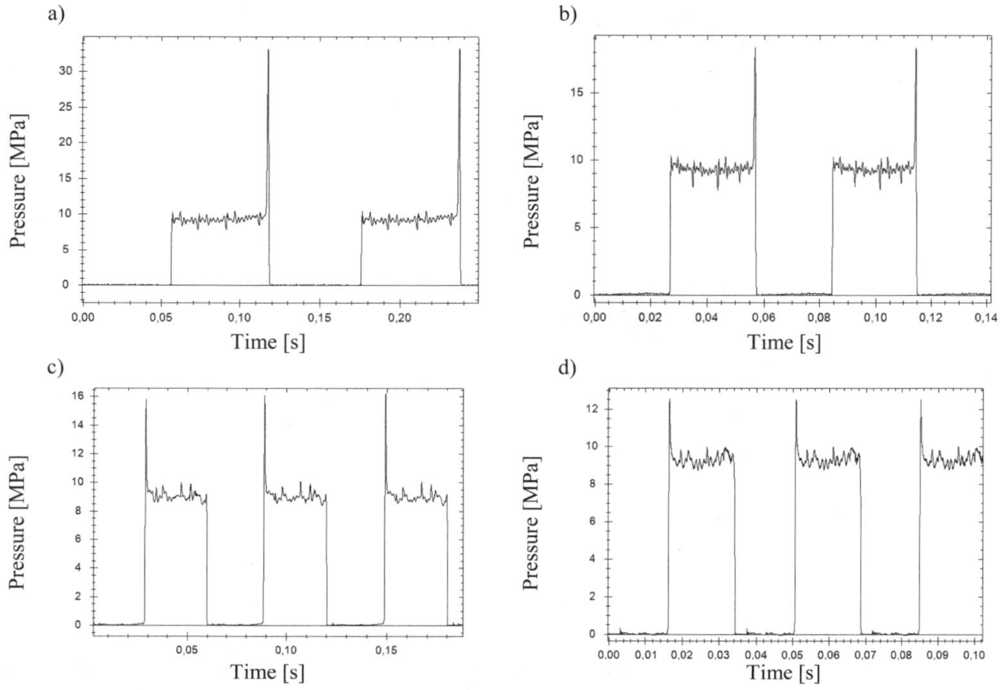

Fig. 11. Pressure variations in pump's chamber (oil temperature - 33°C): a)- rotational speed – 500 rpm, displacement adjustment – 20% ,not compensated, b)- rotational speed – 500 rpm, displacement adjustment – 20%, compensated c)- rotational speed – 1000 rpm, displacement adjustment – 80%,not compensated, d)- rotational speed – 1000 rpm, displacement adjustment – 80%, compensated.

CFD and FEA Simulation

Goals of simulation. Computational Fluid Dynamics (CFD) and Finite Element Analysis (FEA) models were used to facilitate understanding of physical phenomena taking place during the pump cycle operation. They helped to measure parameters not easily available, or impossible to measure. They were useful in visualizing events difficult to grasp with other means. Finally they were used as a valuable tool in analyzing and predicting conditions at a given instant.

The specific task set for simulations was to investigate the sensitivity of pressure in the working chamber on the parameters of the membrane forming the wall of the compensating chamber.

Two important aspects have to be pointed out: the complexity of the model and the interaction between the working fluid and surrounding structure.

Therefore both CFD and FEA models were developed and linked together in order to produce the instrument capable of grasping the coupled phenomena in the high pressure pump.

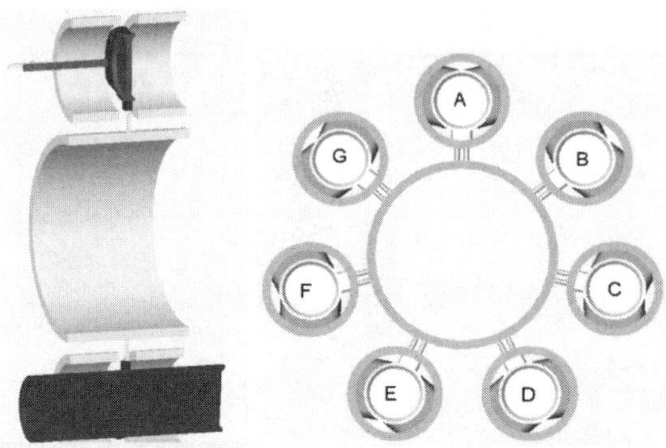

Fig. 12. On the left a simplified CAD model of working fluid features only two out of 7 pump chambers. Components in darker color are moveable. On the right the CAD rendition of the full model is shown. Piston cycles in chambers A through G differ in phases to cover one full revolution of the drive shaft.

Simulation Model Description. The following description shows how the simulation model is being developed and what strategy is assumed to find the correspondence between the membrane behavior and pressure in working chambers.

Owing to the complexity of the pump operation, which comprises dynamic and coupled phenomena, a simplified simulation model was developed. It consists of two working chambers, whose cycles are phased 154.28 degrees away from each other. Additionally a compensation chamber is connected to upper and lower working chambers. For modeling convenience the mesh was set up as if chambers were separated by 180 degrees.

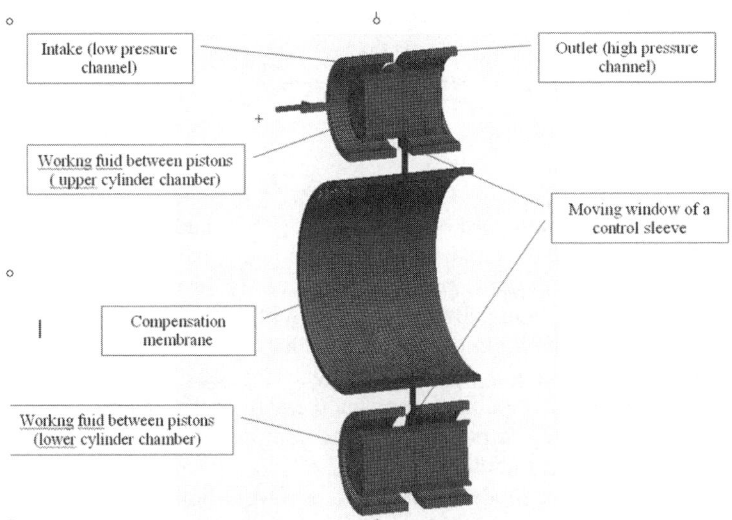

Fig. 13. The CFD model showing all components involved in analysis.

Since fluid volume between chambers changes and a commutating bushing oscillates between high and low pressure zones, so called Moving Dynamic Mesh (MDM) feature was used in the CFD Fluent package to model moving pistons and the commutating bushing. That stipulated imposing motion on moveable zones, as well as controlling cycle phase of opposite chambers. Motion was prescribed to pistons and the bushing via User Defined Function (UDF). UDF was also used to define compressibility of the working fluid by linking oil density to pressure. The pressure-based solver was employed with coupling between oil velocity and pressure.

The CFD model was not equipped with capability to detect cavitation nor was it able to simulate leakages. That is the goal for the next stage of the pump simulator development. In order to save on calculation time, a mirror half of the model defined above was used. The calculation step was set based on the ratio between speed of sound in oil and the length of an element. Fig.13 shows coarse CFD mesh with all significant components of the model.

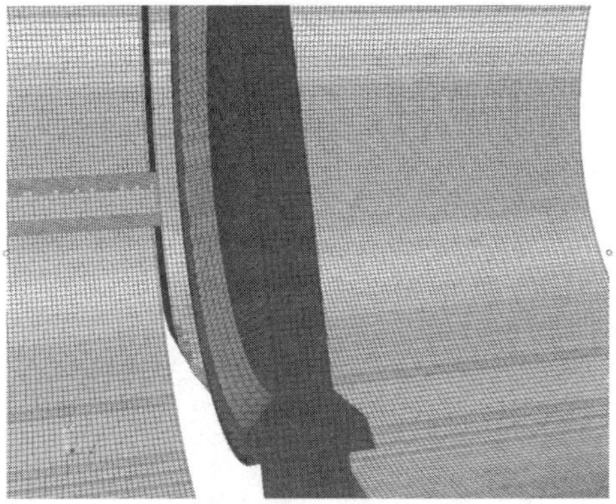

Fig. 14 Refined mesh of the upper working chamber and its related components. Model shown above was created by Fraunhofer Institute SCAI group and was developed in a STAR CCM+ CFD package.

The high pressure setting was fixed at 10 MPa, while the low pressure level was set at 0.2 MPa. More realistic rendering of the CFD mesh shown in Fig.14 was prepared by the SCAI group from Fraunhofer Institute in Bonn. That group is using STAR CCM+ CFD package linked by the Fraunhofer-made MPCCI interfacing module with FEA software package, ABAQUS.

Simulation tasks were defined depending on the configuration of tested objects. There were two major variations in test articles: in the number and size of relief passages and the thickness of the membrane, which impacts the volume of the compensation chamber.

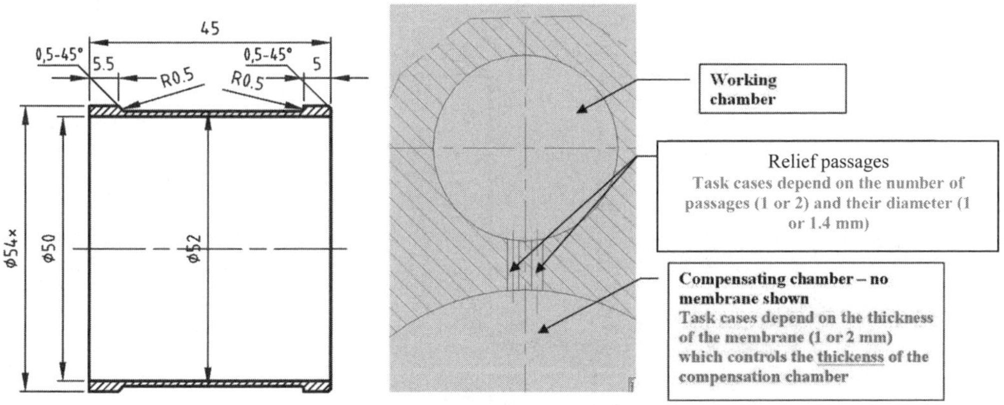

Fig. 15. Task configuration definition depends on the parameters of the relief passages and the membrane. Shown on the left: one of membrane designs used in the model

Fig.16 shows the measured pressure variations of the first analyzed configuration, consisting of a pump with one relief passage of 1 mm diameter. The pump output was set at 100 % capacity with the speed of the drive shaft at 1500 rpm. The implementation of the compensation chamber whose main task was to eliminate large dynamic pressure peaks [4], was said to introduce oscillations encircled in Fig.16. Such distinct oscillations of pressure in the upper chamber are missing from the pressure plot shown in Fig.17 and derived from simulation results. The main reason is the lack of an FSI model in that run. CFD model alone was used to calculate pressure and it managed to recover just one peak at the beginning of the high pressure period (area A in Fig.6). Since the membrane is treated as a rigid body in the CFD package, it precludes the insight into more accurate pressure read out. Another notable detail shown in Fig.17 is a drop of pressure below 0 MPa. Those negative peaks are attributed to cavitation which is likely to occur in small passages. As mentioned above there is a resolve to equip the model with cavitation detection capability.

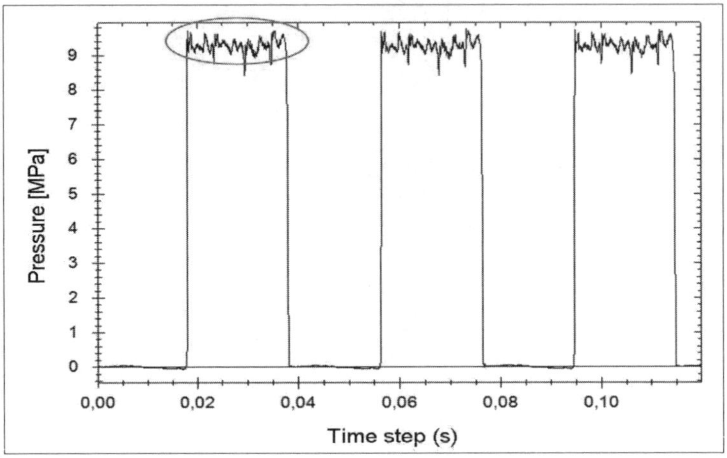

Fig.16. Pressure variations in pump's chamber (rotational speed: 1500 rpm, oil temperature: 33°C, displacement adjustment: 100%). The area encircled in red depicts pressure oscillations attributed to the introduction of the compensation chamber

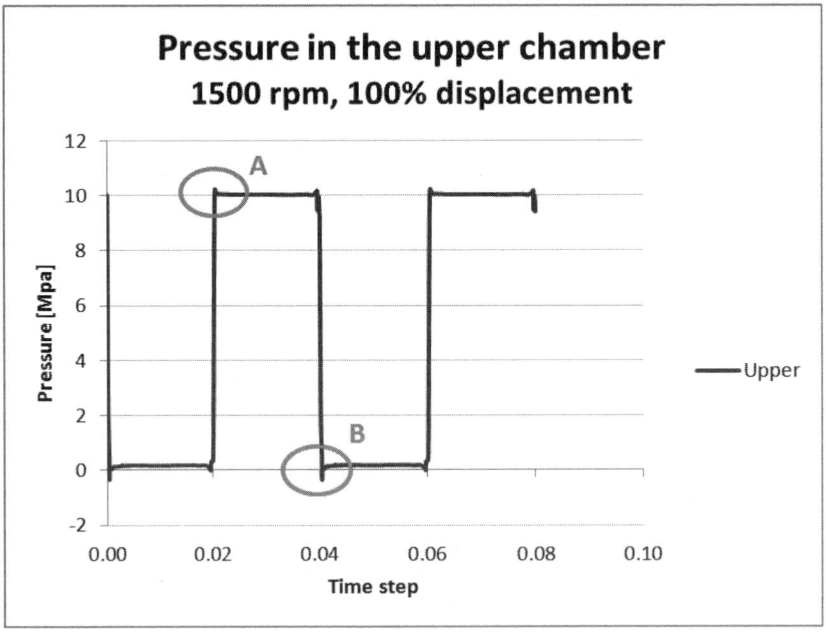

Fig.17. Plot of pressure generated from simulation results. Peak in area A is attributed to the influence of the compensation chamber, peak in area B is linked to cavitation

The assumption was made that the fluid is viscid and compressible. For fluid like hydraulic oil it was necessary to prepare a user-defined routine to calculate density dependent on pressure and the speed of sound dependent on pressure and density variations. Neglecting those dependencies would imply infinite speed of wave propagation and thereby the model inadequacy to replicate phenomena such as rapid change of pressure.
Since:

$$B = dp \,/\, (d\rho \,/\, \rho). \tag{3}$$

Where:
B = Bulk Modulus
dρ = differential change in density of the object
ρ = initial density of the object

Then

$$d\rho = dp \; \rho \,/B. \tag{4}$$

The speed of sound can be calculated from the formula:

$$c = (B \,/\, \rho)^{1/2}. \tag{5}$$

Where:
B = bulk modulus elasticity (Pa)
ρ = initial density of the object

Turbulence was included in the solution and turbulent flow was calculated using the Spalart-Almaras model.

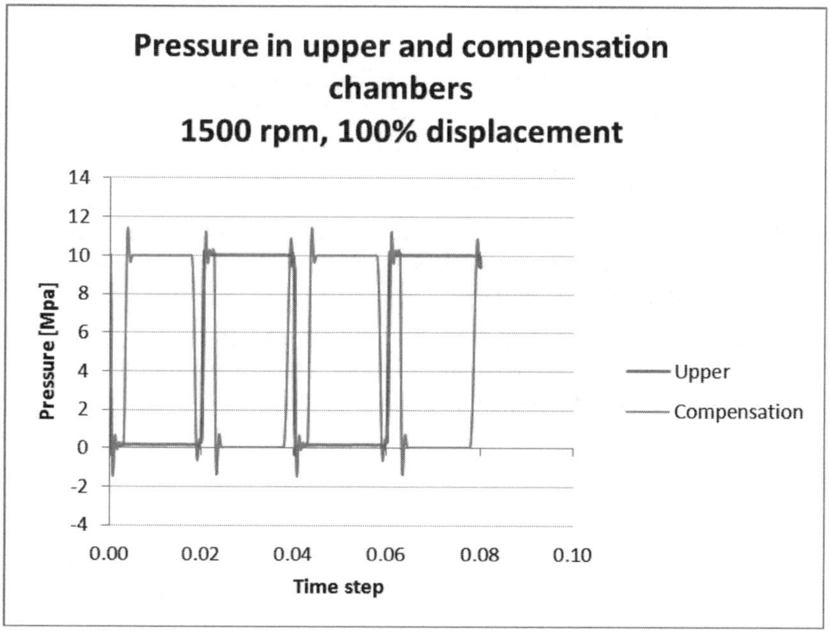

Fig. 18. Superposition of pressure plots from upper and compensation chambers

Fig. 18 shows the superposition of pressure plots related to two distinct points located in the upper chamber and the compensation chamber. Pressure peaks in the membrane seem to be dominating in the pressure switch section of the cycle.

Summary. Prototype tests of the PWK pump equipped with compensation chamber proved it's ability to reduce pressure peaks in wide range of working parameters. To properly design the final version of the compensation chamber proper values of it's volume and elasticity together with shape of relief passage are required. Finding those values in experimental way is time-consuming and expensive. Numerical analysis of pressure variations in pump chambers requires so called Fluid Structure Interaction (FSI) model, which combines specialized codes (CFD and FEA) to get a solution and links them together with an interface (like MPCCI from Fraunhofer). The interface serves as an intermediary through which crucial parameters (like nodal forces) are passed from one code (e.g., FLUENT) to the other code (e.g. ANSYS), and nodal displacements are sent back from ANSYS through MPCCI to FLUENT. The latter method has been employed in the project described herein and is so called a weak coupling. There are two ways to solve the weak coupling problem: staggered – when all solutions are done sequentially or parallel, when both codes are operating simultaneously. Both approaches will be tried.

Reference [5] describes how FSI was used to derive displacement and stresses on the membrane which forms the outside wall of the compensation chamber.

The current status of the simulation project is as follows: FEA models have been completed CFD models are being verified. When that is accomplished the FSI model will be available and tested on all configuration variations of the pump. Results of the FSI simulation will be implemented in the final prototype of the PWK variable displacement pump and verified in laboratory.

References

[1] Osiecki L. Commutation mechanisms of hydraulic axial machines. Monograph, Gdansk University of Technology, 2006.

[2] Osiecki L. Patrosz P. Analysis of dynamic phenomena in the cylinder chamber of the PWK variable displacement pump, in: Klich A., Kozieł A., Palczak E. (Eds.), Badanie, konstrukcja wytwarzanie i eksploatacja układów hydraulicznych, KOMAG, Gliwice, 2010, pp. 185-196.

[3] Osiecki A. Hydrostatic drive of machines, second ed., WNT; Warsaw, 2004.

[4] Osiecki L., Piechna J., Zawistowski T.: Computer analysis of dynamic phenomena in PWK-type variable pumps. Proceedings of International Scientific-Technical Conference, HYDRAULICS AND PNEUMATICS'2009, Wrocław, Poland.

[5] Osiecki L., Piechna J., Zawistowski T, Zylinski B.: Multidisciplinary Approach to the Component Design of a New Axial Hydraulic Pump. Proceedings of the 1st Conference on Multiphysics Simulation, June 2010, Bonn, Germany.

Key Engineering Materials Vol. 490 (2012) pp 45-53

doi:10.4028/www.scientific.net/KEM.490.45

Construction analysis of mini-generator of electric energy using working medium flow as the working environment

Roman Kaczynski[1, a], Izabela Wilczewska[1,b] Boguslaw Hoscilo[1,c]

[1]Manufacturing Engineering Department, Faculty of Mechanical Engineering, Technical University in Bialystok, Bialystok, Poland

[a]rkgraf@pb.edu.pl, [b]i.wilczewska@pb.edu.pl, [c]b.hoscilo@pb.edu.pl

Keywords: mini-generator, electric energy, polymer bearings.

Abstract. Monitoring systems for water distribution and sewerage networks constitutes an indispensable condition for smooth operational management of existing communal infrastructure. The question of an effective use of resources and means to supervise, regulate and control water supply networks requires special care already in the planning stage and also during the implementation process of maintenance equipment. To meet the above demand telemetry and data system to monitor the performance of water mains has been developed making use of GSM/GPRS data transfer technology. One of the fundamental problems of building the system is how to supply energy to the distributed control and measurement devices. A possible solution to these problems could be, as proposed in this paper, a wireless energy supply system using the measuring medium, i.e. the flowing water for energy transport. To this end, a construction of a mini-hydrodynamic generator, i.e. a device transforming the kinetic energy of water into electrical energy using carbon polymers reinforced with carbon fibers has been made, in which the flow of the operational medium is the working environment.

Introduction

The present article attempts to solve two integral sub-tasks related to the construction of mini-generator of electric energy, namely:

- working out the relationship between comparable tribological pairs which employ plastic components applicable in construction of among others the bearings of a mini-generator of electric energy where the working environment is constituted by the working medium;
- investigating the optimum form and construction of mini-turbine.

Application of synthetic materials in sliding bearings on the base of polymers, give wide possibilities to improve technical and economical characteristics in machines and mechanisms. The advantages of its use are: self-lubricating compositions for working node in the water, high durability of components used in chemically active environments (food, water, sea water, etc.). Friction couples, in which polymeric materials are used, are characterized by: low density, low noise, it has vibration damping properties, has good electrical properties and heat insulation properties.

In the article, tribological researches of metal-non-metal couples were executed, where binary scheme of wear in microarea is realized [1], working in water environment. Qualitative and quantitative dependences of tribological pairs work on base of polymers, were determinated. The results were used during designing of real notes of hydro-generator. The work analysis and selection of proper construction of mini-generator is the next step realized in current subject. Nextly the simulations and verification were done, aimed at determine of influence of shape and dimensions of mini turbine blades on efficiency and operation of the device. Important factors affecting the threshold for starting the generator, generated power, speed and noise have been identified.

The meaning and subject scope

Monitoring systems for water distribution and sewerage networks constitutes an indispensable condition for smooth operational management of existing communal infrastructure. The question of an effective use of resources and means to supervise, regulate, and control water supply networks requires special care already in the planning stage and also during the implementation process of maintenance equipment. Considering the large scale of distribution of water and sewage systems, the problem of simultaneous reading of a large number of measurement results gains special importance. The problem can be resolved by the application of an efficacious method for gathering and processing of the measurement data obtained from a random number of distributed nodes in the network by means of GSM/GPRS technology. One of the essential issues in the design of such a system is supplying power to control and measurement instruments in various locations. Presently, this can be achieved by the following means:

- supplying power directly to the measurement points, which is expensive and requires planning permissions and pre-arrangements;
- equipping measurements devices in batteries of high capacity allowing for several years' operation, which is a convenient but expensive solution. Besides, this solution does not ensure constant access to the devices and data, due to the periodical operation of the devices (the devices enter a power-save mode);
- using replaceable and rechargeable batteries, which requires additional effort.

An alternative for the above-mentioned solutions can be a wireless power supply which makes use of the working medium – that is, of the flowing water. For this purpose, construction of hydro-generator was suggested, a device which transforms kinetic energy of flowing water into electric energy. This would provide continuous power supply allowing the built-in batteries of measurement instruments to charge. Moreover, stored up electric energy can be used for the purpose of running measurements and for periodic data transfers. Essentially, the difficulties arising in construction of such devices are: insufficient data concerning the form and size of mini-turbines used in this kind of generators, as well as lack of information related to the problems of insulation and bearing of slide elements, the choice of friction pairs working in the flowing medium and the evaluation of their durability.

The manufacturers of power mini-generators, that is devices generating electric current from mechanical vibrations of surface, temperature differences, light, etc. have recently noted abrupt increase in sales. Mini-generators are used for powering wireless sensors, the products which in last few years have been developing at an astonishing rate. According to Darnell Group's report, the market of mini-generators might be relatively small thus far, but nonetheless it is characterized by an impressive growth rate averaging to 65% per year. Until the year 2010 about 200 million of these devices are to be released [2,3,4]. Companies such as Advanced Linear Devices (USA), EnOcean (Germany) and Perpetuum (Great Britain) hold the greatest share in the market.

Tribological researches methodology

Optimal friction pair should comply the following assumptions:

- work of pair is realized in water environment (tap water);
- friction coefficient should be the smallest;
- contact pressure range is 2,8÷13,5 [MPa];
- sliding velocity range is 0,01÷1,25 [m/s] (maximum rotating velocity of rotor is 1000 [rev/min])

Experimental researches were carried out with the use of two kinds of friction pairs, which materials, used as samples for investigation, were structurally very similar. The samples, the pin diameter equal 3[mm] (Fig.1a), were made of the following composite materials reinforced with carbon fibers:

- Sigma3, the composite material of Saint Petersburg industrial plants, used for lubricating screw propeller shafts and having the following properties (Table 1):

Table 1

Trade name	Density [g/cm3]	Brinell hardness number	Relative stretchablity[%}	Temperature range[°C]
Sigma-3	1,5	100	14	(-40) – (+140)

- Quadrant Epp Ketron® Peek CA30 Polyetheretherketone, 30% carbon fiber with the following properties (Table 2):

Table 2

Trade name	Density [g/cm3]	Brinell hardness number	Relative stretchability[%}	Temperature range[°C]
Ketron Peek CA30	1,41	325	5	do +250

a) b)

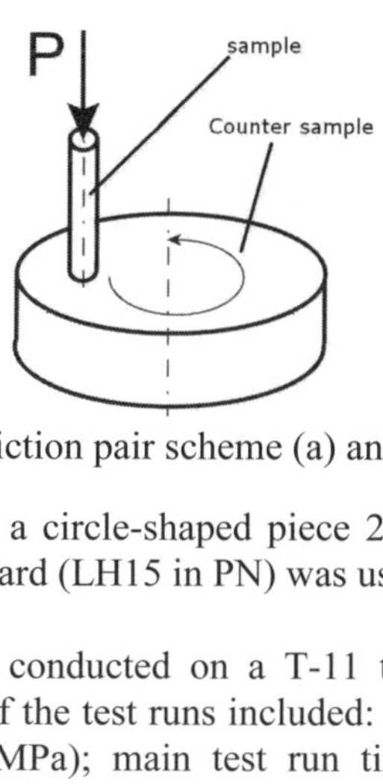

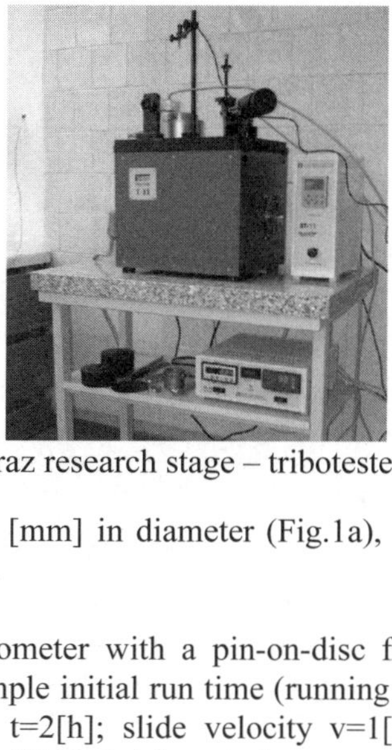

Fig. 1. Friction pair scheme (a) and oraz research stage – tribotester T11 (b)

As a counter-sample, a circle-shaped piece 25,4 [mm] in diameter (Fig.1a), 100Cr6 bearing steel described in EN standard (LH15 in PN) was used.

The researches were conducted on a T-11 tribometer with a pin-on-disc friction (Fig.1b). The constant parameters of the test runs included: sample initial run time (running in) t=10[min] (under the pressure of 0.71MPa); main test run time t=2[h]; slide velocity v=1[m/s]; five values of individual pressure p within the range of 2.8-13.5[MPa]. The whole test procedure was repeated three times for each of the stresses.

Results of tribological researches

The results of experimental analysis were presented in Fig. 2. The graphs present the tribological parameters as a function of contact pressure (2.8; 4.2; 8.5; 11.3 and 13.5 [MPa]) in water environment for the mentioned friction pairs: Sigma3-100Cr6 and Peek CA30-100Cr6.
The microscopic examination and analysis of chemical composition was carried out with the aid of a Hitachi S-3000N Scanning Electron Microscope (equipped with a Quest type EDS X-Ray Microanalysis System). The obtained microscopic pictures of surface layers, as well as the chemical

composition of selected areas of the surface made it possible to carry out an in-depth analysis of friction processes and identification of wear mechanisms occurring in the particular environments (due their large volume the data are not quoted here).

Comparing all the graphs gained through the experimental analysis, it can be noticed that their features for either friction coefficient (Fig.2a), temperature (Fig.2b), or linear wear (Fig.2c) are very similar.

a)

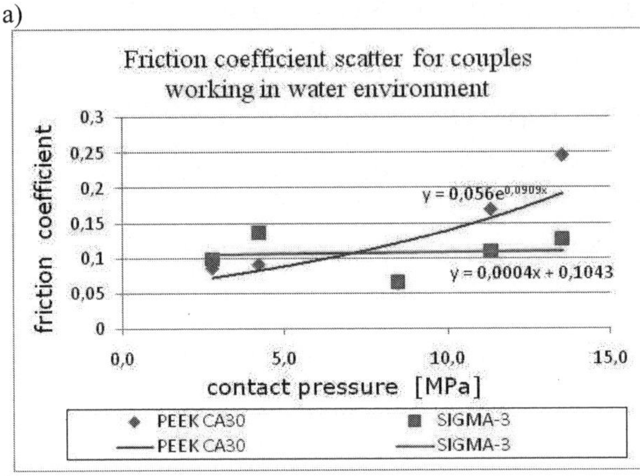

b)

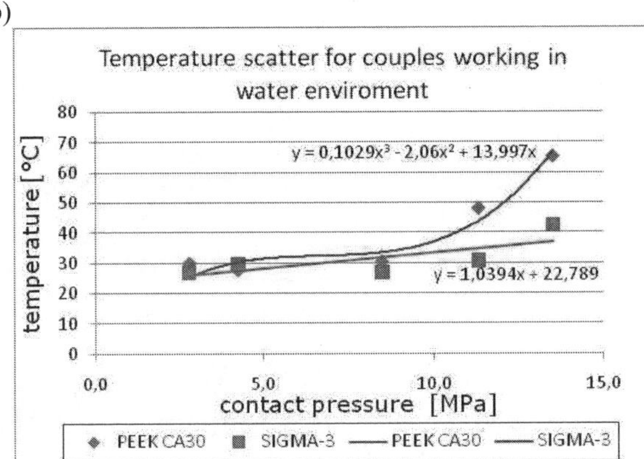

c)

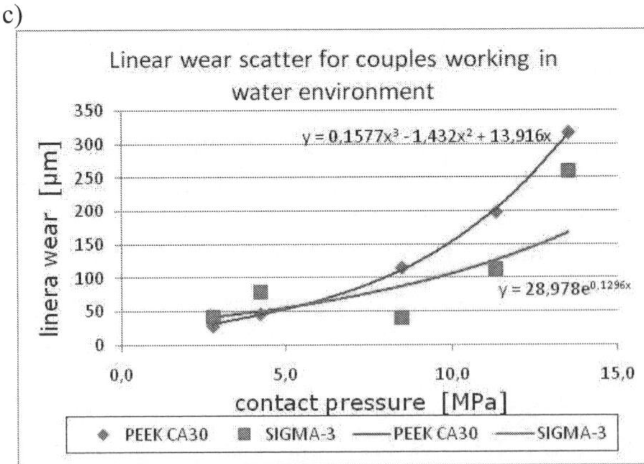

Fig. 2 Friction coefficient (a), temperature (b) and linear wear (c), scatters for friction couples Sigma3-100Cr6 and Peek CA30-100Cr6, in function of contact pressure in water environment [5].

Considering practical application of friction pair and after analysis of gained tribological parameters' values, it is supposed that both friction pairs could be applicated in assumed work conditions: wears of both couples are similar. The friction coefficients' and wears' values of couple Sigma3-100Cr6, are the lowest in the whole range of loads, which suggests using it for the construction of slide elements in the mini-generator.

The test stand for hydrodynamic researches

For further researches a construction of the test stand is needed, for the experimental studies of hydrodynamic generator equipped with a water-driven mini-turbine in the flow channel [6].

The test stand (Fig. 3) consists of a hydraulic system, comprising: a tank of water, two rotary pumps, suction pipe supplying water from the tank to the pump, pressure pipeline with with an integrated straight test section with a diameter of about 80 [mm] and length approximately 3[m], the pipe drain water into the tank.

In a test stand of mini-turbine, built rotary pumps can be connected in parallel or in series with the possibility to work simultaneously with two pumps or only one pump.

In this way, the hydraulic system with pumps in parallel system is able to provide variation in the size of the forced water flow in a very wide range, corresponding to the scope of work of mini-turbine in water system.

However, the series system will provide an opportunity to study mini-turbine in a wide range of water pressure. Water is sucked by the sucking baskets S and then flows by sucking connectors for each pump.

At the exit of each pump, manometers M exist, to measure pressure, while the desired water flow is adjusted by valves Z1, Z4 and Z5.

The volumetric flow intensity will be measured with the rotational flow meter Hydroflow Vortex 1100. Behind flow meter, built-in flow control valve ZR1 is used to regulate the efficiency.

The hydraulic system is equipped with a computerized data acquisition.

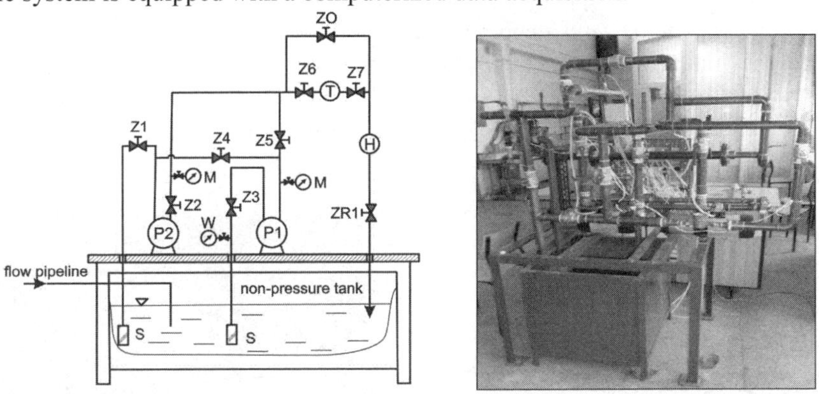

Fig. 3 Hydraulic scheme and a view of the test stand of the generator with mini-turbine: T – mini-turbine with generator, P1, P2 – pumps, W – vacuum gauge, ZO- bypass valve, Z1 - Z7 – cut-off valves, ZR1 – regulation valve, M – manometer, H – liquid flow meter *Hydroflow*.

Range of working parameters' measurements of hydraulic mini-generator with mini-turbine is following:

- the level of the volume flow of water,
- water's temperature and pressure at the supply of tested section,
- rotation speed of mini-turbine,
- start up threshold (minimal volume flow initiating rotation of the rotor),
- current and voltage of generated electricity

Construction of a prototype hydrodynamic mini-generator of electric energy.

The hydro-generators availabled on the market, require substantial flow rates, and their size may prevent or hinder a installation inside the pipeline. Also some difficulties we can find in the methodology and calculation of very small water turbines operating at medium' low speeds. Produced nowadays devices, require stream's velocity more than 1.3 [m/s] (2.5 kn), at a speed of 4 [m/s] (8 kn) they generate power about 100 [W].

On the base of measurements carried out on real water supply systems, it was defined conditions a prototype mini-generator. The analysis indicate that the most advantageous installation will take place on the pipeline with an inside diameter 150 [mm]. The theoretical average kinetic energy of the stream in this case is about 35 [J]. It is also the most commonly used diameter for the installation of measuring devices that communicate in a GSM / GPRS network. The average velocities of flow recorded in the various pipelines were presented in Fig.4.

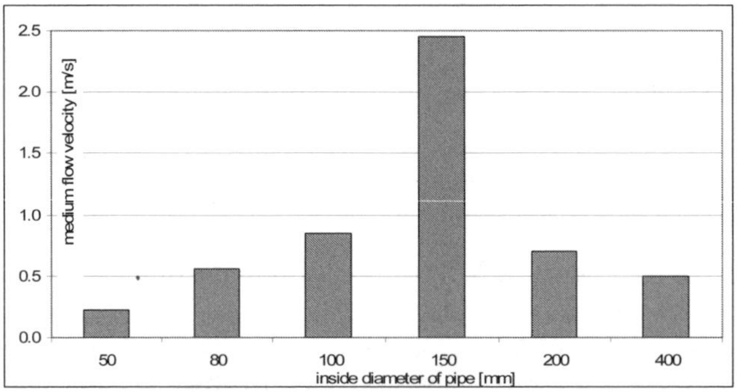

Fig. 4 Average velocity of medium flow depends on diameter of pipe.

On the base of water meter set and stepper motor operating as a generator, the hydro-generator was prepared for the preliminary researches (Fig.5).

The described conception of the project will result in the minimization of efforts related to the construction of the rotor turbine, impeller, and at the same time will allow to focus on the drive transfer system with all the related tribological issues.

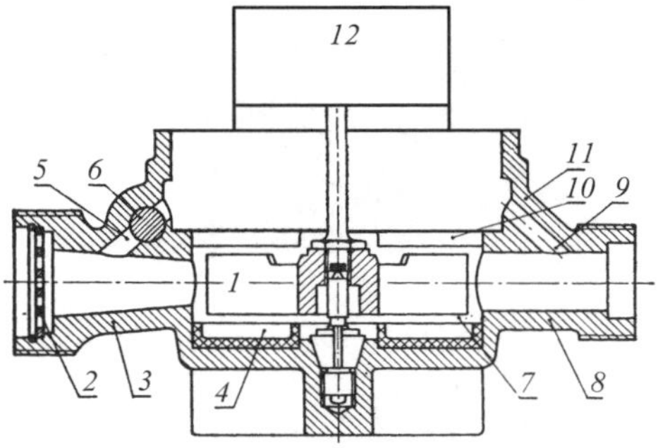

Fig.5 Outline of mini-turbine rotor system with a vane rotor a. vane impeller: 1 – vane rotor ; 2 – filter ; 3 – inflow spout; 4 – swelling plate; 5 – release channel; 6 – control plug; 7 – rotor chamber; 8 – outflow spout; 9 – release channel ; 10 – swelling plate 11 – gear encasement; 12 – direct current generator.

Nextly the optimal way of torque transfer from the turbine to the generator, with simultaneous the generator isolation from the medium was selected. For this purpose, the design of magnetic clutch with the rotor with Nd-Fe-B epoxy bonded magnets was chosen. These magnets can work in water environment.
The rotor with axial location of magnetic elements was proposed as a clutch system. This rotor is prepared for the long-term work in a water environment (in the case of the rotor seal is completely impossible). For the rotor bearings were chosen sliding bearings with the polymer pan. Such a system allows, with little modification, for use directly the rotor as part of an electric machine.

Several approaches being studied, the measurements of obtained electrical values were conducted depending on the speed of the rotor of the stepper motor.
From the observation, the use of multiphase stepper motors as generators of electrical energy, may require application of an electronic system, which adjust received multiphase process to the needs of battery's sets' charging

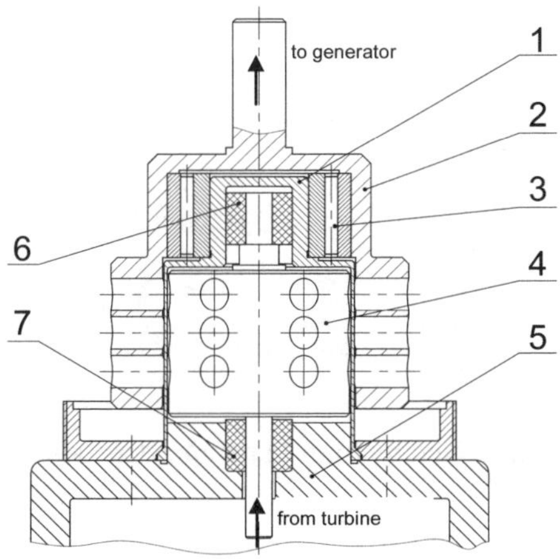

Fig.6 Water proof magnetic clutch: 1–water proof element; 2-dry outside rotor; 3-needle bearing; 4-wet permanent magnet rotor (Nd-Fe-B); 5-turbine enclosure; 6,7-self-lubricating slide bearing.

The role of such a system can fulfill the matrix converter, described e.g. in [7,8]. The test results of prototype of mini-generator's were given in Fig 7 and 8. The results are related to the test flow in the pipeline with a diameter of 12 [mm].

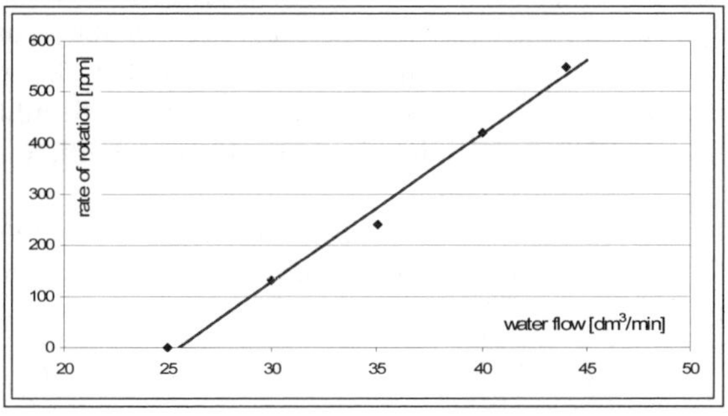

Fig. 7. Rate of rotation of water miniturbine vs water flow.

The minimum value of flow velocity that run tested mini-generator is about 3 [m/s]. The rotation speed of generator's axis and the effective voltage and efficient intensity, were measured, gradually increasing the flow. The obtained results were converted into the value of average power (Fig 8).

As a result of measurements, a maximum average power for the prototype of mini-generator at assumed values of the flow, were obtained. It was found that mini-generator with a flow rate of about 4.5 [m/s], generates power with a value about 5 [W], what corresponds demand of measuring devices.

Efficiency of the turbine-clutch-generator presented in Fig. 5 is about 30%, at a flow rate value equal 4.5 [m/s]. In order to increase efficiency and because of the nature of work and possibility of achieving flow rates not exceeding 2.5 [m/s], turbine geometry similar to the Kaplan turbines (or turbine propeller), using whole cross-section of the pipeline, in subsequent prototypes will be used.

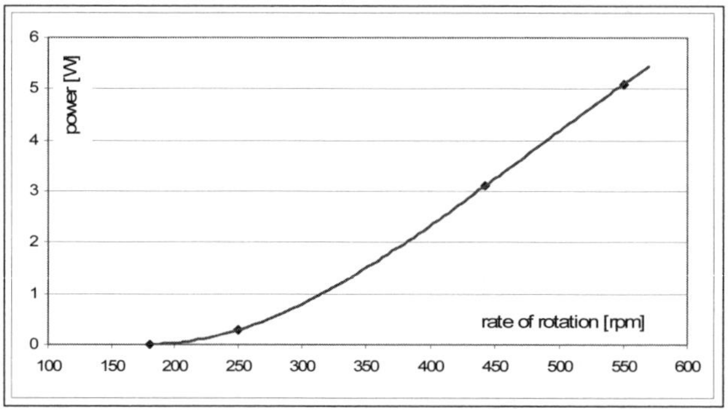

Fig. 8 Electric power generated by mini-generator depends on rotor rate of rotation.

Measurable effects

The construction assuming the position of all devices' unit in one housing and the placement of a whole construction in a string of pipeline, linking it permanently using flange connection, is also considered. The maximum power from the turbine and the smallest losses in the flow, can be obtained using the whole profile of pipe. However, this solution significantly reduces the possibility of use such a solution to a newly built or refurbished networking and relatively is costly and labor-intensive assembly.

Also, it is assumed that the constructed mini-generator will provide sufficient power to charge batteries powering the control and measurement instruments. This device will transform the energy of the medium flowing through a pipeline into electric energy. The essential element of the device is a mini-turbine designed to achieve the maximum efficiency in the broadest possible range of medium flow speed at minimized flow interruptions. Carrying out further research and simulations concerning the influence of geometry and size of turbine's vanes will allow for their rational and effective realization, reducing the costs related to practical execution and analysis of several prototypes with turbines of different geometry

To sum up, a measurable outcome of the present project, should be an economical effect resulting from the construction of a power supply system providing electric energy to distributed control and measurement instruments.

References

[1] M.B. Rubin, W.E. Bakhareva: Bearings in the ship technology, Reference book: Ship building, 1987r., s. 201.

[2] Darnell Group Report: Distributed Generation Growth Is Increasingly Vigorous and Consistent Compared with Past Experience, Darnell Publishes New Report on Power Electronics for Distributed and Cogeneration April 21, 2006

[3] Darnell Group Report: Energy Harvesting & Micro Batteries: Market Forces and Demand Characteristics, Third Edition, November 2009

[4] Pronotex Annual report and accounts 2006

[5] I. Wilczewska, R. Kaczynski, Identification of friction and wear parameters of polymers strengthened by carbon fibres in different environments", Acta Mechanica et Automatica, 2009, vol.3.

[6] R. Kaczynski: Minigenerator of electric energy using working medium flow as the working environment for tribological nodes. Electro-energy News, Vol 3, 2011, s.11

[7] T. Sobczyk: Analysis of voltage transfer by a 9x3 matrix convert with simultaneous frequency changing; Technical magazine 1-E/2008, Technical University in Cracow Publisher, Cracow 2008, p. 79-93.

[8] J. Tadeusz, T.J. Sobczyk, Small thermal power station with higher efficiency, Technical University in Cracow, Institute of Electromechanical Energy Transition, Cracow, 2007

Key Engineering Materials Vol. 490 (2012) pp 54-63
© (2012) Trans Tech Publications, Switzerland
doi:10.4028/www.scientific.net/KEM.490.54

Crack propagation in zirconia MgO-PSZ ceramic material

Waldemar Karaszewski

Technical University of Gdansk, ul. Narutowicza 11/12, 80-233 Gdansk, Poland

walkaras@pg.gda.pl

Keywords: crack propagation, zirconia, rolling contact fatigue

Abstract. The properties of ceramic materials such as elevated hardness, high temperature capability, low coefficient of thermal expansion are of interest for rolling element materials. Widely used ceramic materials in engineering applications are silicon nitride, zirconia and alumina. The paper presents an experimental study of the fatigue life of MgO-PSZ ceramic material in lubricated contact with defined types of cracks. A ceramic angular contact bearing operation was simulated using a four-ball machine. Zirconia balls were artificially damaged with pre-cracks. Ringformed pre-cracks were propagated on the ball surface by a blunt impact load using a tungsten carbide ball under five various impacts. Research of the presented paper shows that fatigue failure under rolling contact loading is markedly sensitive on ringformed pre-cracks made by a different impact energy (drop height). Propagation of surface pre-cracks takes place under a fatigue load. Primary and secondary surface crack propagations are described.

Introduction

High quality ceramic materials have shown advantages for rolling element bearing applications. The main properties of ceramic materials are high stiffness, high hardness, high temperature capability, low coefficient of thermal expansion and a lower density compared with steel. In the 1960s, alumina ceramic bearings were tested [1]. The first information about silicon nitride bearing was presented in the 1970s [2]. Nowadays a lot of research work has been done investigating silicon nitride bearings [3-9,16-18]. In particular, silicon nitride hybrid bearings (silicon nitride ball and steel races) have the ability to withstand high speeds and high loads.

Quality control of the ceramic ball surfaces has reached a satisfactory level. However, surface cracks are found on ceramic balls [10]. Therefore, it is important to conduct research to investigate the influence of these cracks on fatigue life performance. Surface cracks may exist on ceramic balls due to HIP manufacturing pressing faults (porosity, inclusions) or impact during transportation or exploitation (Fig.1, Fig.2). The most common types of surface cracks are cone and ring types caused by Hertzian stress field [2-6]. The surface cracks have a large impact on the fatigue life of ceramic balls under cyclic stress. Propagation of surface cracks under cyclic stress decreases the fatigue life rapidly [4, 5, 7, 8].

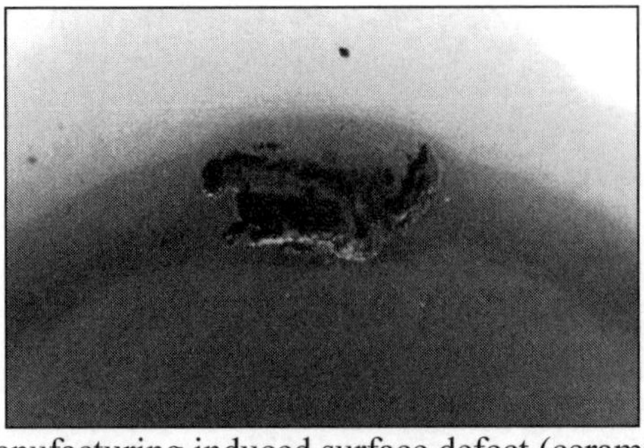

Fig.1. A manufacturing induced surface defect (ceramic zirconia)

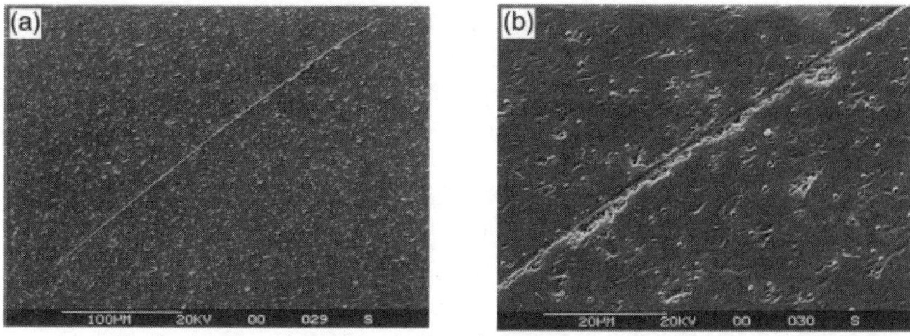

Fig.2. A manufacturing induced surface crack (silicon nitride) [2]

It is estimated that up to 50% of the cost of ceramic bearings may be attributed to the inspection and finishing of the ceramic elements [6]. The best way of cost reduction is the possible use of ceramic balls with defined surface crack defects (the applied load should be below threshold).

Zirconia is the next potential engineering material for bearing applications. There is not much information about crack propagation in zirconia materials [11, 12] and there is less data about failure modes. The main purpose of the research described in this paper is a study of the influence of ringformed pre-cracks made by different impact energy on the fatigue failure of MgO-PSZ. The paper also contains information about some manufacturing faults found in zirconia materials.

Experimental methods

The four-ball machine. A four-ball machine, shown in Fig.3, was used to test ceramic balls. The machine consists of elements which simulate an angular contact of ball-bearing: upper ball (1), lower balls (2), steel cup (3). The assembly is loaded via a piston (5) below the steel cup. The upper zirconia ball is assembled to a drive shaft (4) and contacts three lower balls. Lower balls are silicon nitride or steel rolling elements. Different lower ball materials were chosen to receive different Hertzian stresses. Upper and lower balls are immersed in lubricating oil. The upper ball is the test specimen. Four-ball machines have been used to examine rolling contact by various researchers [3-5, 8].

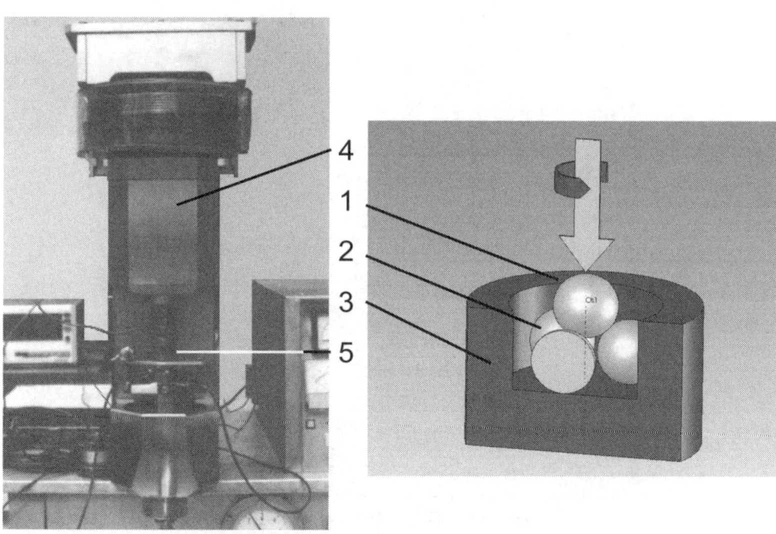

Fig. 3. A four-ball testing machine
(1-upper ball, 2-lower balls, 3-steel cup, 4-spindle/drive shaft, 5-load piston)

Pre-cracked zirconia balls. The tested zirconia balls were manufactured as 12.7mm diameter balls. Table 1 describes properties of MgO-PSZ material used in the research. Zirconia balls were manufactured by Comballs Materials Co. The roughness (R_z) of zirconia balls is 0.08 μm, the ball diameter tolerance is ± 2.5 μm and the ball roundness tolerance is 0.6 μm. Theses data are based on the manufacturer's information.

Table 1. Zirconia ceramic properties (MgO-PSZ)

	MgO-PSZ
Chemistry	10molMgO+ ZrO_2
Compressive strength [MPa]	1900
Max. service temperature [^{O}C]	1100
Young modulus [GPa]	200
Hardness [HV]	1020
Density [g/cm^3]	5,73

The ring cracks were created on the zirconia balls surface by impact loads. This method was chosen to create similar cracks which can be created during falls of the balls or exploitation as an effect of impact load. A zirconia ball was placed in a ball holder and tungsten carbide ball (12.7 mm diameter) was placed on top of the zirconia ball (Fig. 4). The position of the zirconia ball holder was designed to introduce the ring crack in a zirconia ball in place of a 4-ball contact.

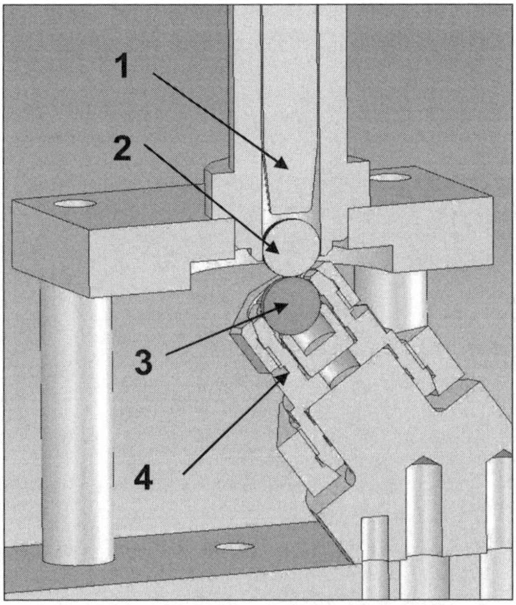

Fig. 4. A ring crack device maker
1 – dropped weight, 2 – tungsten carbide ball, 3 – zirconia ceramic ball,
4 –upper ball holder placed onto drive tribometer shaft

The tolerances of tungsten carbide balls are the same as those of the zirconia balls. Table 2 describes the properties of tungsten carbide material. Typical ring cracks were produced by a 400 g impact load, dropped from five different heights: 150 mm, 200 mm, 300 mm, 500 mm, 750 mm. These drop heights refer to five different impact energy levels: 0.6J, 0.8J, 1.2J , 2J, 3J (denoted K1, K2, K3, K4, K5 respectively). A microscopic analysis reveals that there are no cracks for only first two impact loads. The Hertzian cracks were observed only for the three highest drop heights. Fig.5 presents the artificially induced K3 crack. The depth of the ring pre-cracks are 54 μm, 96 μm, 160 μm for K3, K4 and K5 cracks respectively.

Table 2. Tungsten carbide properties

Chemistry	93,5 - 94,5 % WC 5,5 – 6,5 % Co
Compressive strength [MPa]	5200
Tensile strength [MPa]	1520
Maximum useful temperature [^{O}C]	430
Young modulus [GPa]	640
Hardness [HV]	2200
Density [g/cm^3]	13,2

a) b)

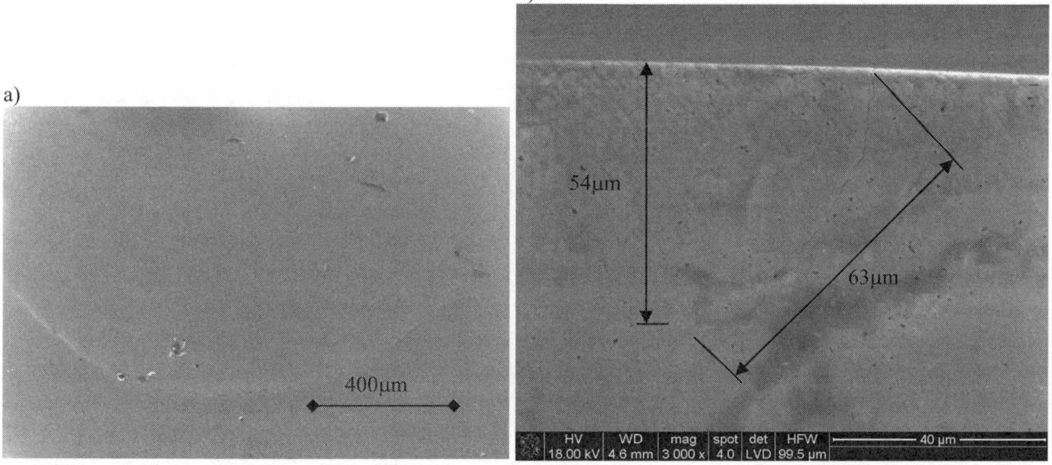

Figure 5. An artificially induced K3 crack: surface (a) and cross-section (b) view of zirconia ball (MgO-PSZ)

Testing procedure. All tests were conducted at 750 N load and a spindle speed of 1500 rpm (Fig.6). The linearly increasing load from zero to a maximum load prevents vibrations at the beginning of the tests. The rolling velocity was 0.3m/second and the Hertzian stresses were 4.7 GPa and 4.2 GPa for MgO-PSZ/Si_3N_4 couples and MgO-PSZ/steel couples respectively. Each test was repeated four times. Table 3 describes the properties of silicon nitride and bearing steel properties. Table 4 describes the contact ball surface properties.

A base lubricant (Tab. 5) without any special additives was chosen to eliminate any possible chemical reaction with zirconia materials and to take only properties of the ceramic material into consideration.

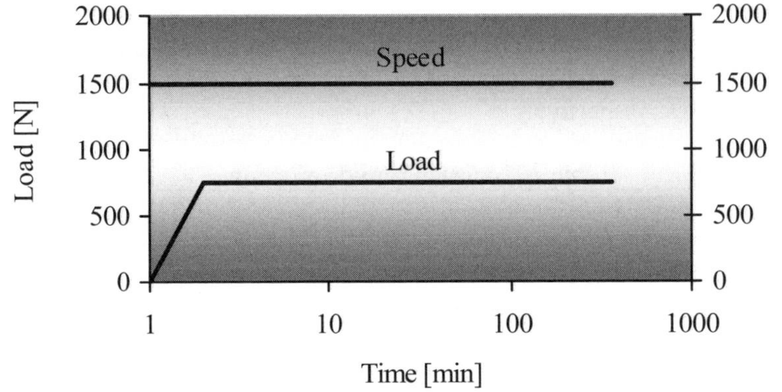

Fig. 6. Testing parameters

Table 3. Lower balls material properties

	Si_3N_4	Steel 1.3505 (DIN)
Chemistry	99,9 % Si_3N_4	0.95-1.1% C 0.25-0.45% Mn 0.15-0.35% Si 1.3-1.65% Cr max. 0.025% P max. 0.025% S max. 0.3% Ni
Compressive strength [MPa]	2353	880
Max. useful temperature [^{O}C]	1400	180
Young modulus [GPa]	324	210
Hardness [HV]	1500	700
Density [g/cm^3]	3,2	7,8

Table 4. Contact ball surface properties

Couples	Equivalent modulus of elasticity [GPa]	Equivalent radius of curvature [mm]	Contact area radius [mm]	Hertzian stresses [GPa]
$ZrO_2(MgO)/Si_3N_4$	132	3.175	0.194	4.7
$ZrO_2(MgO)$/steel	111	3.175	0.206	4.2

Table 5. Base lubricant properties

	Base lubricant
Trade name	T-9
Producer	Nynas
Properties:	
Kinematic viscosity ($40^{O}C$) [mm^2/s]	9.13
Sulfur content [mass %]	0.01
Aromatic hydrocarbons content [mass %]	9.0

The time for failure was recorded and the tests were stopped at a high vibration level. A specific sound at high vibration level appeared at the moment of surface failure. Tests were suspended after 360 minutes if there was no high vibration level. Ceramic ball surfaces were cleaned and examined by electron microscopy.

Experimental results and discussion

Figure 7 shows experimental results for various pre-cracked rings made by three different impact energies (K3, K4, K5). The ring cracks propagate under cyclic load. There isn't big difference in average time to failure between MgO-PSZ/Si_3N_4 and and MgO-PSZ/steel couples. There aren't also differences in time to failure for cracks K3 and K4. The largest impact energy (K5) causes very rapid crack propagation under a cyclic load. The analysis of variations in time for failure for K3, K4 and K5 impact energy shows that MgO-PSZ/Si_3N_4 couples have larger variations of time for failure than MgO-PSZ/steel couples.

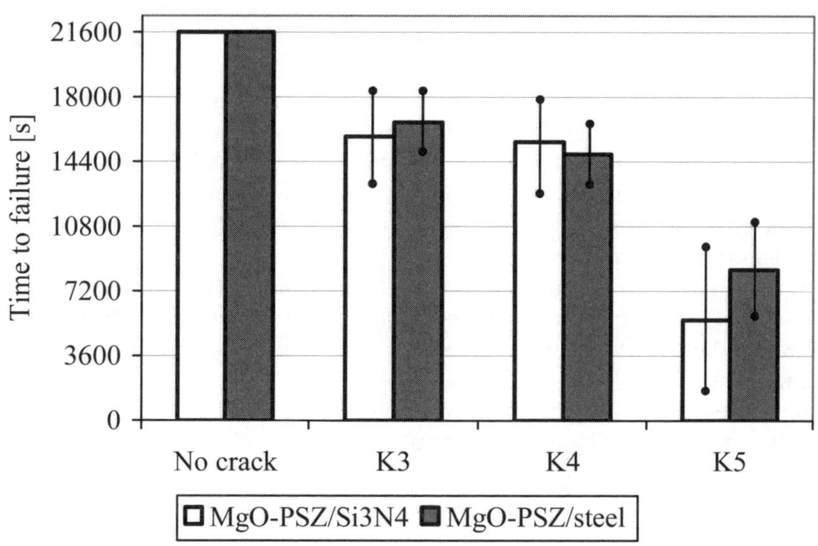

Fig. 7. The influence of drop height (impact energy) on time to failure (MgO-PSZ zirconia balls)

To understand the mechanism of ring crack propagation, experimental observations of the spalling processes are very important. Experimental research results along with results of other researchers [4, 5, 7, 8, 13, 14, 16-18] allow to understand how the crack propagates. Fig. 8 shows the scheme of crack propagation from the pre-crack. Fig. 8a shows the scheme of the cross-section of a typical ring pre-crack It is difficult to observe pre-cracks on the surface using electron microscopy (Fig. 5). Some researchers demonstrate this crack by using a penetrant dye in ultraviolet light [3,4]. Typical ring-crack sections show an initial, sub-surface vertical or ring crack shapes which propagate conically away from the contact area [8, 15].

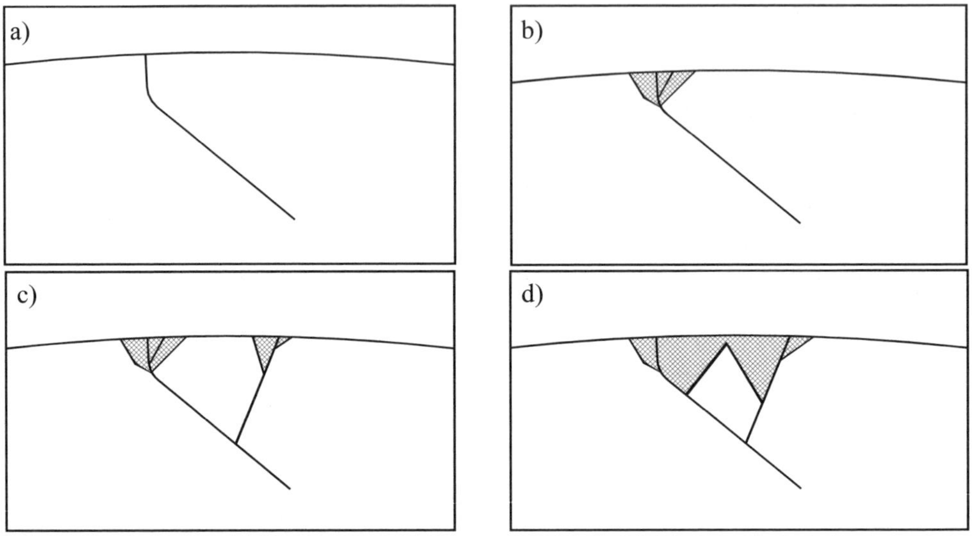

Fig. 8. Stages of crack propagation (MgO-PSZ zirconia balls)
a) pre-crack, b) primary propagation, c) secondary propagation, d) final stage propagation

Pre-cracks propagate under a cyclic load. Propagation can be divided into primary (Fig. 8b) and secondary (Fig.8c) propagations. These cracks are presented on the zirconia ball surface (Fig. 9). The cracks propagate in the middle of a wear track. Primary propagation takes place from the surface to the original pre-crack. This phenomenon originates with the beginning of material removal from the ball surface. Fig.10 presents two sections of the same zirconia ball. The primary propagation changes direction (Fig. 10c) and this is connected with structure defects in the material structure. A secondary crack propagates from the primary crack towards the surface. This is visible in Fig. 10c and Fig. 10d. A secondary crack reaches the surface and forms spalling in the form of another ring crack as in the primary propagation. When the secondary crack reaches the ball surface large amount of material is removed. The beginning of the formation of the next ring crack on the ball surface is shown by an arrow in Fig.10a.

During the final stage of propagation the material is removed (Fig.8d, Fig.11) from the surface area created during the primary and secondary stages of propagation. The ring cracks which propagate towards the surface from the secondary and final cracks, form the damaged area, as a letter "W" (Fig. 11c).

Fig.11a illustrates also the higher surface roughness of the wear track. This phenomenon is connected with surface damage by the presence of wear particles.

Experimental observations show also that there isn't any differences in crack propagation modes for MgO-PSZ/Si_3N_4 and MgO-PSZ/steel couples.

Conclusions

The cracks propagate under rolling contact cyclic stresses. Failure modes in lubricated rolling contact of ceramic materials have been investigated by numerous researchers [2-7, 9, 13, 14, 16, 17, 18] but the research was mainly connected with silicon nitride.

In this paper, the influence of crack propagation on the rolling contact fatigue behavior in MgO-PSZ ceramic zirconia has been studied experimentally. Based on this work, the following conclusions may be formulated:

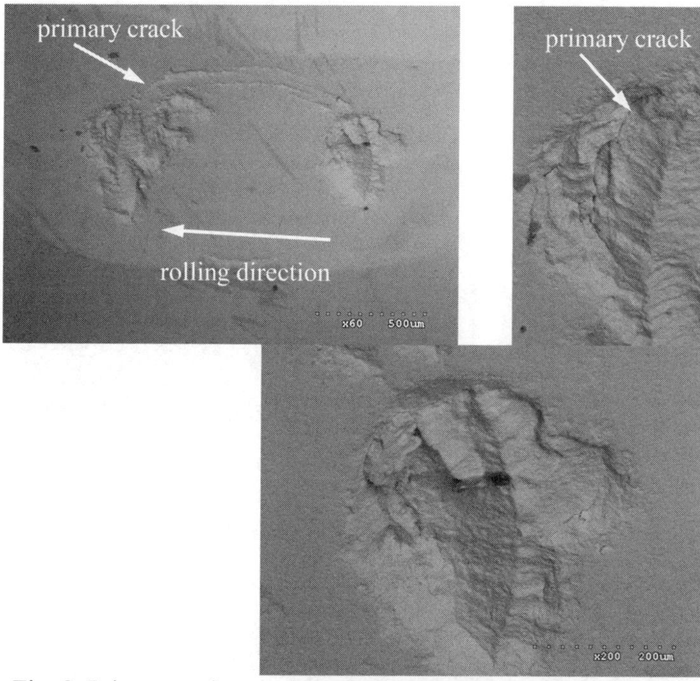

Fig. 9. Primary and secondary crack propagation of a zirconia ball (MgO-PSZ/Si_3N_4 couples)

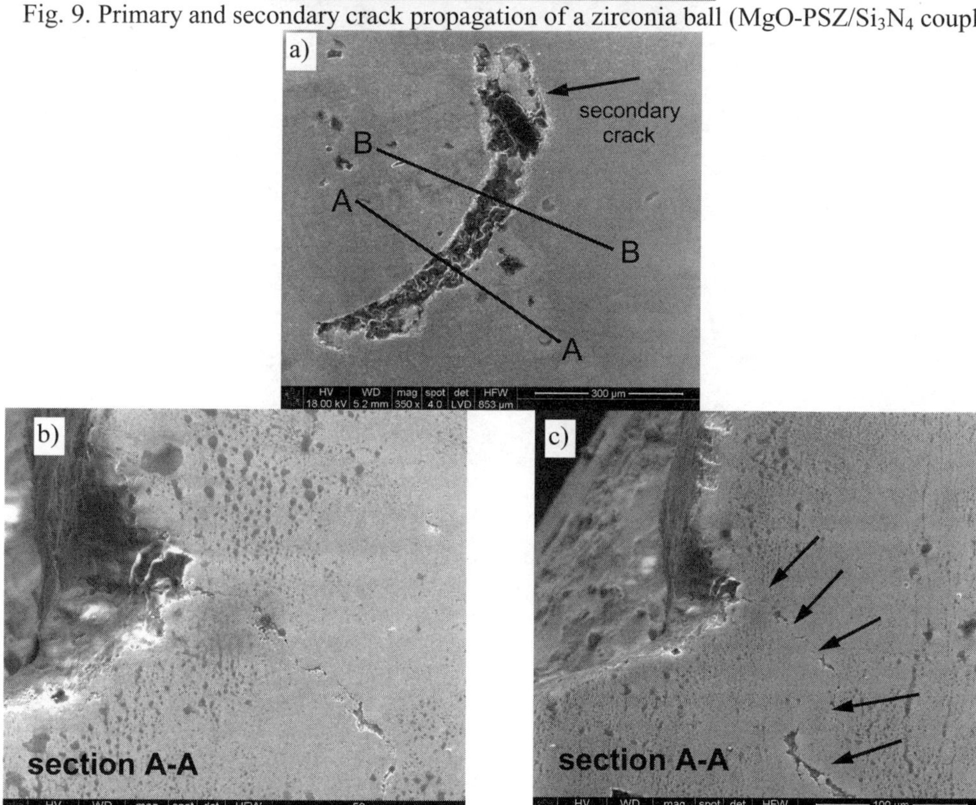

Fig. 10. Subsurface crack propagation of ceramic MgO-PSZ: ball surface (a), section A-A (b,c)

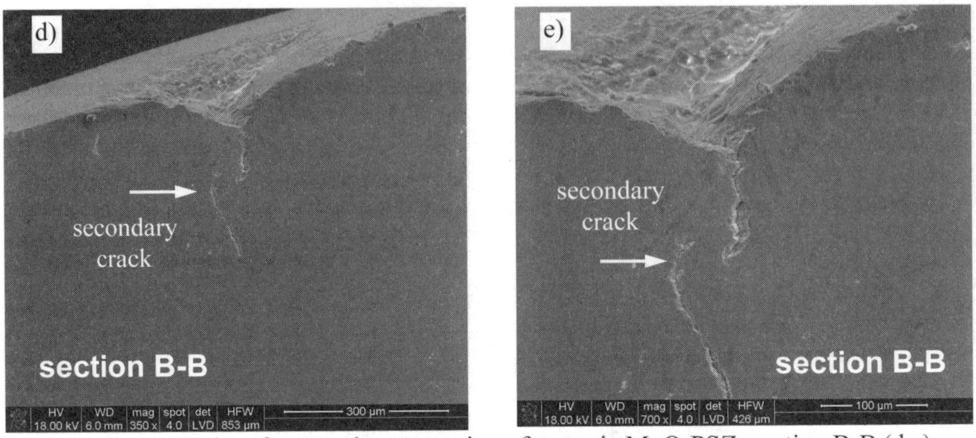

Fig. 10. Subsurface crack propagation of ceramic MgO-PSZ: section B-B (d,e)

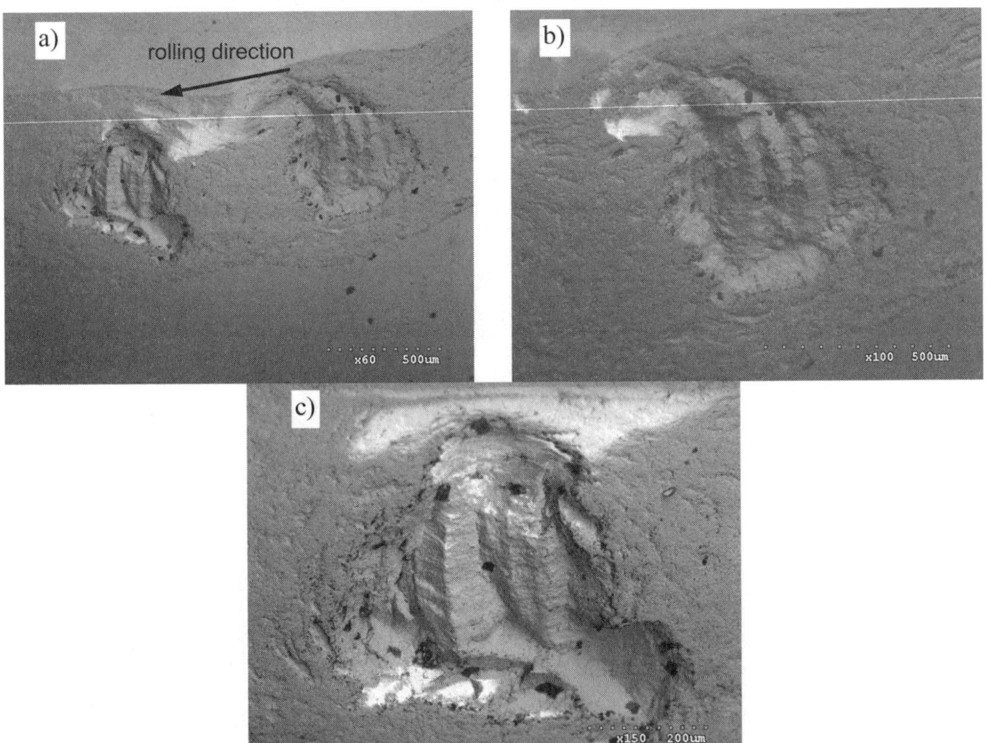

Figure 11. The final stage of crack propagation (MgO-PSZ/Si_3N_4 couples)

- Fatigue life, measured by number of cycles to failure, is influenced by such factors as: the size of ring pre-cracks introduced by different impact energies and shapes of the pre-cracks.
- The ring crack propagation can be divided into primary, secondary and final stage propagation.
- Secondary cracks propagate from the primary crack towards the surface.
- Failure modes of crack propagation in MgO-PSZ zirconia material are the same for MgO-PSZ/ Si_3N_4 and MgO-PSZ/steel couples.

This study was funded by the Polish Ministry of Science and Higher Education under contract no. 4TO7C03528.

References

[1] E.V. Zaretsky, W.J. Anderson, Rolling-contact fatigue studies with four tool steels and crystallized glass ceramic. Trans ASME J Basic Eng, 83, vol. 4 (1961),

[2] H.K. Lorosch, J. Vey, R. Weigan, E. Gugel, H. Kessel, Fatigue strength of silicon nitride for high-speed rolling bearings, ASME Paper 1979:79-GT-68.

[3] Y. Wang, M. Hadfield, Rolling contact failure modes of lubricated silicon nitride in relation to ring crack defects. Wear 225-229 (1999) 1284-1292.

[4] Y. Wang, M. Hadfield, Failure modes of ceramic rolling elements with surface crack defects, Wear 256 (2004) 208-219.

[5] M. Hadfield, T.A. Stolarski, R.T. Cundill, S. Horton, Failure modes of ceramic elements with ring crack defects, Tribology International 26 (1993) 157-164.

[6] C. Cirel, T.A. Stolarski, Factors affecting the wear of ceramic balls, Wear 172 (1994) 191-196.

[7] W. Karaszewski, The influence of oil additives on spread cracks in silicon nitride, Tribology International 41 (2008) 889-895.

[8] Y. Wang, M. Hadfield, Failure modes of ceramic rolling elements with surface crack defects, Wear, 256 (2004) 208-219.

[9] L. Wang, R.W. Snidle, L. Gu, Rolling contact silicon nitride bearing technology: a review of recent research, Wear 246 (2000) 159-173.

[10] R.T. Cundill, Rolling element bearings into 21st century, IMechE, Seminar, London, November, 1990.

[11] E.P. Rajiv, I. Amita, S.K. Seshardi, Tribological properties of cobalt-partially stabilized zirconia (PSZ) composites in dry sliding conditions, Wear 189 (1995), 100-106.

[12] P. Gautier, K. Kato, Wear mechanisms of silicon nitride, partially stabilized ziconia and alumina in unlubricating sliding against steel, Wear 162-164 (1993), 305-313.

[13] Z.A. Ghar, H. Hadfield, S. Tobe, Y. Wang, Ceramic rolling elements with ring crack defects – a residual stress approach, Material Science and Engineering A, 404 (2005), 221-226.

[14] Y. Wang, M. Hadfield, The influence of ring crack location on the rolling contact fatigue failure of lubricated silicon nitride: experimental studies, Wear 243 (2000), 157-166.

[15] J.P. Tillett, Fracture of glass by spherical intender, Proceedings of Physical Society B 69 (1956) 47-54.

[16] W. Karaszewski, Surface cracks of ceramic rolling elements, Acta Mechanica et Automatica, vol. 3, nr 1(7), 2009, 68-70.

[17] W. Karaszewski, Influence of ring cracks on fatigue in silicon nitride, Tribologia, R. 35 vol. 3 (2004) 159-168.

[18] W. Karaszewski, Limitation of fatigue life of ceramic couples due to surface ring cracks, Tribologia, R. 40 vol. 2 (2009), 29-38.

Key Engineering Materials Vol. 490 (2012) pp 64-75

doi:10.4028/www.scientific.net/KEM.490.64

Designing a pumping system of gear pumps featuring the involute internal gearing

Damian SLODCZYK[1, a], Jarosław STRYCZEK[1,b]

[1]Wroclaw University of Technology

Institute of Machine Design and Operation

Łukasiewicza 7/9

50-371 Wrocław

Poland

[a]damian.slodczyk@pwr.wroc.pl,

[b]jaroslaw.stryczek@pwr.wroc.pl

Keywords: gear pump, pumping system, gears with internal involute gearing, axial clearance compensation

Abstract. On the basis of the analysed references it has been shown that an important problem in designing hydraulic machines is to develop a method of designing a pumping system of pumps with the internal gearing, which would integrate designing gears and compenstion elements. The authors of the paper has defined a design algorithm that consists of three basic stages: designing a concept of the system, designing the gear subsystem, and designing the compensation subsystem. The design process at particular stages have been described, and the results have been presented in engineering drawings. An original computer program 'Internal Pumping Set' which enhances the process of designing the pumping system has been created. An application of the program has been presented, and a design of an exemplary pumping system made.

Introduction

As stated in [1, 2], gear pumps make the most frequently used group of energy generators in the hydraulic drive and control systems. The most technologically advanced designs in this group are gear pumps featuring the internal involute gearing. Papers [3, 4, 12] indicate the following design and operational advantages of the pumps:

- compact structure, small size, high delivery/mass ratio q/m,
- high working pressure p while maintaining high delivery q, low ratio of delivery pulsation δ, high volumetric efficiency η_v and total efficiency η_c,
- low noise level.

The key factor providing the advantages is a proper design of the basic pump system with the internal gearing, which the pumping system makes. A schematic design of the system has been shown in Fig.1 (according to [5]). The pumping system includes:

- gear subsystem (1, 2) featuring the internal involute gearing, separated with an insert (3),
- subsystem of compensation plates (4,5).

The system is located in the body (6), and it works as follows: when the gear subsystem (1, 2) is driven to rotate, it sucks the working fluid up from the inlet zone (*IN*) and by means of the intertooth space, charges the fluid into the outlet zone (*OUT*), thus providing the displacement of the pump.

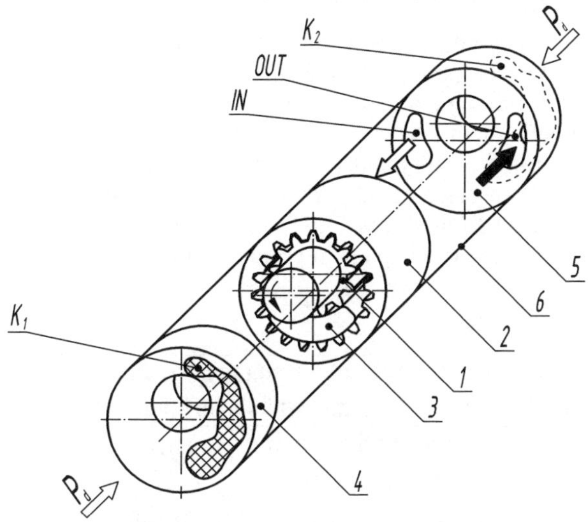

Fig. 1 Engineering drawing illustrating the pumping system of the gear pump featuring the internal invloute gearing
1 – external gear, 2 – internal gear, 3 – separation insert,
4, 5 – compensation plates, 6 – body of the pump,
K_1, K_2 –compensation surfaces, P_d – force pressing the compensation plates,
IN – inlet zone, OUT – outlet zone

Separation of the two zones is secured by the separation insert (3). The gear subsystem collaborates with the subsystem of the compensation plates (4, 5). On their sides there are compensation surfaces (K_1, K_2) created, to which the working fluid at a high pressure is delivered. It results in pressing them by forces (P_d) to the front surfaces of the gears and minimising the axial clearances that are the main source of the internal leakage. The compensation plates subsystem provides the internal tightness of the pump and enables achieving high working pressures p as well as volumetric efficiency η_v.

The theoretical basis of designing the gear subsystem with consideration of geometry, kinematics, and hydraulics has been presented in [6] and in [1, 2]. The practical design solutions for the gears have been shown in [12]. Whereas in [4] ways of compensating the axial clearances in the internal gearing pump are discussed. The efficiency of the axial clearances compensation has been proved also in [7, 8] for the case of the gerotor pump, the design and operation principle of which is similar to the one featuring the internal involute gearing.

What can be concluded from the presented analysis is that there are known methods of designing particular subsystems, namely the gear subsystem and the compensation elements subsystem. In the literature on the subject, the synthetic methods of designing the entire system are not presented. Simultaneously, it seems to be important, because in the light of the description and Fig. 1, there is a clear interaction between the gear subsystem and the compensation elements. In this situation, as the objectives of this paper the following have been considered:

- developing a method for designing the pumping system of gear pumps featuring the internal involute gearing, which would synthesize the gear subsystem and the subsystem of the compensation elements,
- making a computer program that would enhance the design method,
- verifying the method by designing an exemplary pumping system.

Algorithm of the pumping system design method

Fig. 2 illustrates a block diagram of the algorithm of the method of designing the pumping system of the internal involute gear pumps. The algorithm includes three basic stages:

- stage 1, when the concept of the pumping system is developed,
- stage 2, when the gear subsystem's design is conducted,
- stage 3, when the compensation subsystem's design is conducted.

The design process goes from stage 1, through stage 2, up to stage 3, and within each of the stages some specific tasks are performed (see Fig. 2). The algorithm also illustrates the possibility of introducing modifications while the design is in progress through feedback between the stages.

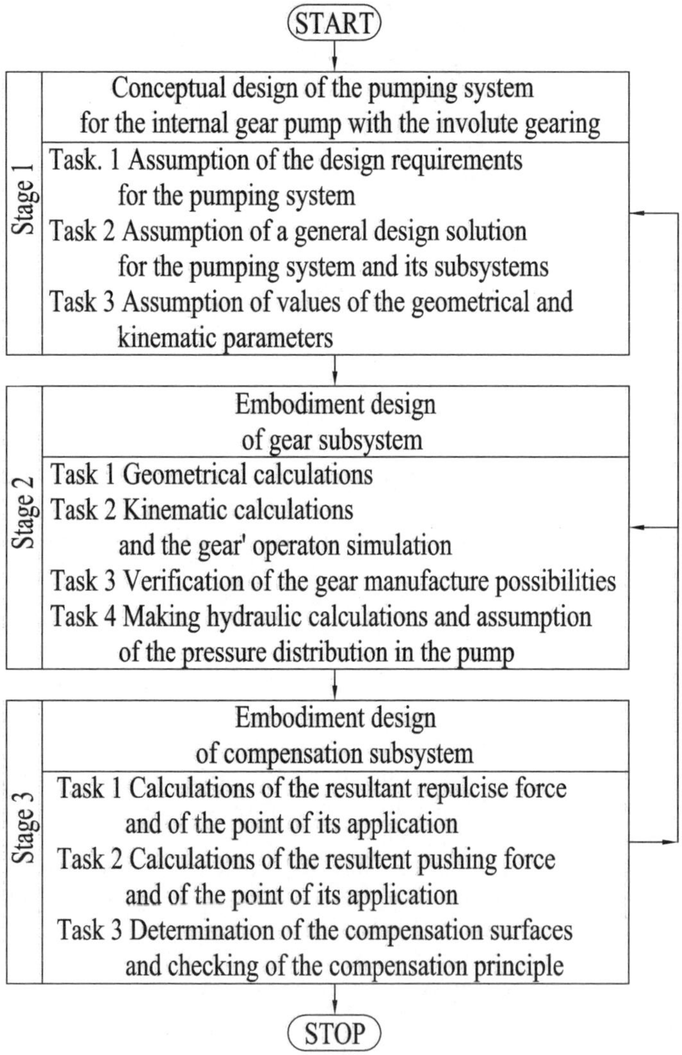

Fig. 2 Algorithm of designing the pumping system

Developing a concept of the pumping system

According to the algorithm shown in Fig. 2, the first task to complete at this stage is to define the fundamental design requirements for the pumping system. The requirements are the following:

- providing as small diameter of the pumping system as possible,
- providing high delivery q and low ratio of delivery pulsation δ,
- providing high working pressure p, high volumetric efficiency η_v and total efficiency η_c,

In order to meet the requirements, first the general design solution for the pumping system needs to be assumed, which is carried out in task 2.

The concept work is started from the analysis of the gear subsystem, a design of which has been illustrated in Fig. 3a and b.

The figures show that for the susbsystem considered can be various numbers of teeth z_1 of the driving wheel, and teeth z_2 of the driven wheel, thus changing tooth difference z_2-z_1. Normally, small numbers of teeth z_1 and z_2 are taken into consideration, namely over a dozen, as well as a small tooth difference z_2-z_1 of a few teeth, which has been presented in Fig. 3a. This way, a small external diameter D of the subsystem is achieved, and simultaneously, room for the gear separation insert is made. Due to the small number of teeth z_1 and z_2 , as well as the small tooth difference z_2-z_1, the tooth interference beyond the meshing area [9] is to be taken into account. In order to eliminate the interference, it is necessary to reduce the height of the teeth (tooth height ratio y<1), to apply a wider angle of the tooth profile α, and above all, to carry out the angular correction [9].

Finally, making a selection of the abovementioned toothing parameters, and then the correction, expected should be a change in tooth contact ratio ε, which, consequently, ought to be equal to a value from the range $\varepsilon = 1.1 \div 1.3$. It secures the meshing fluency, separation of the inlet zone (*IN*) from the outlet zone (*OUT*), as well as a respectively little capacity of the closed intertooth area.

Furthermore, following [6] it can be observed that the assumption of the small numbers of teeth z_1, z_2 causes the reduction in the pump's delivery. In this situation, the set delivery is provided by selecting higher values of m module and tooth width b. At the same time, as presented in [6], the reduction of tooth difference z_2-z_1 results in the reduction of delivery pulsation δ as well as in the lowering of the noise level during the pump's operation.

In the considered gear system (Fig. 3a and b), various location of the insert separating the inlet zone (*IN*) from the outlet zone (*OUT*) of the pump can be assumed. The insert can be positioned centrally (*c*) in relation to both zones, or it can be shifted (*o*) towards the outlet zone. Thus, the pressure distribution in the pump can be modified.

Continuing the concept work one should consider a design solution for the compensation plates subsystem, which has been presented in Fig. 3c-g. The plates can collaborate with the fronts of the gears, moving in the pump's body owing to pushing forces P_d (Fig. 3c and d) or getting elastically deformed [10] under the influence of forces P_d (Fig. 3e). Additionally, it is possible to execute the compensation by one plate, namely the so called one-sided compensation (Fig. 3c) or by two plates, that is the two-sided compensation (Fig. 3d).

Also, various profiles of the compensation plates can be taken into account. There can be full plates (Fig. 3f), including both the inlet zone (IN) and the outlet zone (OUT), or incomplete plates (Fig. 3g), including only the outlet zone (OUT).

What stems from Fig. 3c, d, f, and g is, that the diameter of the compensation plates is equal to the external diameter D of the gear subsystem. Respectively, when diameter D is smaller, then it is easier to pair the front surfaces of the gears and the compensation plates. Thicker plates C_1 (rigid) can be more easily and more precisely moved in the pump's body under the influence of pressure forces P_d, both in the one-sided and the two-sided compensation (Fig. 3c-d). Less thick (flexible) plates C_2, however, are better in the case of the elastic elements compensation (Fig. 3e).

On the full compensation plates, though, (Fig. 3f), it is easier to shape the compensation surfaces than on the incomplete plates (Fig. 3g).

The catalogue of the concept solutions presented in Fig. 3 stands for only a part of a wider catalogue which can be further developed and completed. By means of such a catalogue, and combining various elements of gears and compensation plates one can achieve different concept solutions of the pumping system. For instance, by combining solutions presented in Fig. 3a, d, f one achieves the system shown in Fig. 1.

Another important task at the concept stage is to assume values of the geometrical, kinematic, and hydraulic parameters for the selected pumping system.

In the group of the geometrical parameters one should specify:

- tooth number z_1 and z_2, as well as tooth difference z_2-z_1,
- tooth height ratio y,
- nominal tooth profile angle α_0,
- module m.

Considering kinematic parameters one should determine tooth contact ratio ε for the designed gearing.

In the group of the hydraulic parameters one should assume:

- specific delivery (per revolution) q and delivery pulsation δ of the pumping system,
- working pressure p.

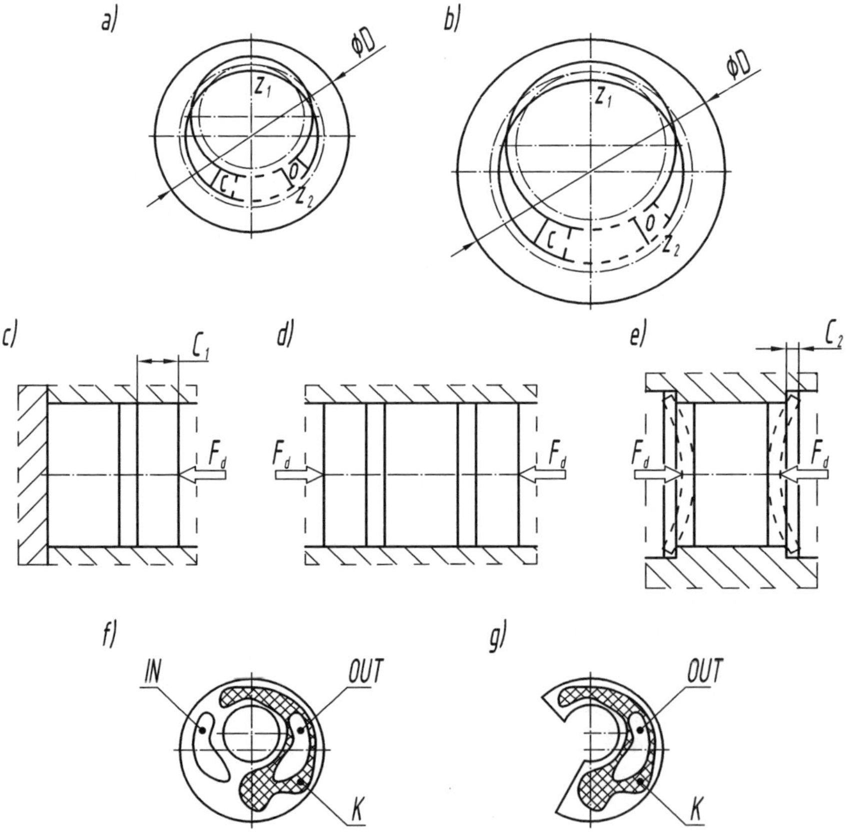

Fig. 3 Concept solutions for elements of the gear pump featuring the internal involute gearing
a, b) gear subsystem solutions
c-g) compensation plates subsystem solutions

Designing the gear subsystem

The first task, which is illustrated by the algorithm presented in Fig. 2, at this stage is to carry out the geometrical calculations for the gear subsystem assumed earlier in a form of a concept solution.

For the calculations, the generally known formulae presented in the gear literature [9, 11] have been used, to which values of the geometrical parameters assumed at the previous stage are used. Important steps in the process of this task's completion are the following:

- conducting the gearing correction P and determining coefficients x_1, x_2,
- calculating diameters d_{a1}, d_{a2}, d_{f1}, d_{f2}, d_{w1}, d_{w2} of the gears, which include correction coefficients x_1, x_2,
- checking thickness at the top of teeth Sa_1, S_{a2} in order to make sure that the sharpening at their top as a result of the correction does not occur,
- assuming external diameter D of the gear subsystem.

Consequently, as shown in Fig. 4, the geometrical form of the gears is achieved.

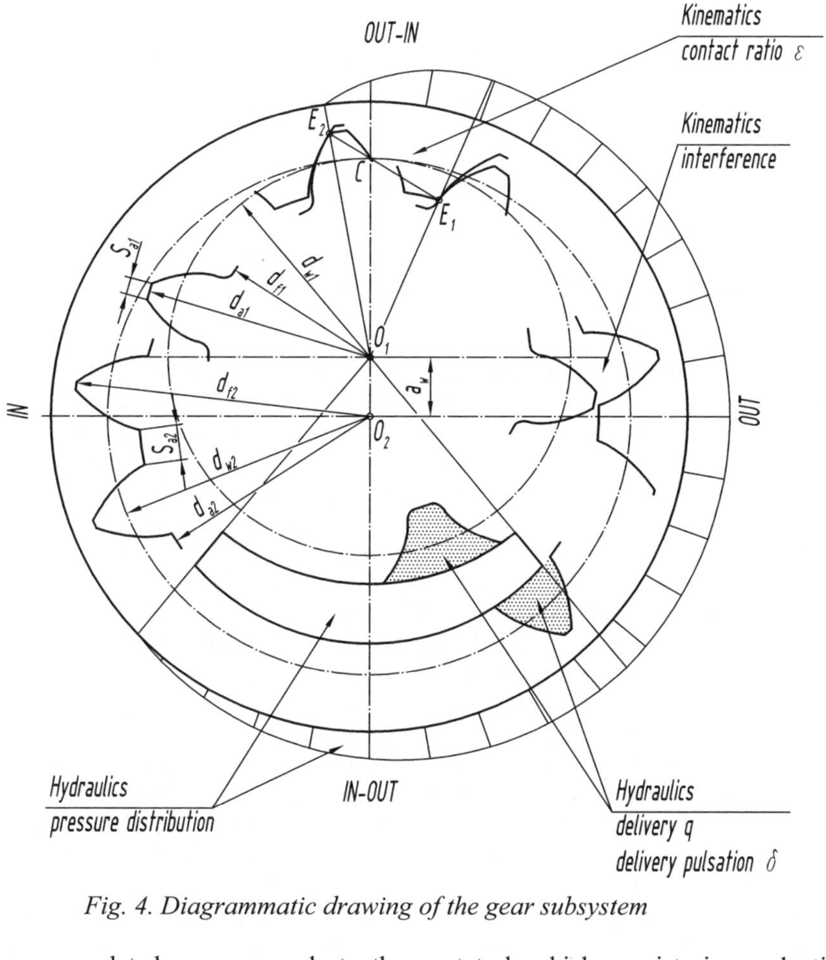

Fig. 4. Diagrammatic drawing of the gear subsystem

With the above completed, one proceeds to the next task which consists in conducting the kinematic calculations. In this case too, the formulae given in the literature [9, 11] are applied, and the following performed:

- checking the meshing and determining the real contact ratio ε as well as comparing it to the value assumed in the assumptions which say that it ought to fall in the range of $\varepsilon = 1.1 \div 1.3$,
- checking if the tooth interference beyond the gearing does not occur.

The carried out kinematic analysis makes it possible to state whether the meshing will be proper, which is also shown in Fig. 4.

The third task is to verify the possibility of making the gears. The objective should be to make it possible to manufacture the gears by means of Fellows gear cutter which still is the basic method of manufacturing internal gears. It is also possible to make gears by means of broaching or by WEDM (Wire Electrical Discharge Machining).

The final, fourth task to complete is to conduct the hydraulic calculations consisting in the determination of:

- specific pump delivery q, i.e. the delivery per one revolution of the pump's shaft,
- pump delivery pulsation coefficient δ.

For calculating the quantities used are the formulae given in [6]. It is proved there that the quantities depend on the parameters of the toothing and the gearing assumed at stage 1, task 2, which can be described by means of the following general dependencies:

$$q = f(z_1, z_2, y_1, y_2, x_1, x_2, m, b, \alpha_0, \alpha_w) \quad (1)$$

$$\delta = f(z_1, z_2, y_1, y_2, x_1, x_2, \alpha_0, \alpha_w) \quad (2)$$

The results of the hydraulic calculations have also been presented symbolically in Fig. 4. What needs to be specified at the end of this task is the position of the insert separating the gears, which makes it possible to determine the internal pressure distribution in the pump, namely to determine four characteristic zones: the inlet zone (*IN*), the pressure increase zone (*IN-OUT*), the outlet zone (*OUT*), the pressure decrease zone (*OUT-IN*). The shift of the insert towards the outlet zone (*OUT*), which has been depicted in Fig. 4 allows the reduction of this area of the gear subsystem which is under the influence of the high pressure, and consequently, the reduction of the compensation area.

Designing the compensation subsystem

The operation principle of the hydraulic compensation of the axial clearance has been presented in Fig. 5. The compensation consists in maintaining constant thickness of clearances h made between the fronts of the gears (1, 2) and the fronts of the compensation plates (4).

The changing working pressure p in the pump generates the changing repulsive forces P_{od}, which, on their part, cause the shift, or deformation, of the compensation plates and the change in axial clearances' thickness h. In order to maintain constant thickness h of the clearances, it is necessary to push the compensation plates with forces P_d which are generated by the activity of the working fluid delivered onto the outer surface of the plates. Papers [1, 7, 8] contain the conditions of the balance of the compensation plates, which are formulated as:

- pushing force P_d should be equal or slightly greater than than repulsive force P_{od},
- coordinates (x_d, y_d) of the pushing force application point should be identical with coordinates (x_{od}, y_{od}) of the repulsive force application point.

The conditions have been presented in Fig. 5 – sections *B-B* i *C-C*.

Taking into account the abovementioned conditions, the first task to complete at stage 3, ought to be calculating the resultant repulsive force Pod as well as its application point (xod, yod).

As shown in Fig. 5, section A-A, for that purpose, the pressure distribution in the pump assumed at the previous stage is used, i.e. in the pressure increase zone (*IN-OUT*), in the outlet zone (*OUT*) as well as in the pressure decrease zone (*OUT-IN*). By multiplying the pressure value in a given zone by the value of the surface worked on by the pressure, partial repulsive forces P_{od1}, P_{od2}, P_{od3} in the zones are generated. By summing the partial forces, achieved is the resultant repulsive force P_{od} which has been shown in Fig. 5, section *B-B*. Coordinates (x_{od}, y_{od}) of the application point of repulsive force P_{od} are determined basing on the general rules of the technical mechanics.

The next step in the design process is completion of task 2, when the value of pushing (compensation) force P_d is assumed, which is illustrated in Fig. 5, section *C-C*. According to the guidelines provided in [8], it ought to be:

$$P_d = (1.05 \div 1.3) P_{od} \quad (3)$$

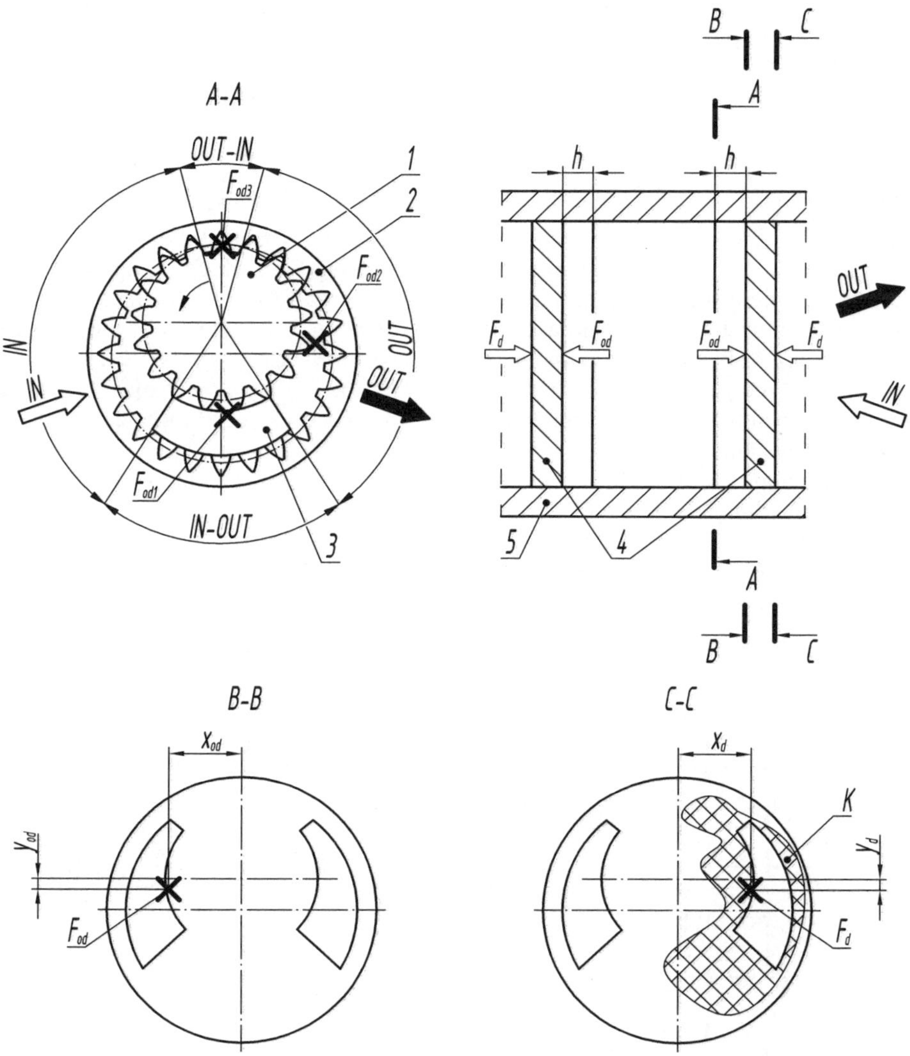

Fig. 5 Diagrammatic drawing of the compensation plates subsystem
1,2 – gears, 3 – separation insert, 4 – compensation plates, 5 – body of the pump,
h – axial clearance, K – compensation area,
P_{od} – resultant repulsive force, P_d – resultant pushing force,
x_{od}, y_{od} – repulsive force coordinates, x_d, y_d – pushing force coordinates

The pushing force surplus is needed to overcome the friction forces that are generated while the compensation plates (4) are moving in the body of the pump (5). Having assumed the value of force P_d, the shape of the compensation area is defined, which has been depicted in Fig. 5, section *C-C*, as (*K*).

Assuming that the whole area is worked on by the pump delivery pressure *p*, the value of the area is defined as:

$$K = P_d / p \tag{4}$$

At the end, as task 3 of this stage, the centre of gravity of the compensation pump (x_d, y_d) is determined, in which applied is the resultant pushing force P_d. As it has been stated above, it is important to aim at making coordinates (x_{od}, y_{od}) of repulsive force P_{od} identical with coordinates (x_d, y_d) of pushing force P_d. It is executed by selecting an appropriate profile and size of the compensation area (*K*).

Computer program 'Internal Pumping Set' and an example of its application

The assumed algorithm along with the theoretical design basis that follows it were used to develop a special computer program named 'Internal Pumping Set', which is supposed to enhance the process of designing the pumping system of the gear pump featuring the internal gearing. The third stages included in the algorithm have their equivalents in three blocks of the program. As a result of the activity of block 1, the pumping system concept is achieved, as a consequence of the activity of block 2, the gear subsystem is produced, whereas the activity of block 3 results in the making of the compensation plates subsystem. The results are given in a numerical form as well as in a form of diagrams and animations.

Using the program allowed the designing of a prototypical pumping system of the pump featuring the internal involute gearing.

At the first design stage of the process, various design solutions presented in the abridged catalogue of solutions in Fig. 3 were analysed, and for further design, a general solution was selected and shown in Fig. 6.

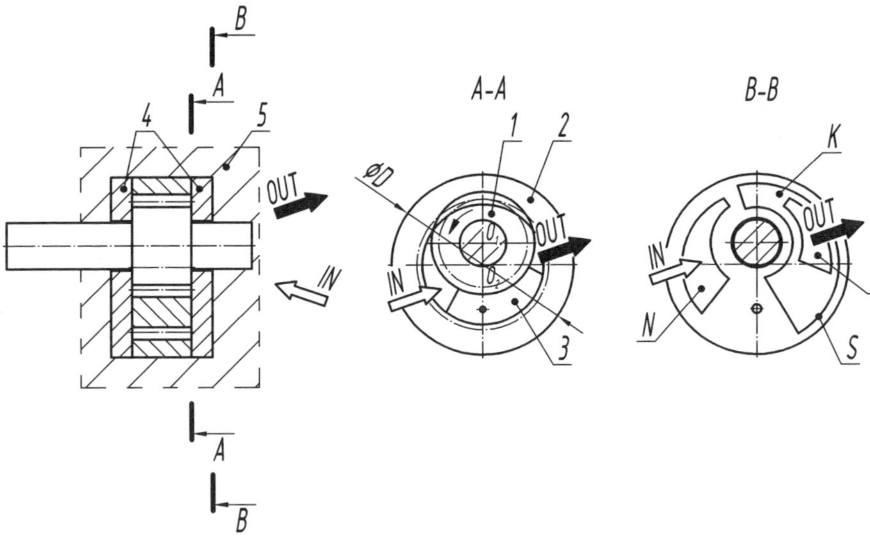

Fig. 6 A concept of the pumping system generated by means of Inernal Pumping Set computer program
1, 2 – gear system, 3 – separation insert, 4 – compensation plates, 5 – body of the pump, K – compensation area, S – sealing ring, N – kidney hole

The gear subsystem consists of gears (1, 2) featuring a small numer of teeth z_1, z_2 and a small tooth difference z_2-z_1=3. This will provide a relatively small internal diameter *D* of the subsystem. The separation insert (3) ought to be shifted towards the outlet zone (*OUT*), thus enlarging the inlet zone (*IN*), and simultaneously reducing the area of the high pressure operation on the outlet side (*OUT*).

The compensation subsystem is made by the full plates (4) that are pushed towards the front surfaces of the gears (1, 2) from both sides. This is called two-sided compensation. On the outer sides, the compensation plates (*K*) limited by the sealing rings (*S*) are shaped. The inlet and the outlet to/from the gear subsystem is through the kidney holes (*N*) made in the compensation plates.

The following values of the geometrical parameters have been assumed: $z_1 = 10$, $z_2 = 13$, $m = 4$, $y = 1$, $\alpha_0 = 20$ [°]. It has also been added that the tooth contact ratio $\varepsilon > 1$. The following hydraulic parameters have been assumed:

$q = 10$ [cm3/rev.], $p_t = 16$ [MPa].

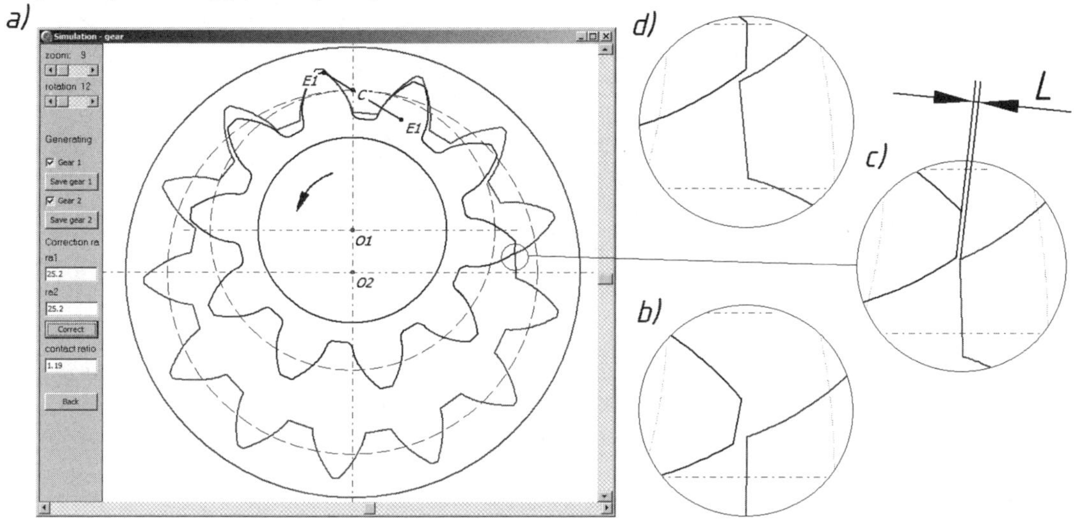

Fig. 7 The gear subsystem designed by means of the Internal Pumping Set program
a) geometrical form of the gear subsystem,
b, c, d) elements of the gears' motion simulation showing the absence of interference in the gearing

Using the Internal Pumping Set program the authors of the article have designed the gear subsystem shown in Fig. 7. Respectively, Fig. 7a illustrates the geometrical form of the gears. By means of the program, the simulation of the gears' motion was carried out. It was stated that the meshing is performed properly, and the tooth contact ratio is $\varepsilon = 1.19$. In the diagram, contact line $E_1E_2 = \varepsilon\ p_b$ is marked. As a result of the simulation of the motion, the absence of the tooth interference beyond the gearing area was proved. It has been presented respectively in Fig. 7b, c, d, where it is to see that the teeth move in relation to ech other with a clearance (*L*).

At the final stage of the design process, the Internal Pumping Set program was used to design the compensation subsystem which is shown in Fig. 8. The figure illustrates that the previously designed gear subsystem (1) is separated with the insert (2) and limited from the side by the compensation plate (3). In the plates, holes (4, 5) have been designed. The holes (or ports) stand for the inlet (IN) and the outlet (OUT). Around the outlet port (OUT), the compensation area has been created, which is limited by the sealing ring (S).

By animating so designed a pumping system, it is possible to verify the collaboration of the gears and the compensation plates.

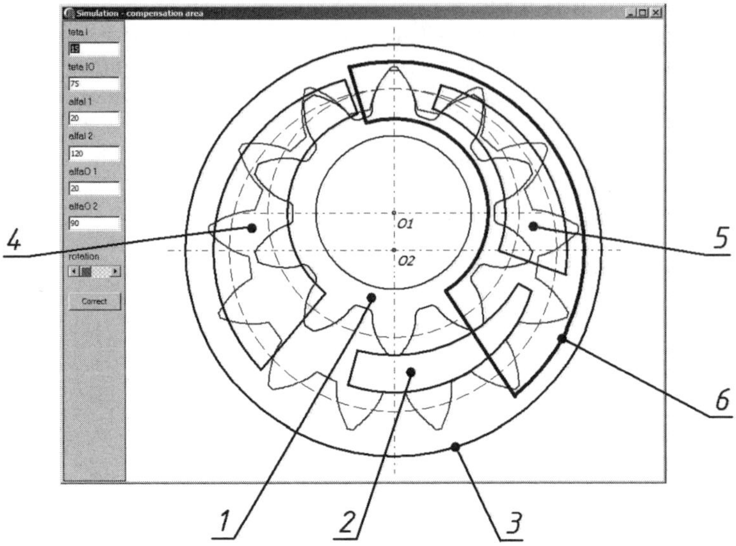

Fig. 8 The compensation subsystem designed by means of the Internal Pumping Set program
1 – gear subsystem, 2 – separation insert, 3 – compensation plate,
4, 5 – inlet and outlet ports , 6 – sealing ring

Summary

According to the aims formulated in the introduction of the paper, a method of designing the pumping system of the gear pumps featuring the internal involute gearing, as well as a special computer program Internal Pumping Set, enhancing the method have been developed.

The authors of the article have proved the applicability of both the design method and the computer program by designing an exemplary pumping system shown in Fig. 7 and 8.

Using the program one can optimise the design solution for the pumping system according to various criteria, both geometrical, kinematc and hydraulic.

The algorithm of the design method shown in Fig. 2 is open in its character. It can be improved and developed, e.g. within the range of stage 1, by developing the concept solutions, or within the range of stage 2, by developing the hydraulic calculations. Respectively, the Internal Pumping Set program itself can be improved and developed.

References

[1] S. Stryczek, Hydrostatic Drive, WNT Warszawa, 1995,

[2] J. Ivantysyn, M. Ivantysyn, Hydrostatiche Pumpen und Motoren. Konstruktion und Berechnung, Vogel, 1993,

[3] C. Dorflinger, E. Reinke, Innenzahnradpumpen: Einstufig auf 345 bar, O+P, number 10/1997,

[4] R. Pippes, C. Mayer, Gap-compensated high-pressure internal gear pumps, 3rd International Fluid Power Conference, Aachen, 2002,

[5] D. Slodczyk, S. Stryczek, Computer program for the design of hydraulic machines' pumping set with internal involute gearing, 6th International FPNI-PhD Sympsium, Purdue, 2010,

[6] J. Stryczek, Hydraulic Machines' Gears, Wrocław, 2007,

http://www.dbc.wroc.pl/dlibra/docmetdata?id=2045&from=&dirids=1,

[7] J. Stryczek, Gerotor- und Orbitmachinen, O+P, number 6/2005,

[8] S. Bednarczyk, J. Stryczek, Axial clearance compensation in the gerotor pump, Proceedings of the 20th International Conference on Hydraulics and Pneumatics, Prague, 2008,

[9] L. Müller, Design of Gear Transmissions, WNT Warszawa, 1996,

[10] S. Bednarczyk, J. Stryczek, Axial clearance compensation in gear displacement machines, Proceedings of the 3rd FPNI-PhD Symposium on Fluid Power, Terrassa, 2004,

[11] K. Ochęduszko, Gears, WNT Warszawa, 2009,

[12] Eckerle pump catalogue, www.eckerle.com, Voith oumo catalogue, www.voithturbo.com.

Key Engineering Materials Vol. 490 (2012) pp 76-82
© (2012) Trans Tech Publications, Switzerland
doi:10.4028/www.scientific.net/KEM.490.76

Designing of high-speed machine shafts of carbon composites with highly nonlinear characteristics

Zbigniew Dąbrowski[1, a], Piotr Deuszkiewicz[2, b]
[1] Institute of Machine Design Fundamentals, Warsaw University of Technology, Narbutta 84, 02-524 Warszawa, Poland
[2] Institute of Machine Design Fundamentals, Warsaw University of Technology, Narbutta 84, 02-524 Warszawa, Poland
[a] zdabrow@simr.pw.edu.pl, [b] uszy@simr.pw.edu.pl

Keywords: rotating system dynamics, non-linear vibrations, carbon-epoxy composite

Abstract. Carbon composites (carbon-aramide) constitute modern "Hi-Tech" materials of an extremely favourable ratio of load to mass transfer. Thereby the idea of using them for long transmission shafts seems to be very attractive, specially in aviation and sailing. Unfortunately structures made of carbon-aramide composites are characterised by a strong anisotropy and in effect by non-linearity of elastic characteristics, due to which they have significantly different dynamic influences than steel structures. To this end there is a need of application of different calculation algorithms than the ones widely used in Construction Engineering. It requires taking into account the coupling between transverse and torsional vibrations, which essentially projects onto the possibility of the critical states occurrence. This is indispensable since shafts made of expensive composite materials are profitable in places where their low elementary mass is essential (long high-speed transmission systems). The paper presents the theoretical discussion of this problem, empirical investigations of nonlinear characteristics as well as the proposition of designing methods.

Introduction

Structural calculations of steel shafts (or more generally metal shafts) constitute the topics of the classic lecture on fundamentals of machine construction and it would seem that all related problems have been solved. A generally applied idea of available algorithms is brought to the calculation of diameters from the kinetostatic strength condition, which determines the complex state of stresses on the basis of the Huber's hypothesis (mainly by calculating the so-called equivalent moment of inertia) and then performing verifying calculations concerning deflections. Dynamic problems are usually limited to determinations of critical velocities and sometimes natural frequencies of torsional vibrations. When a sophisticated structure requires more precise calculations it can be done by the accurate combination of moments (with their variability taken into account) instead of amplitudes [2] and by a more detailed dynamic analysis. The problem of fatigue strength becomes then of a substantial significance. Algorithms which start the calculation train from satisfying the dynamic assumptions, within critical states, have – so far – not been and probably will never be widely accepted [2], since modern computers are so fast that it allows to analyse dynamic properties simply as a function of changed dimensions – within the allowable limits – thus by systematic searching method.

The situation is completely different in case of shafts made of carbon or carbon-aramid composites. Such composite is – in the first place – of an unisotropic structure not satisfying, at a deformation in any direction, the Hooke's law and secondly is not a material – in our hitherto understanding – since it obtains material properties only after being shaped (!). Thus, it is not possible to transfer the results obtained for an ideal sample (usually in a flat beam form) into the machine shaft working under difficult (complex) load conditions. At the moment it can be stated, that several published theoretical papers concerning mechanics of composites as well as empirical investigations leading to mathematical models [3, 7] are still within the laboratory tests boundary. When investigating dynamics of high-speed systems with composite elements we are coming across several new effects requiring rather complicated examinations [3, 5]. Not without meaning, for the development of our knowledge, is confidentiality of producers since they are not willing to disclose the technological details of production and the obtained exploitation parameters. Composite driving systems are extremely attractive in aviation and navy structures of a military character, since apart from a favourable power to strength ratio, their resistance to physical fields also counts (especially their non-magnetism). This probably explains the fact that, e.g. information of the Swedish navy on starting building the power transmission system of mine destroyer with the main shaft of carbon composite stopped immediately after the ship launching and it is not known whether the concept was really realised.

Problem characteristic

Two main calculation problems exist at shafts constructing of carbon composites. The first concerns a composition of loads acting in various directions including the ones unfavourable for composites. In this case the achievements of the 'classic' theory of laminates can be helpful and the problem can be analysed with the so-so accuracy, however a lot is still to be done. The example of the complex state of stress in which addition of a small axial force caused the construction failure is given in paper [4]. The second problem concerns dynamics, two examples are given below.

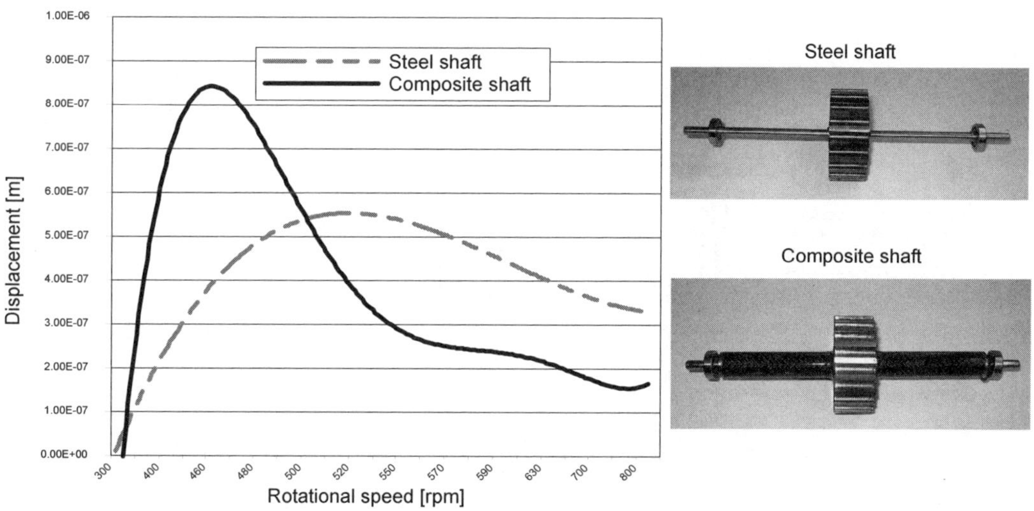

Fig. 1. Amplitude-frequency characteristic of steel shaft and a carbon-epoxy composite shaft.

Comparison of the amplitude-frequency characteristic of steel and composite shafts of the same static bending strength and the same static stiffness is presented in Figure 1. The wheel of a relatively high mass was placed in the middle of the shaft. The power spectrum of the impulse response of the shaft mounted in an outrigger way, without any mass inside, is illustrated in Figure 2. It can be easily noticed that in the first case – in spite of large differences – the effect can be described by the mathematical model of a similar structure as for the steel shaft, but in the second case the generation of the presented spectrum requires a complex nonlinear system. As it is well known, the steel element – mounted in the similar fashion – vibrates with at least the third (fourth) natural frequency and the amplitudes of consecutive harmonics are decreasing fast.

Empirical investigations

In an attempt to explain the effect seen in Figure 2 a series of experimental tests of carbon-epoxy shafts in various reinforcement configurations (Table 1) were carried out. The shaft in the reinforcement system 0÷90° versus the geometric axis of the shaft was selected for further considerations, as being the most representative.

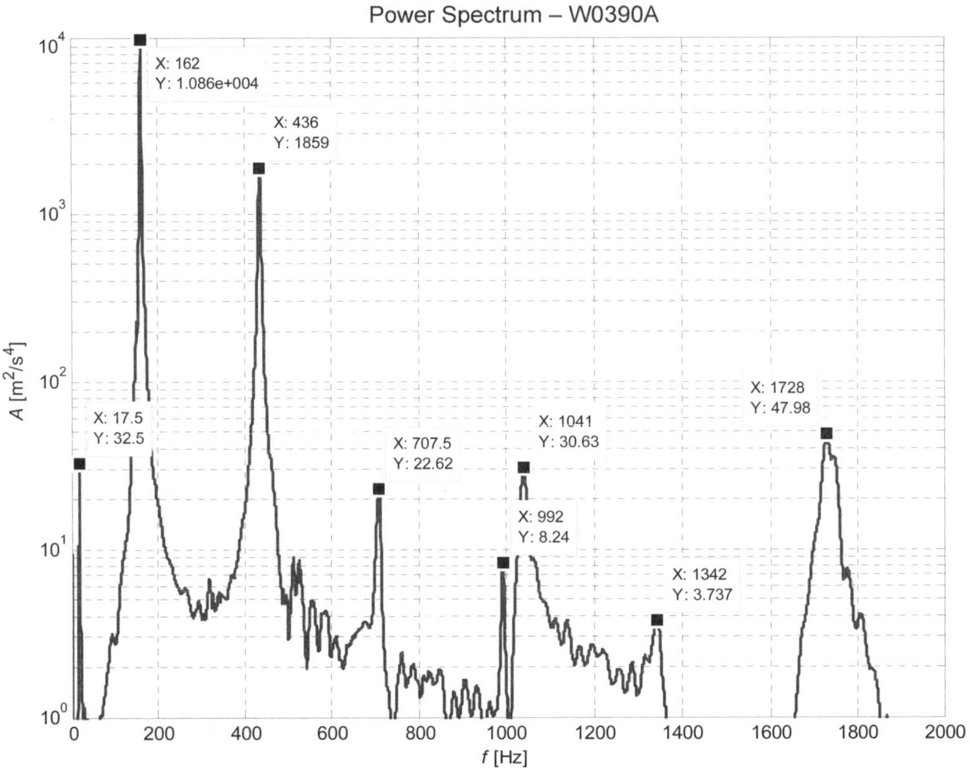

Fig. 2. Impulse response of the carbon-epoxy composite shaft.

Table 1. Parameters of the investigated carbon-epoxy shafts.

Sample / Parameters	W01/W02	W03	W04	W05
Reinforcement system	±45°	0÷90°	±45°	±45°
Kind of fabric	Sleeve Ø26 [mm]	Fabric	Sleeve Ø26 [mm]	Sleeve Ø25 [mm]
Number of layers	6	4	4	4
Mass [g]	150	135	115	115
Outside diameter [mm]	~ 26	~ 25.5	~ 25.0	~ 24.5
Wall thickness [mm]	~2	~ 1.5	~ 1.5	~ 1.5
Total length [mm]	950	950	950	950
Active length [mm]	900	900	900	900

The example of the examination results of static characteristic, the equivalent Young module calculated each time from the simple linear model and impulse response are presented in Figure 3. Static tests were performed several times and the results should be considered reproducible. The given characteristics were approximated by the fifth-degree curve. The fact that the static deflection characteristic is nonlinear and degressive (although without a visible hysteresis), but this non-linearity is rather weak (!), is drawing attention. Similar characteristics are linearly approximated in several applications and satisfactory qualitative and quantitative effects are obtained. Thus, what is the essence of the problem and how to model such cases to obtain identifiable models?

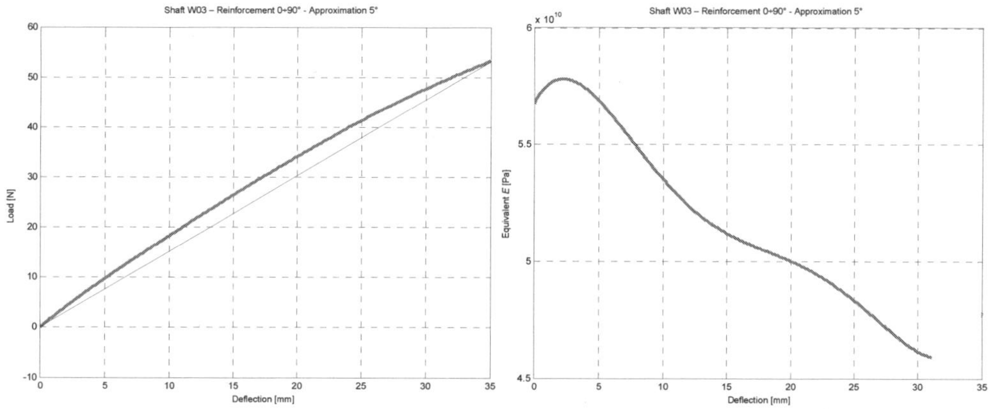

Fig. 3. Examples of elastic characteristic and the equivalent Young module (approximation by the fifth-degree curve).

Attempts to explain the problem

Let us try to explain the effect on the grounds of the theory of non-linear differential equations. For a better clarity of considerations (and the possibility of "non-numerical" discussion) we will first discuss the simplest discrete model of the rotating shaft with a mass concentrated in the middle.

Such model is described by the known equation system [2]:

$$\begin{aligned} m\ddot{h}+kh &= me\left(\ddot{\varphi}\sin\varphi+\dot{\varphi}^2\cos\varphi\right)\\ m\ddot{v}+kv &= me\left(-\ddot{\varphi}\cos\varphi+\dot{\varphi}^2\sin\varphi\right)\\ I\ddot{\varphi} &= ke\left(v\cos\varphi-h\sin\varphi\right)+M_1-M_2 \end{aligned} \tag{1}$$

Where h and v determine – respectively – displacements in the planes perpendicular to the constant reference system, φ – angle of rotation, M_1 and M_2 – driving and anti-torque moments and e – mass eccentricity. The system of equations is non-linear and contains coupling between transverse and torsional vibrations. However, this coupling is very weak for small deflections and the general practice of system linearisation by assuming the stationary motion ($\ddot{\varphi}=0$) is fully justified and leads to results compatible with observations with a sufficient accuracy. The situation will completely change when to the first two equations of the system (1) we will introduce non-linear characteristics of elasticity and damping. The occurrence of even the weakest non-linear disturbance in each equation does not allow for its simple linearisation. Thus, system (1) will be reduced to the form:

$$\begin{aligned} m\ddot{h}+kh+f\left(h,\dot{h}\right) &= me\left(\ddot{\varphi}\sin\varphi+\dot{\varphi}^2\cos\varphi\right)+f_h\left(\varphi,\dot{\varphi}\right)\\ m\ddot{v}+kv+f\left(v,\dot{v}\right) &= me\left(-\ddot{\varphi}\cos\varphi+\dot{\varphi}^2\sin\varphi\right)+f_v\left(\varphi,\dot{\varphi}\right)\\ I\ddot{\varphi}+f\left(\varphi,\dot{\varphi}\right) &= ke\left(v\cos\varphi-h\sin\varphi\right)+f\left(v,\dot{v},h,\dot{h}\right)+M_1-M_2 \end{aligned} \tag{2}$$

This is the system of differential equations "totally" non-linear and insoluble in the accurate way. However, it is possible to try obtaining the expected solutions on the bases of analytical approximate methods. At first, we will reduce the system to the so-called normal coordinates (which means such coordinates each of which vibrates with first frequency of natural vibrations). This leads to uncoupling of equations in a linear system, while in a non-linear system only left members of equations will uncouple. The sequence of proceedings is given below.

Omitting non-linear elements in the system (1) and reducing it to the dimensionless form (by dividing each of equations by the inertia), we perform the linear transformation:

$$h, v, \varphi \rightarrow \xi_i : \begin{cases} h = \lambda_{11}\xi_1 + \lambda_{21}\xi_2 + \lambda_{31}\xi_3 \\ v = \lambda_{21}\xi_1 + \ldots \\ \varphi = \lambda_{31}\xi_1 + \ldots \end{cases} \tag{3}$$

where: λ_{ij} – form coefficients of the linearised system.

In variables ξ_i the system obtains the form:

$$\ddot{\xi}_i + \omega_{0i}^2 \xi = \Phi_K \left(\xi_1 \ldots \xi_i, \dot{\xi}_1 \ldots \dot{\xi}_i, \Omega t \right) \tag{4}$$

where Φ_K is the transformation of the right members and the non-linear functions of equation (2) (in accordance to (3)), and Ω is the fundamental frequency of the forcing expansion in the Fourier series. It is possible to provide the general form of solution for such system [1] on the bases of analytical approximate methods.

$$\xi_j = a_j(t)\cos\psi_j(t) + \frac{1}{(2\pi)^{1+n}} \sum_n \sum_m \frac{\exp i(m\Omega + \sum_k \psi_k n_k)}{\omega_{ok}^2 - (m\Omega^2 + \sum_k n_k \omega_{ok}^2)} \cdot$$

$$\cdot \underbrace{\int_0^{2\pi}\int_0^{2\pi}\int_0^{2\pi} \ldots \int_0^{2\pi}}_{n+1} f_k \exp\left\{(-i)[m\Omega t + \sum_k n_k \psi_k]\right\} d\psi_1 \ldots d\psi_n dt \tag{5}$$

$$\forall m,\, n,\, k \in \left\{ N - (m; n; k \;:\; \omega_{ok}^2 - (m\Omega^2 + \sum_k n_k \omega_{ok}^2) = 0 \right\}$$

Accurate calculations of function f_K parameters are, of course, highly time consuming and in the age of the developed digital technique at least unprofitable, but the discussion of the form (5) allows to draw essential qualitative conclusions:

1. The number of possible resonance responses is the same as the number of zero values, which can occur in the denominator of sum (5).
2. The necessary condition for the resonance response occurrence is the non-zero value of Φ_K function. To make it possible for higher frequencies, the nonlinear characteristics must be approximated by a relatively long expansion in power series (for approximation of $x_i + \varepsilon x^3$ the third harmonics occur, at most!).

3. Values of coefficients occurring in Φ_K functions decrease fast with an increase of inertia coefficients (in a dimensionless system these coefficients are in denominators and further transformations of variables are linear transformations).

Thus, the final conclusion is as follows. In order to obtain the spectrum presented in Figure 2, the elastic characteristics must be functions of higher orders and the shaft has to be of a small inertia.

On the bases of the given above assumptions the accuracy of the measured static characteristics was improved and the dependencies presented in Figure 4 were obtained in the reproducible way.

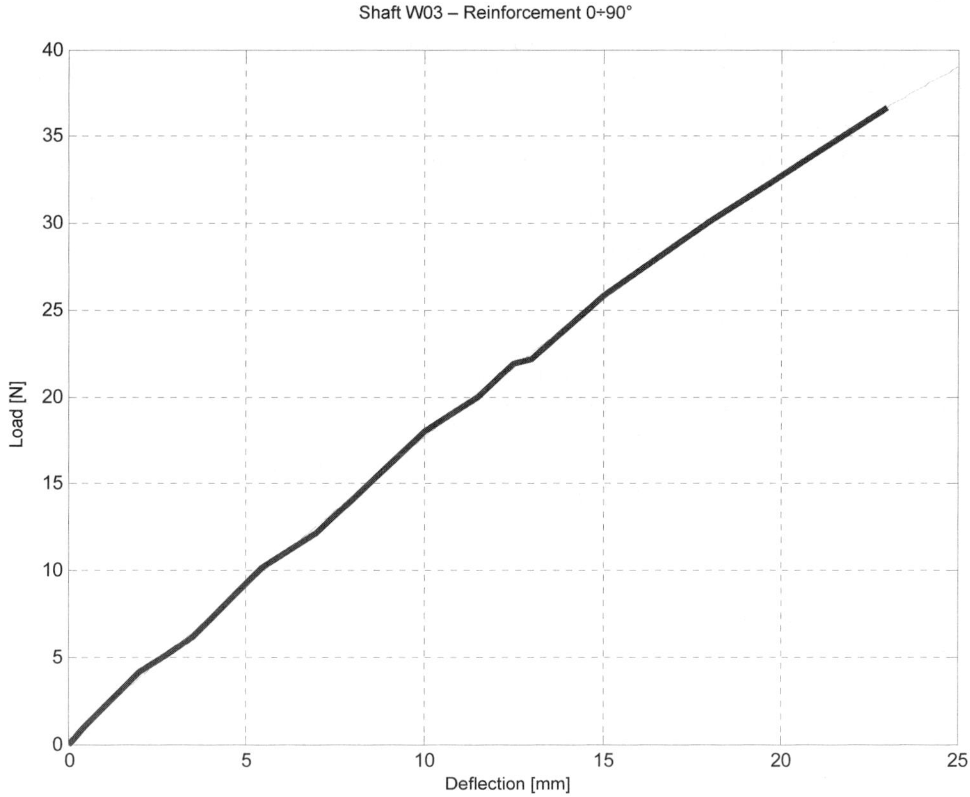

Fig. 4. "Accurate" elastic characteristic of the composite shaft.

Such dependence, being the result of the technological imperfection of the composite product, allows explaining the observed effect.

Conclusions

The following conclusions can be drawn on the bases of the performed considerations.

High-speed machine shafts made of carbon-aramide composites can enter into several critical states, in which successive spectral dominants do not occur for the successive multiplication factors of the fundamental frequency. The amplitude of the first critical state does not have to be (and usually is not) the highest. The amplitude-frequency characteristic resembling the metal shaft one, can be obtained only by a drastic increase of the shaft mass (but after all this is not the purpose of such structures!).

Simulation investigations are difficult and require the preliminary structural identification of the model. Regardless of whether the discrete or continuous model would constitute the grounds for digital calculations, it is necessary to estimate non-linear disturbances with a high accuracy and their approximation by the multinomial of the appropriate degree. In practice, it is only possible (for the time being) to identify the model on the basis of the prototype investigations and to use it for further optimisations. The authors do not see the possibility of the a'priori development of the reliable mathematical model. It is hoped that, a certain correction of this situation will be provided by improved production technologies.

References

[1] Batko W., Dabrowski Z., Engel Z., Kicinski J., Weyna S.: Modern Methods of Vibroacoustic Diagnostic – part I (in polish), Institute for Sustainable Technologies – National Research Institute, Warsaw – Radom, 2005.

[2] Dabrowski Z.: Machine Shafts (in polish), Polish Scientific Publishers PWN, Warsaw, 1999.

[3] Dabrowski Z., Dziurdz J., Skorski W.W.: Vibration of Carbon Fibre Composites Masts, (in polish), Institute for Sustainable Technologies – National Research Institute, Warsaw – Radom, 2007.

[4] Deuszkiewicz P.: Diagnostic Ability Research on Composite Shafts (in polish), XXXIV Polish Symposium on Machine Diagnostic, 2007.

[5] Kurnik W. (red.), Tylikowski A., Starczewski Z., Przybylowicz P.: Nonconventional Materials in Diagnostic and Active Vibration Reduction – DIADYN series (in polish), Institute for Sustainable Technologies – National Research Institute, Warsaw – Radom, 2008.

[6] Pakowski R., Deuszkiewicz P.: Dynamic Research on Nontypical Powertrain Systems (in polish), XXX Symposium on Marine Power Plants – SymSO, Gdynia, 2009.

[7] Kurnik W. (red.), Tylikowski A.: Mechanics of laminating elements (in polish), Warsaw University of Technology Publishers, Warsaw, 1997.

Key Engineering Materials Vol. 490 (2012) pp 83-89
© (2012) Trans Tech Publications, Switzerland
doi:10.4028/www.scientific.net/KEM.490.83

Development of Bevel Gear Motion Transmission Graphs with FEM

Jacek Pacana

Rzeszow University of Technology, The Faculty of Mechanical Engineering and Aeronautics

pacanaj@prz.edu.pl

Keywords: bevel gears, FEM, transmission errors

Abstract. The article presents the course of the calculations and conclusions from the analysis of a bevel gear motion transmission using FEM. The motion transmission graphs show the angular variation of the driven gear in the case of a driving pinion that rotates with a constant angular velocity. In contrary to the classical Tooth Contact Analysis, which is carried out on stiff bodies, FEM analysis includes deformations of mating gears. Thus, it brings more realistic information on dynamic behavior and noise characteristic of analyzed gear drive.

Introduction

Computer numerical calculations currently play a fundamental role in the process of mechanical engineering. At many stages of design of machine parts, various types of analysis and computer simulations are used to determine their size, strength, durability and dynamic parameters resulting from the nature of the work. Among the numerical methods, the finite element method (FEM) is the most popular one. Thanks to many commercial programmes which implement this method, the solution to many design difficulties is much easier and the results allow for a clear and precise interpretation of them. However, to obtain the correct results of desired accuracy, it is necessary to properly prepare a computational model for FEM calculations [1,5,6,11].

Thus, the main task of a designer must be proper preparation of both the computational model, its discretization, definition of the initial and the boundary conditions and relevant analysis of the results and their correct interpretation.

The FEM analysis of gear drives is a specific one due to the complex geometry of the mating tooth flanks and multi-tooth meshing.

The main aim of the FEM analysis of a bevel gear was to define its motion transmission graphs and examine what is the impact of mesh density and the size of the calculation time step on the accuracy of the calculations .

Preparation of the computational model

This paper presents an analysis of the spatial models of gears constituting the bevel gear loaded with a constant torque. Prepared in an external CAD system, solid models of both a pinion and a gear wheel were then imported to the FEM program ABAQUS. All the steps necessary to prepare and carry out the calculations were performed with use of this software. Models with created finite elements mesh, prepared for numerical calculations, are shown on picture 1. Complex geometry of the gears and nonlinear nature of contact analysis results in huge demands in computer calculation capacity.

Therefore, it is essential to build such a computational model that will provide both sufficiently accurate results and economic use of computing resources of a computer. First of all, only active parts of mating gears are used, including five contacting teeth of each gear, and appropriate boundary conditions are applied. Also the geometry is simplified by deleting some dress-up features, which do not affect the stiffness of the original model and have no impact on the system kinematics [2,10].

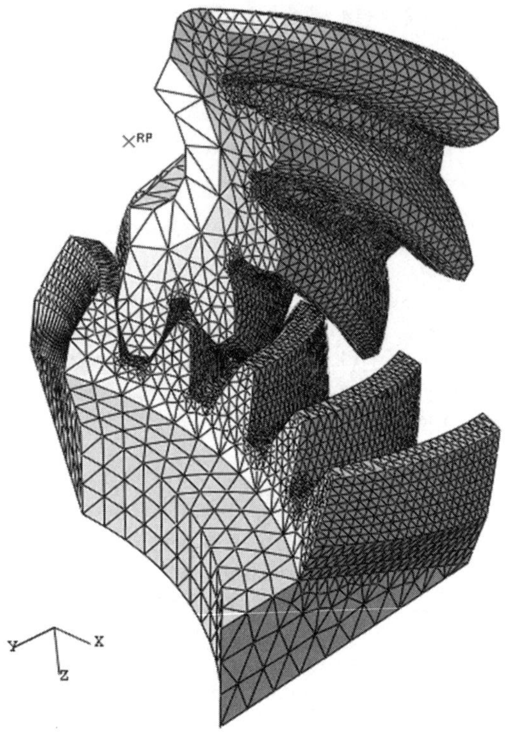

Fig. 1. View of computational models of bevel gears prepared for calculations in Abaqus, divided into finite elements of 2mm

Five different computational models of the same bevel gear have been prepared, with various minimal sizes of finite elements located in most sensitive areas: 5, 3, 2, 1, 0.5 [mm].

The analysis covers the whole range of the gear meshing cycle operating under the constant value of applied torque T = 400 [Nm]. The entire calculation is divided into 100 time steps, each step taking the pinion rotation angle of 0.02 [rad]. For some selected models, with small sizes of finite elements, also calculations with a division for 1000 time steps were carried out, then a single step had the value of 0.002 [rad]. After verification of all entered data and geometric boundary conditions, calculations were made, in result of which correct solutions for all the computational models were obtained.

Processing and analysis of results

The interpretation of the FEM calculation results obtained on the basis of previously prepared models was conducted with use of ABAQUS postprocessor. Due to large number of computational models prepared and low cognitive value for some of the results obtained, only those results will be presented, in the following part of the paper which clearly show the results of the analysis. The von Misses stress distribution for the pinion and the gear are presented in Fig. 2 for the mesh of 0.5 [mm] size. It shows the situation of a single tooth pair contact.

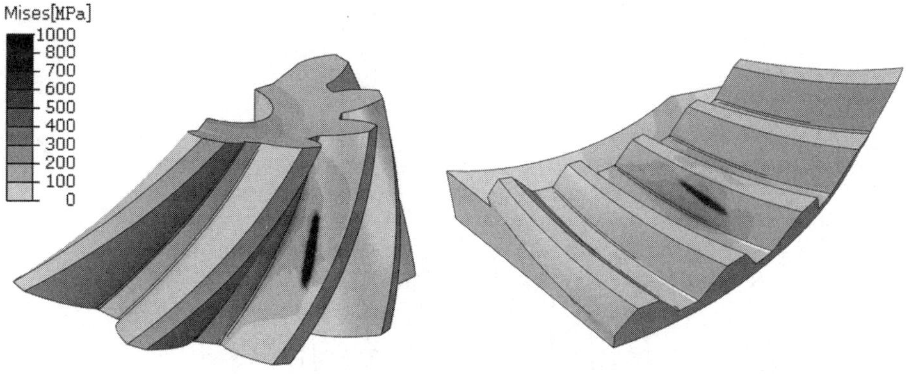

Fig. 2. The von Misses stress distribution for the models of mesh size 0.5 [mm] for: a) the pinion and b) the gear

For all studied computational models the highest stress occurred in the middle of the tooth width for both the pinion and the gear. This stress distribution is a consequence of proper tooth geometry of the wheels and results also from the nature of their cooperation. In the area of tooth contact the maximum stress reaches the value of about 900 [MPa] and the maximum bending stress at the root of the tooth of both the gear and the pinion is about 250 [MPa].

Based on observations of results for all studied computational models, it can be concluded that with increasing density of the finite element mesh, the contact stress distribution on mating tooth surfaces is more structured and it takes a known form of an ellipse [4,8,12].

Motion transmission graphs

The main purpose of the analysis was to determine the motion transmission graphs for a bevel gear under load. The motion transmission graphs show the angular variation of the driven gear in the case of a driving pinion that rotates with a constant angular velocity. Contrary to the classical Tooth Contact Analysis, which is carried out on stiff bodies, FEM analysis includes deformations of mating gears. Thus, it brings more realistic information on dynamic behaviour and noise characteristic of analyzed gear drive.

Motion transmission graphs together with bearing contact determined for both drive and coast side of tooth are essential for assessing the accuracy of bevel gears [14,15].

In order to define the function of motion transmission, both pinion and gear rotation angle values ($\varphi^{(1)}$, $\varphi^{(2)}$) must be determined for all time steps covering the whole range of the tooth meshing cycle. The rotation angles are measured as rotations of characteristic points chosen in such areas of pinion and gear models that do not experience any deformation.

Transmission error $\Delta\varphi^{(2)}$ for a given value of the pinion rotation angle is represented as:

$$\Delta\varphi^{(2)}(\varphi^{(1)}) = \varphi^{(2)}(\varphi^{(1)}) - u_{21}\varphi^{(1)} \qquad (1)$$

where: u_{21} – gear ratio = z2/z1

The motion transmission graphs for all five models with different mesh densities were elaborated. For FEM models created with finite elements of larger size, obtained graphs are ambiguous and difficult to interpret like, for example, the motion transmission graph obtained for the model with the element of 3 [mm] size shown in Fig. 3.

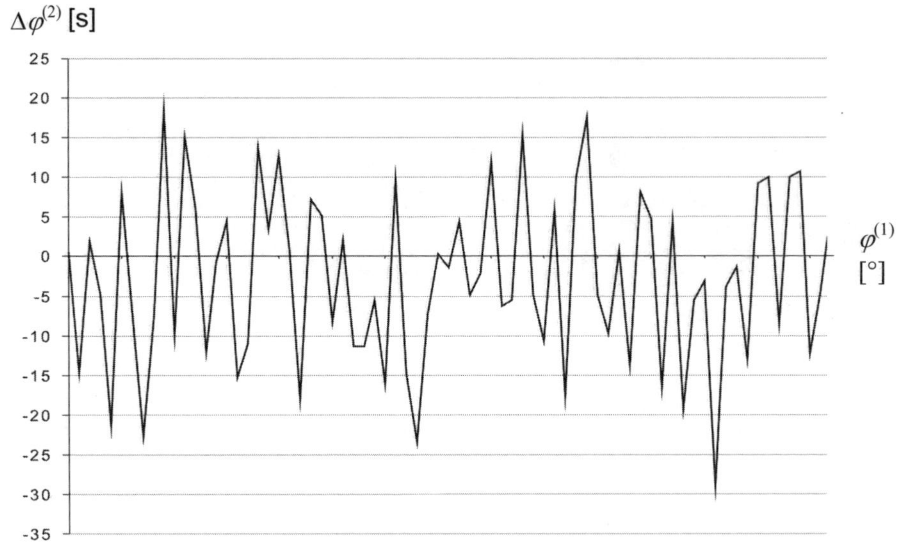

Fig. 3. Motion transmission graph obtained for the model with the element of 3 [mm] size.

For the FEM model with the element of 0.5 [mm], the graph is much clearer (Fig. 4). It resembles the graphs obtained for real bevel gears on testing machines.

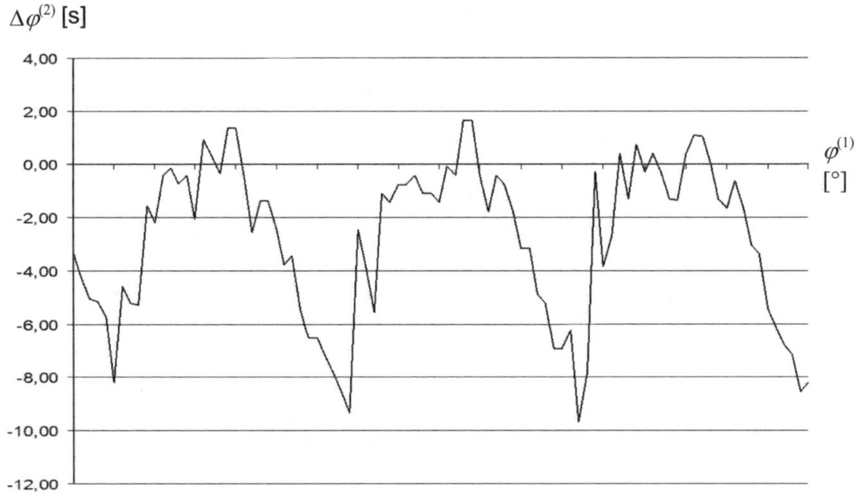

Fig. 4. Motion transmission graph obtained for the model with the element of 0.5 [mm]

The resulting graphs are not smooth because they contain inaccuracies in the calculated angle of rotation and, in particular, geometry errors resulting from replacement of the smooth outline of theoretical side of the tooth surface with discrete finite elements.

Also an attempt was made to designate the graph for a smaller time step value in terms of the rotation angle of pinion equal to 0.002 [rad]. However, the graph obtained, as presented in Fig. 5, is also irregular and unclear. Thus, it can be concluded that the quality of the results obtained in the analysis of bevel gear transmission errors depends on the discrete value of pinion rotation angle (calculation time step), not directly but in relation to the size of the finite element mesh used for discretization of computational models.

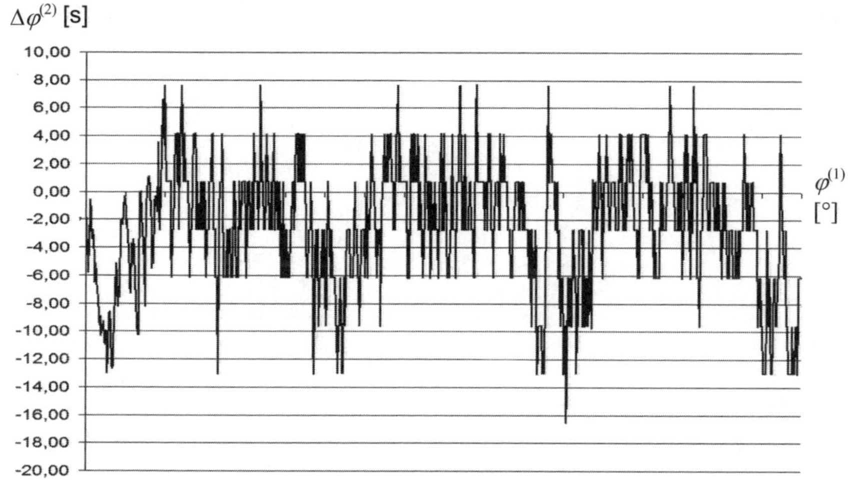

Fig. 5. Motion transmission graph obtained for the model with the element of 0.5 [mm] and the time step equal to 0.002 [rad]

In order to compare results obtained from the calculations on basis of the FEM method to theoretical values, the above graphs have been smoothed out. The resulting graphs for the FEM model with the element of 0.5 [mm] and both sizes of time steps taken are presented in Figures 6 and 7.

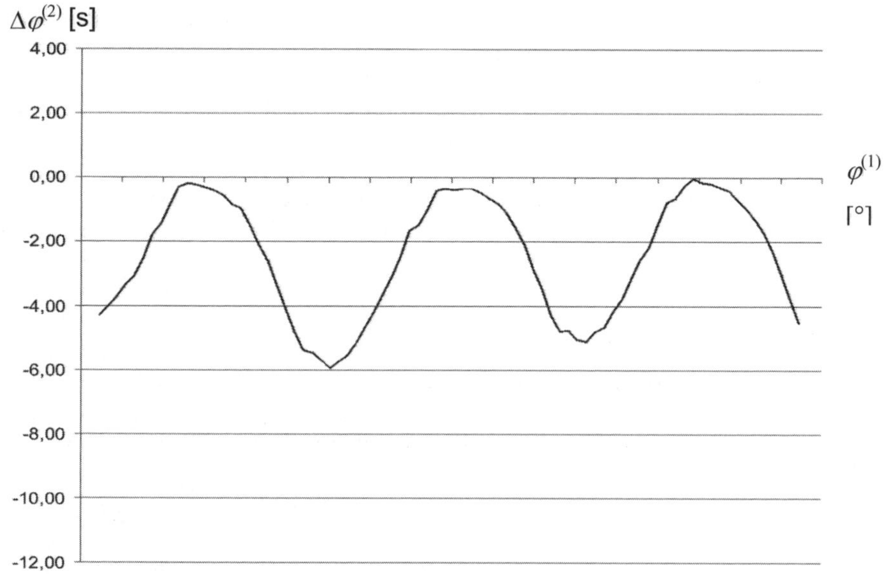

Fig. 6. Smoothed motion transmission graph obtained for the model with the element of 0.5 [mm] and the time step equal to 0.02 [rad]

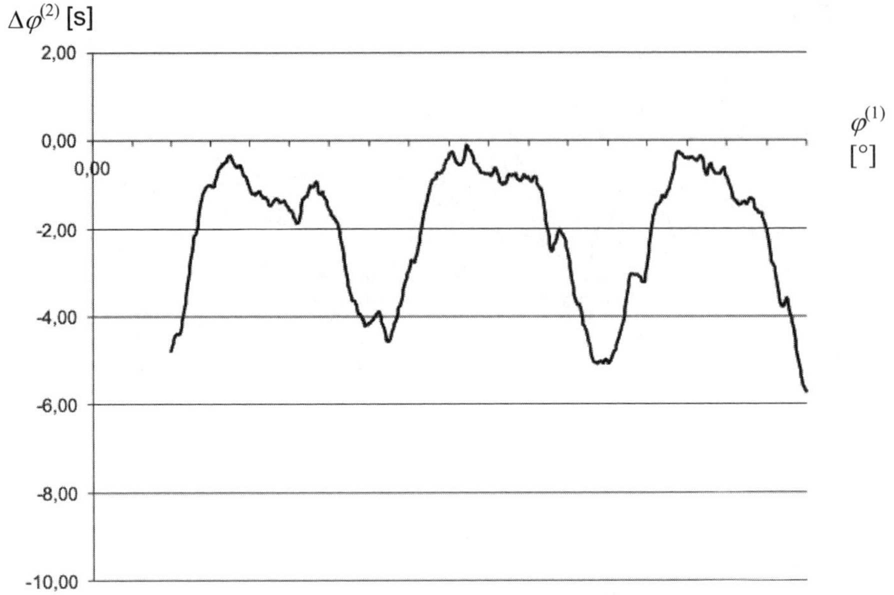

Fig. 7. Smoothed motion transmission graph obtained for the model with the element of 0.5 [mm] and the time step equal to 0.002 [rad]

The graph obtained for the calculation step equal to 0.02 [rad] is most similar to an inverted parabola and its shape is similar to that obtained with TCA method. The graph obtained for calculations performed with the time step of 0.002 [rad] reveals a discrepancy as compared to the previous one.

In this case, the graphs are clearly flattened at the apexes. This flattening is the result of elastic deformations of mating teeth which appears in the contact area. Similar results were obtained in experimental investigations of gear drives [3,13].

Additional disruption of graphs are caused by bending of the teeth under load. This phenomenon is noticeable only in models with the time step of calculations of 0.002 [rad], which justifies the use of large numbers of time steps.

Conclusions

The performed calculations confirmed the need for use of computational models with the greatest possible density of the finite element mesh in critical areas of the gears.
This has an impact both on the correctness of the values of stress and the accuracy of their distribution. However, due to a recurring characteristic pattern of stress distribution, models with lower-density mesh are recommended to be used in the initial calculations.

As part of the analysis, motion transmission graphs were created showing the instantaneous acceleration and deceleration to constant gear pinion rotation. Smoothed figure presented in Figure 6 confirmed the cyclical nature of speed variation of the gear. With the calculations carried out for a very small time step unit (Fig. 7), additional non-uniformity can be observed (flattening of the graph) due to load gear.

On the basis of uneven traffic graphs of bevel gear, obtained in result of computer simulation, it can be concluded that the value of the maximum displacement of traffic does not exceed 6 seconds of arc. Therefore they fall within the acceptable range of 10 seconds [7, 9]. The values obtained allow to assess the quality of the work of highly analyzed bevel gear whose work will be steady and gentle in nature, without a noticeable step change speed.

Acknowledgments

Financial support of Structural Funds in the Operational Programme - Innovative Economy (IE OP) financed from the European Regional Development Fund - Project No POIG.0101.02-00-015/08 is gratefully acknowledged.

References

[1] Bąk R., Burczyński T. (2001). *Strength of materials with elements of computer approach.* Warsaw, Poland: Wydawnictwa Naukowo-Techniczne.

[2] Budzik G., Pacana J. (2008). *Analysis of MES solution correctness depending on type and number of finite elements used.* Kosiče: Acta Mechanica Slovaca, 3-A/2008, PRO-TECH-MA Ročnik 12.

[3] Gosselin C. (1999). *Computation and measurement of the kinematical motion error of actual hypoid gears under load.* Paris: 4th World Congress on Gearing and Power Transmission.

[4] Handschuh R.F., Bibel G.D. (1999). *Experimental and Analytical Study of Aerospace Spiral Bevel Gear Tooth Fillet Stresses.* Transactions of the ASME, Journal of Mechanical Design, 121.

[5] Kawalec A., Wiktor J. (2001). *Computer simulation of spur gear generation - part of general CAE system for gear design and manufacturing.* Warsaw: 2nd International Conference on Advances in Production Engineering.

[6] Kleiber M. (1995). *Computer methods of solids mechanics.* Warsaw: PWN

[7] Litvin F.L. (1968, 1994). *Theory of Gearing.* Moscow: Nauka, second edition 1968, also, Englewood Cliffs, NJ: Prentice-Hill, 1994

[8] Marciniec A. (2001). *Analysis of engagement and cooperation trace of bevel gears with circular-arched tooth line.* Scientific Bulletins of Rzeszów Uniwersity of Technology, 188, Mechanics, 57.

[9] Marciniec A. (2002). *Synthesis and analysis of engagement of bevel gears with circular-arched tooth line.* Rzeszów: Oficyna Wydawnicza Politechniki Rzeszowskiej.

[10] Pacana J., Budzik G., Kudasik T. (2008). *Defining bending stress in gears with finite elements method.* Kosiče: Acta Mechanica Slovaca 3-A/2008, PRO-TECH-MA Ročnik 12.

[11] Rusiński E., Czmochowski J., Smolnicki T. (2000). *Advanced method of finite elements in load-bearing structures.* Wrocław: Oficyna Wydawnicza Politechniki Wrocławskiej.

[12] Sałaciński T. (2003). *The basis of shaping and cooperation simulation of bevel gears teeth surface.* Warsaw: Oficyna Wydawnicza Politechniki Warszawskiej, Mechanika, 200.

[13] Skawiński P., Siemiński P. (2008). *Study of cooperation trace and generation of moving diagrams of spiral bevel gears in CAD software.* Scientific Bulletins of Rzeszów Uniwersity of Technology, 259, Mechanics, 75.

[14] Weck M., Stadtfeld H. (1984). *Optimization of spiral bevel gears.* Industrie Anzeiger, 48/1984.

[15] Wójcik Z. (2004). *Bevel gears in Gleason system, construction and technology.* Rzeszów: Oficyna Wydawnicza Politechniki Rzeszowskiej.

Key Engineering Materials Vol. 490 (2012) pp 90-96
© (2012) Trans Tech Publications, Switzerland
doi:10.4028/www.scientific.net/KEM.490.90

Dynamics Conscious Approach to Tribometer Design and Tribo-testing

Jacek I.LUBINSKI[1,a], Szymon GRYMEK[1,b], Krzysztof DRUET[1,c], Henryk OLSZEWSKI[2,d]

[1] Mechanical Engineering Faculty, Gdansk University of Technology, ul.G.Narutowicza 11/12, 80-233, Gdansk, POLAND

[2]Ocean Engineering and Ship Technology, Gdansk University of Technology, ul.G.Narutowicza 11/12, 80-233, Gdansk, POLAND

[a]jacek.lubinski@pg.gda.pl, [b]sgrymek@pg.gda.pl, [c]kdruet@pg.gda.pl, [d]h.olszewski1@wp.pl

Keywords: tribometer, dynamics, vibration, tribometrology, repeatability, reproducibility

Abstract. In the paper findings are discussed on the issue of dynamic characteristics of a tribometer as a factor influencing the result of the tribological experiment. An advanced approach to tribo-testing is attempted with integrated dynamic analysis of the tribometer and the sliding pair. The fundamental idea is explored of the tribometer being regarded as any machine in which friction is inflicted with all the resulting consequences such as e.g. wear, vibration, fatigue etc. The research described is concentrated around friction induced vibration and it's potential consequences to the results obtained in a tribological experiment. General methodology used for the task described is presented.

Influence of machine dynamics and vibration on friction in tribological experiment.

There are numerous examples of the influence of friction induced vibrations and dynamic characteristics of the machine upon friction and wear in sliding contacts. Among most important factors are the properties of sliding materials and lubricant, load/velocity input and contact geometry, but also dynamic characteristics of the machine (tester) which influenced the friction process. In some cases it a variation in test results was observed as a consequence of alteration of the stiffness of the support system for the sliding contact, even when no measurable vibration was present [1, 2].

Kato et al. [6] conducted tests on how wear is influenced by friction induced vibration and rigidity of support of the passive specimen in direction tangential to sliding velocity. Tests were carried out on various metallic materials with the use of an inertial tribometer in a flat – on – cylinder contact configuration. Friction induced vibration was found to influence the wear of sliding pair, also a decrease in specimen support rigidity resulted in increased wear scatter.

Aronov et al. [2, 3] presented experimental results from unlubricated steel – on steel tests carried out on a pin – on – disc tribometer. A mathematical model was created with the account taken for of 3 degrees of freedom of the pin system. Disc was assumed to represent a massive, rigid body in a constant velocity motion. The random irregularities on disc surface were observed to inflict harmonic vibrations of the pin at frequencies close to natural frequencies of vibration of the system. Vibration amplitude decreased with the running in of the disc surface and the increase of load. Most significant form of vibration was the tilting of the vertical axis of the pin; The change of sliding surfaces contact from conformal to concentrated caused abrupt increase of measured friction.

In later paper [3] wear increase was observed as a result of the increase of contact system stiffness normal to the friction surface. Wear changes correspond linearly to stiffness changes mentioned. It was proposed, that increased natural frequencies of the system contribute to faster accumulation of fatigue cycles in the sliding surface, thus accelerating wear.

Aronov et al. [1] also discovered a wear increasing effect of the increase in sliding system stiffness in pin – on – disc tests while investigating mild wear, which was also increased with the increase of load. Stiffness increase effect was present for both critical directions: tangential and normal to the sliding surface.

Grudzinski et al. [5] emphasized the necessity of correct representation of friction forces when conducting dynamic modelling of a mechanical system, especially when subjected to alternating load and velocity. In theoretical mechanics it is a running problem to account for friction in the system in a way, which would, in effect yield usable simulation results

Any motion perpendicular to the sliding contact interface will strongly influence friction. A non constant force acting in the direction added to a constant force component can result in temporary contact losses or normal force fluctuation, which, according to results obtained by Tolstoi [10] and Martins et al. [9] may result in faked changes of observed friction coefficient of 10 to 30%; results obtained in own research by authors of this paper confirm the finding [7, 8], but an observed friction coefficient decrease amounted, in extreme cases, to 90%, as compared to non vibrating conditions.

In numerous works the issue of stick – slip vibration accompanying low velocity sliding. Van De Velde et al. [12] embarked on the problem by modelling a test stand as a 2 mass system. In simulation the friction process was not influenced by changes in dynamics of the machine, but the vibration of the machine can interfere with the measurement of sliding velocity, thus introducing error. For large disproportion between the most and the least rigid component in the system an opposite circulation can be observed in the measured dynamic friction coefficient, contrary to the anticipated by a model calculations.

For better understanding of friction dynamics special tribometers are designed e.g. a pin – on – disc machine [11], designed for easy delimitation of the dynamic characteristics of all components by experimental methods and for simplicity of dynamic interactions in the system. Interchangeable components for controlled alteration of stiffness were introduced into the system. This allows for changes in natural frequencies of the passive specimen holder. The mathematical model was tuned with the help of experimental data. The authors ,Tworzydlo et. al., proposed, that the excitation of steady friction induced vibration is only possible if there is a negative gradient of friction coefficient as a function of sliding velocity. In the discussed work both stick – slip and quasiharmonic vibrations were observed. Coupled normal and tangential vibrations were observed with surface separations occurring at certain input parameter combinations. Also tilting of the pin was observed resulting in concentrated contact occurring with very high friction coefficient (even in excess of 1) and rapid deterioration of the disc surface.

As a classical example of a wide, multi – factor fundamental tribological research programme the VAMAS can be recalled [4] as an attempt to achieve repeatability and reproducibility in sliding friction experiments. The aim was to properly identify test parameters vital to the assumed goal. Czichos et. al. [4] presented a summary of test results. Steel and alumina specimens were tested in standard conditions. Repeatability of test results on same test rig in one laboratory was satisfactory – maximum scatter of friction coefficient ±13% and wear ±14%. Comparison of results from various laboratories revealed less favourable results of ±20% and ±38%, respectively. The main reason indicated was the variation in shapes and contact geometry of specimens in different test rigs. In most tests no vibration measurement was available, so it can only be speculated on the basis of measured friction coefficient scatter, that in a given test there could be a vibration factor involved.

Own research results on test rig dynamics influence friction induced vibration and tribological experiment results

Authors have conducted extensive research on the subject of friction induced vibration and the influence of dynamic configuration of the test rig upon the result of the tribological experiment [7, 8, 14]. In Fig 1 a schematic is presented of a typical geometry of a sliding contact used in PT type tribometers for friction and wear tests. Initially a simple model was used to model the dynamic behaviour of the system, comprising of 2 masses, with one of the specimens treated as being part of a rigid system of large inertia. In several test series the tribometer was utilised in various mass/inertia and test system support stiffness. In Table 1 the natural frequencies of the test set – up are given available in the presented test program. The changes were achieved by replacing components designed for the purpose of executing changes of inertia and stiffness of support of the test set up. The research programme was centred around the analysis of the influence of purpose made changes to the dynamics of the tribometer on the course of experiments on sliding friction conducted under repeatable conditions. Sintered alumina specimens were used and distilled water or paraffin oil were used as lubricants. Experiments were conducted with constant velocity/variable load or variable velocity/constant load conditions.

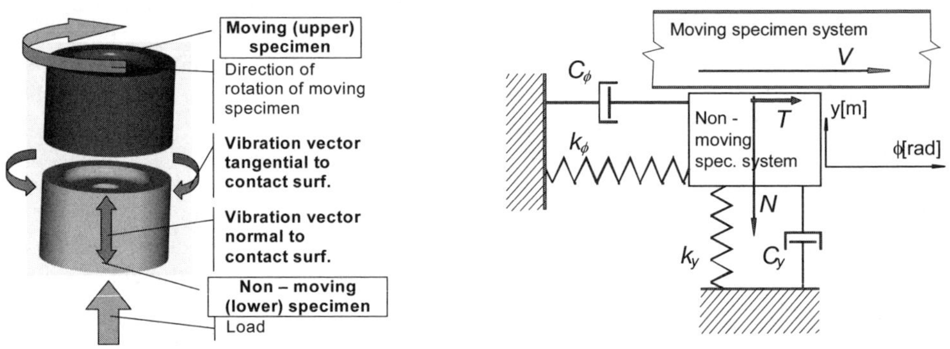

Fig. 1 Schematic of the experimental contact geometry in a PT tribometer (left) and a dynamic model of the tribometer reduced to a two mass system

Table 1 Natural frequencies of vibration of the 2 mass model of the test set up of the PT tribometer available in test with the use of inertial and stiffness altering components.

tangential to contact surf. $\mathbf{f_n}$ [Hz]	normal to contact surf. $\mathbf{f_s}$ [Hz]
77	1402
150	
193	1875
248	

The vibration modes analysed in the experimental programme were the ones representing two crucial directions of motion in the system: tangential and normal to the friction interface. An array of combinations of inertia/stiffness were tested and experimental data comparative analysis then conducted. As one of most useful parameters the primary component frequency of the spectrum of vibration recorded. If no vibration was present during the test a zero frequency was displayed in a graph.

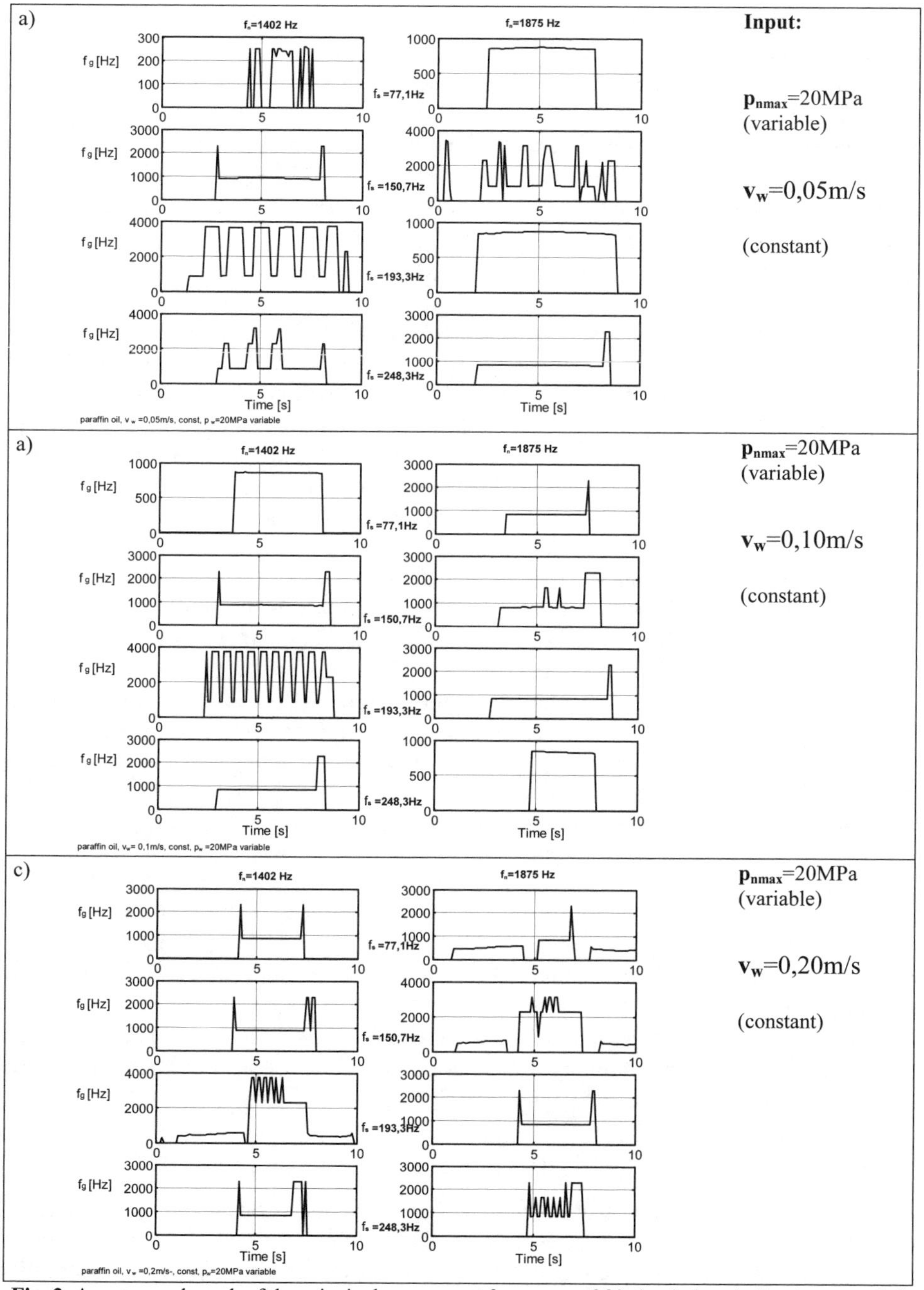

Fig. 2. A compound graph of the principal component frequency of friction induced vibration recorded in tests carried out for 8 different dynamic configurations of the test set – up with paraffin oil lubrication and variable load ($\mathbf{p_{nmax}}$=20MPa), and 3 constant sliding velocities ($\mathbf{v_w}$: 0,05 m/s **(a)**, 0,10 m/s **(b)**, 0,20 m/s **(c)**)

In Fig. 2 a set of graphs is presented. The aim of tests was to better understand the influence of dynamic configuration of the rig on the experimental results and sliding friction induced vibration. The tests presented were carried out with paraffin oil lubrication at constant sliding velocity (v_w=0,05 m/s (a), 0,10 m/s (b), 0,20 m/s (c)) and variable contact pressure (p_{nmax}=20 MPa).

At v_w=0,05 m/s and the normal natural frequency of f_n=1402 Hz it is visible, that the range of input load, within which vibration was observed, is broadened with the increase of the tangential frequency of vibration of the test set up (f_s) from 77,1 Hz, through 150,7 Hz, to 193,3 Hz. Increase in f_s results in vibration being excited at lower surface pressure. Further increase of the f_s frequency, up to 248,3 Hz, results in a narrowing of the range of load with vibration in the test, but the effect is weak. The vibration classified as low frequency, with main harmonic frequency f_g≈250 Hz, was only observed in the set of tests carried out at v_w=0,05 m/s (f_s=77,1 Hz, f_n=1402 Hz), presented in Fig. 2 a. The most common principal harmonic frequency observed was approximately f_g≈850 Hz, beside it two typical frequencies are present f_g≈2125 Hz and 3725 Hz.

At higher sliding velocity (Fig. 2 **b** – $\mathbf{v_w}$=0,10 m/s) the increase of $\mathbf{f_s}$ from 77,1 Hz to 193,3 Hz is accompanied by widening of the band of $\mathbf{p_n}$ in which vibration is observed. At both $\mathbf{v_w}$=0,05 m/s and $\mathbf{v_w}$=0,10 m/s principal component frequency of vibration above 850Hz are dominating the process. A noticeable presence of vibration at principal frequencies below 850Hz could be observed in tests at sliding frequencies of $\mathbf{v_w}$=0,20 m/s (Fig. 2 **c**). The 'low frequency' vibration did not occur only in three tests out of eight presented; the cases correspond to natural frequency of vibration $\mathbf{f_s}$=77,1 Hz i 150,7 Hz (for $\mathbf{f_n}$=1875 Hz – Fig. 2 right column) and $\mathbf{f_s}$=193,3 Hz. When analysing tests carried out with $\mathbf{f_s}$=77,1 Hz and 150,7 Hz it can be stated, that, regardless of the normal natural frequency $\mathbf{f_n}$, the increase of $\mathbf{f_s}$ results in a broadened band of contact pressure within which the 'high frequency' vibration is observed.

At the configuration resulting in the f_n=1875 Hz the dependence of the load band in which the 'high frequency' vibration is observed from the f_s value seems identical for all the sliding velocities in the programme. For f_n=1402 Hz the tendency of the changes in the picture of the friction induced vibrations described for $\mathbf{v_w}$=0,05 m/s and 0,10 m/s does not seem to be observed also at $\mathbf{v_w}$=0,20 m/s. When the configuration is changed from $\mathbf{f_s}$ from 77,1 Hz to 150,7 Hz the band of load accompanied by vibration is broadened, but no changes in the frequency pattern of the observed vibration are recorded. In tests with of $\mathbf{f_s}$ =193,3 Hz the 'low frequency' can be observed and a distinct narrowing of the contact pressure band in which the 'high frequency' vibration is recorded. It can be speculated, that tests could be carried out with configuration changes made at smaller intervals will reveal interactions, should they exist, between the 'high' and 'low' frequency friction excited vibration. Such finding could be very interesting for the friction measurement, because in the tests presented the influence of the 'low frequency' vibration on the friction coefficient was a decrease of up to 90% as compared to runs with no recorded vibration or with 'high frequency' vibration, which, on the other hand, had no noticeable effect on the friction coefficient measured.

The course of vibration spectrum and amplitude changes are very similar for both, the tangential and normal vibrations. The maximum amplitude of normal vibration at 'high frequency' amounted to about 0.7 μm and at 'low frequency' approximately 2.5 μm.

Final remarks

From publications available and from own experience a general conclusion can be drawn, that it seems like an inevitable development in experimental technology for use in tribological experiment to introduce a profound dynamic analysis of the test stand as a means of improved usability of test results. Current level of the development of both the design tools (3d CAD), computer modelling tools for predictive dynamic modelling a machine at the design stage and the capacity of data gathering systems allow for a postulate to recognise the dynamic analysis of the tribological experiment as a support tool to be used on a regular basis in tribology. The benefits of the introduction of the analysis of the dynamic behaviour of a tribometer performing a test and an in – test vibration measurement, with it's post – experimental analysis, are expected to be improvements in accuracy, repeatability and reproducibility of tribological experiments. Also, with a certain amount of experience accumulated, it seems possible to develop pre – experimental routines enabling the dynamic response characteristics of test rigs to be tuned to meet the requirements of a certain type of tribological experiment; Such routines could, for example, help prevent the onset of friction induced vibration during the course of a sliding friction and accompanying wear experiments. The alteration of mass/inertia and stiffness distribution or application of controlled damping to certain components of the mechanical system of a tribometer coupled with vibration analysis assisted data processing could add a new quality to the tribological experiment.

A tribometer, being a machine, should thus be treated as a dynamic system with friction on the design stage as well as while designing the tribological experiment for the benefit of research on friction and wear, but also for improved applicability of experimental results in technology. Such approach should result in added value to products developed with the help of new, dynamics enhanced, tribological research practice.

Currently authors are conducting research on the development and experimental verification of the dynamic models of the PT series tribometers for further use as a component of a standard experimental procedure associated with the class of tribological test rigs.

ACKNOWLEDGEMENTS

The financial support from the *Polish Ministry for Science and Higher Education* in the form of research grant No N N 504 411 437 is hereby gratefully acknowledged.

Summary

If you follow the "checklist" your paper will conform to the requirements of the publisher and facilitate a problem-free publication process.

References

[1] Aronov V., D'Souza A.F., Kalpakijan S., Shareef I., (1983) Experimental Investigation of the Effect of System Rigidity on Wear and Friction - Induced Vibrations, Trans. of the ASME, Vol. 105, pp.206 - 211

[2] Aronov V., D'Souza A.F., Kalpakijan S., Shareef I., (1984) Interactions Among Friction, Wear, and System Stiffness - Part 1: Vibrations Induced by Dry Friction, Trans. Of the ASME - Journal of Tribology, Vol. 106, pp. 59 - 64

[3] Aronov V., D'Souza A.F., Kalpakijan S., Shareef I., (1984) Interactions Among Friction, Wear, and System Stiffness - Part 2: Wear Model, Trans. Of the ASME - Journal of Tribology, Vol. 106, pp. 65 – 69

[4] Czichos H., Becker S., Lexow J., (1987) Multilaboratory tribotesting: results from the Versailles Advanced Materials and Standards Programme on Wear Test Methods, Wear 144, pp.109-130

[5] Grudziński K., Konowalski K., Kostek R., (2000) Modelowanie kontaktu tarciowego w układzie dynamicznym, s.9 – 16 Proc. of the Int. Conf. „Friction 2000" Modelling and Simulation of the Friction Phenomena in the Physical and Technical Systems, Warsaw University of Technology, Warsaw, 23-24 Nov. 2000

[6] Kato K., Iwabuchi A., Kayaba T., (1982) The effects of friction induced vibration on friction and wear, Wear Vol. 80, pp. 307 – 320

[7] Lubinski T, Druet K., Lubinski J.I. (1997) Dynamic features of a tribometer affect friction and wear test results, Proc. 1st World Tribology Congress, London, 1997.

[8] Łubiński T., Druet K., Łubiński J.I., (1996) Doświadczalne badania drgań wzbudzanych tarciem w styku materiałów ceramicznych, Krajowa Konf. Nauk.-Techn., „Inżynieria Łożyskowania", '96, Gdańsk 4-5 czerwca 1996.

[9] Martins J.A.C, Oden J.T., Simoes F.M.F., (1990) A study of static and kinetic friction, International Journal of Engineering Science , Vol. 28, 29-92

[10] Tolstoi D.M., (1967) Significance of the normal degree of freedom and natural vibraitons in contact friction, Wear, Vol.10, pp.193 – 213

[11] Tworzydlo W.W., Hamzech O.N., Zaton W., Judek T.J., (1999) Friction induced oscillations of a pin – on – disk slider: analytical and experimental studies, Wear 236, pp.9 – 23

[12] Van De Velde F., De Baets P., (1998) The relation between friction force and relative speed during the slip-phase of a stick-slip cycle, Wear No. 219, pp.220-226

[13] Van De Velde F., De Baets P., Degrieck J., (1998) The friction force during stick – slip with velocity reversal, Wear 216, pp. 138 – 149

[14] Łubiński J.I., Druet K., Łubiński T., (2001) Variable dynamics test set-up for friction induced vibrations investiga-tion, W: [CD-ROM] Presentations/Papers, Authors and Keywords. 2nd World TRIBOLOGY Congress. Vienna, Austria, 03-07 September 2001. Vienna: Austr.Tribology Soc.**2001 ISBN 3-90-1657-09-6

Key Engineering Materials Vol. 490 (2012) pp 97-105
© (2012) Trans Tech Publications, Switzerland
doi:10.4028/www.scientific.net/KEM.490.97

Experimental Investigation of Prototype Water-Lubricated Compliant Foil Bearings

Artur Olszewski[1, a], Michał Wodtke[1,b] and Piotr Hryniewicz[2,3, c]

[1]Gdansk University of Technology, ul. Narutowicza 11/12 80-233 Gdansk, Poland

[2] Polish Academy of Sciences, ul. Fiszera 14, 80-952 Gdansk, Poland

[3] Current address: Widener University, Mechanical Engineering, Chester, PA 19013, USA

[a]aolszews@pg.gda.pl, [b]mwodtke@pg.gda.pl, [c]phryniewicz@mail.widener.edu

Keywords: foil bearing, water lubrication, experimental tests, test rig

Abstract. First gas-lubricated compliant foil bearings (CFBs) were built in the 1950s. Due to their significant advantages, such as oil-free operation, good tolerance to bearing misalignment and very low maintenance, they have been penetrating the bearing applications for high speed compressors, air-cycle machines and gas turbines. The work presented here investigates a novel idea of water-lubricated compliant foil bearings, which could be used in applications where environmentally friendly lubrication is desired, for example in hydroelectric turbines or water pumps. Experimental results collected for three prototype water-lubricated foil journal bearings are presented. The tests were conducted under steady radial load and with the sliding speed varied incrementally. A sequence of design improvements is presented, with the best bearing demonstrating friction coefficient of about 0.01 at the sliding speed of about 4 m/s and the radial load of about 300 kPa. Encountered difficulties, research methodology and the testing equipment are also described.

Introduction

Higher than ever environmental awareness forces engineers to search for safe, environmentally friendly and reliable solutions in many areas, including bearing systems. This is especially true in hydroelectric power plants, large water pumps, marine drive trains and other applications where equipment breakdown may lead to water contamination. Unfortunately, relatively low viscosity of water (relative to oil) causes that water-lubricated hydrodynamic bearings have low load carrying capacity and are very sensitive to any rotor misalignment. However, it seems that these shortcomings can, at least to some extent, be alleviated by replacing traditional rigid bearings with compliant foil bearings.

Compliant foil bearing (CFB) technology is relatively new and has been limited almost entirely to gas-lubricated bearings, with one notable exception of the foil bearings lubricated with cryogenic fluids [1]. Due to their advantages (e.g., high damping coefficients, good tolerance to misalignment and relatively high load capacity) gas CFBs found their first commercial applications in high-speed turbomachinery (air-cycle machines, gas turbines and compressors) replacing more traditional oil-lubricated bearing solutions [2]. The initial successful applications triggered substantial research and development effort worldwide, pioneered mostly in the USA and France. However, nearly all published literature investigates air-lubricated bearings ([3], [4]) and has focused mostly in two areas: development of theoretical models that would aid bearing design [5], and experimental validation (or feasibility studies) focusing mostly on dynamic behavior of high-speed rotors supported on CFBs [6]. Published experimental work indicates that testing of CFB systems is relatively difficult, largely because the measured quantities (such as temperatures at selected bearing locations and frictional torque) must be evaluated during high rotational speeds and oftentimes at high temperatures as well. Moreover, the implemented sensors may significantly alter the determined characteristics of the investigated bearing system [7].

In search of new bearing solutions, the authors proposed and developed a new concept and design methodology of water-lubricated journal CFBs. Three prototypes (differing mostly in the design of their elastic support) were manufactured and tested, with the goal of verifying their proper operation and determining their operating characteristics under steady load and steady speed conditions. The experiments were performed using test stand PG2-1Ł (described below) capable of measuring steady state characteristics of hydrodynamic journal bearings. The experimental results were also used to evaluate the design methodology and manufacturing methods used to build the prototype bearings.

Design constraints

It was decided early on that the prototype foil bearings would be tested in the existing journal bearing test stand PG2-1Ł, modified for water lubrication. This decision defined the envelope of bearing overall dimensions and the operating conditions. Because the load direction in the test stand remains constant, the bearing angular extent of about 180° was chosen, which greatly simplified assembly (no need to remove the rotor) and lowered the cost of the prototypes. The journal diameter was selected as 98 mm, with the same bearing length ($L/D = 1$). The details of bearing design were presented in Reference [8] and the manufacturing methods used followed mostly the methods presented in [9]. The top foil was made out of precipitation hardened stainless steel coated with commercially available low-friction coating (based on PTFE and graphite), which was selected based on a comparative wear testing of several candidate coatings [10].

Design effort

Design effort of the prototype water-lubricated CFB was preceded by several numerical studies using specially developed for this purpose numerical package, capable of predicting steady state characteristics of journal CFBs. The numerical model consisted of two coupled modules: one simulated the flow and pressure distribution in the hydrodynamic film and the other calculated the deflection of the bearing structure. Detailed geometry of the elastic support and especially the geometry of the bump foil was optimized in order to maximize the bearing load capacity at a given minimum film thickness. Selected results of the performed numerical study were presented in References [8] and [11]. The results of the numerical study were used to design first prototype water-lubricated CFB. The bearing was then manufactured and its performance evaluated experimentally.

Testing methodology

Experiments were performed utilizing journal bearing test rig PG2-1Ł available at Gdansk University of Technology. The test rig can determine static characteristics of hydrodynamic journal bearings with diameters up to about 110 mm. A schematic of the test rig is presented in Figure 1.

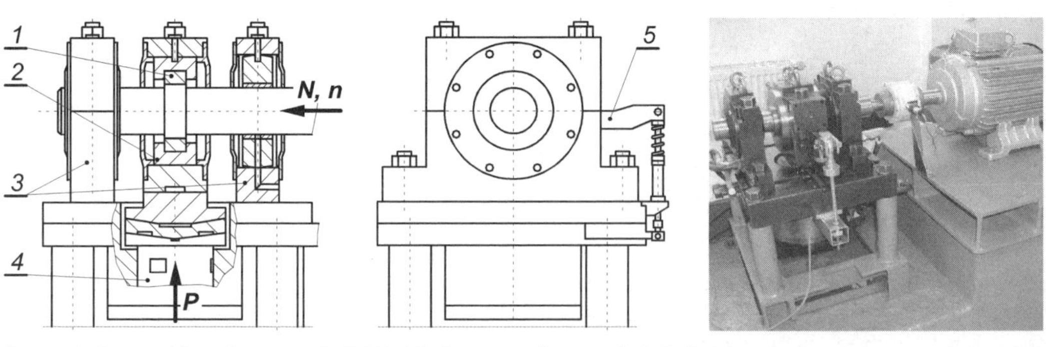

Figure 1. Journal bearing test rig PG2-1Ł (see text for symbol definitions).

The tested bearing (1) is mounted in the test head (2), which is located between two oil-lubricated hydrodynamic bearings (3) supporting the rotor. The bearing load is applied by a calibrated hydraulic piston (4) by supplying oil under a known pressure to the space under the piston. The

tested bearing was completely submerged in water, with the total water volume of about 0.8 liters. The rotor was driven by an electric motor with the nominal power of 11 kW and the rotational speed of 0 – 4000 rpm controlled by an inverter. A closed-loop controller ensured high repeatability of testing conditions, especially for the start / stop operation. The measured data was recorded with a PC-based, fully automated data acquisition and data logging system.
Reliable evaluation of water-lubricated hydrodynamic bearings can be done primarily by very precise measurement of the frictional torque in the tested bearing. Measurement of other indicators, such as the film thickness and hydrodynamic pressure, is quite difficult and seldom performed. The measured frictional torque is a good indicator of the type of the developed lubricating film (full film or mixed lubrication) and allows determination of operating conditions at which full hydrodynamic film is developed. Since the test head (2) is supported by a hydrostatic spherical bearing, the frictional torque in the tested bearing can be precisely measured, even under substantial applied radial loads, using a calibrated elastic beam (5) and a non-contact displacement probe.

Prototype I. Schematic view and a photograph of the first designed and built prototype is presented in Figure 2.

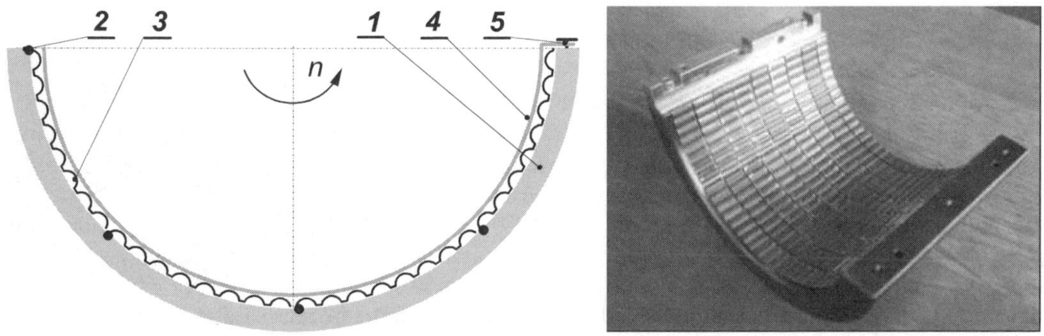

Figure 2. Schematic and photograph of Prototype I with top foil removed (see text for symbol definitions).

In this prototype the elastic support consisted of 36 bump foil segments (3) positioned in 9 rows of 4. Each bump foil segment had 8 bumps (with the geometry defined in Table 1) and was attached at its left end (2) to the rigid steel half-sleeve support (1). Since water-lubricated bearings operate at low temperatures, gluing with high-strength industrial glue was attempted. The gluing operation was performed in a specially designed holder, which ensured proper positioning and firm contact between the glued parts. Top foil (4) was attached at the trailing edge with screws (5).

Table 1. Basic dimensions of the bump foil segments and the top foil in Prototypes I and II (bumps numbered sequentially from the fixed end).

bump number	R [mm]	s [mm]	h [mm]	t1 [mm]	t2 [mm]	
1, 8	2.50	3.85	0.6	0.125	0.135	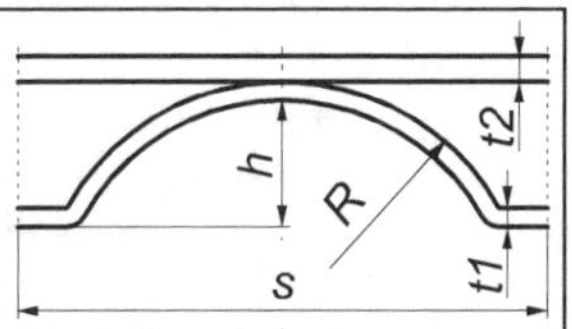
2 to 7	3.00	4.40				

Prototype I performed poorly in testing right from the start. After a few minutes of operation many of the bump foils detached from the steel support and testing was abandoned. It was concluded that gluing does not provide sufficient bonding strength. As a consequence of the bearing failure, the coating on the top foil was worn through in multiple locations and the exposed hardened steel damaged the surface of the journal as well.

Prototype II. Because of the setbacks experienced with Prototype I, it was decided to build a second, significantly redesigned, second prototype. Two important modifications were introduced (Figure 3). The first modification was the new method of attaching bump foils (3) to the half-sleeve (1), which eliminated the need of gluing. In the new design the bump foil sections (3) were mechanically attached to steel blocks (2), which in turn were placed in axial groves machined in the steel half-sleeve (1). The bump foil geometry remained unchanged (Table 1), which allowed the use of the existing hardware in the manufacturing process. However, due to the introduced modification, the locations of bump foil attachment changed for half of the bump foils (compare Figure 2 and Figure 3), resulting most likely in a significantly altered stiffness distribution along the angular direction [11]. The second important change, which was introduced, was replacement of the original steel top foil (coated with low-friction polymeric coating) with a thicker top foil made entirely of a commercially available low-friction polymeric foil (4). The way the top foil was attached to the half-sleeve (1) remained unchanged (5). The new top foil eliminated the danger of damaging the journal if the lubricating film of water is suddenly ruptured and a high-speed rub occurred.

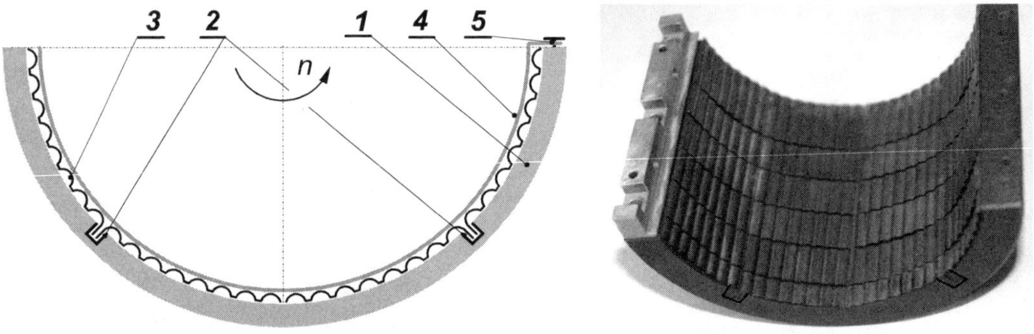

Figure 3. Schematic and photograph of Prototype II with top foil removed (see text for symbol definitions).

Experimental results of Prototype II. Initial tests confirmed that the new method of attaching the bump foils worked well. The foils were attached reliably and, as an additional benefit, could be replaced easily, if needed. During the first few minutes of bearing operation at 1 kN load (Figure 4) a significant reduction of the frictional torque was observed, indicating that the bearing top foil underwent a wear-in process.

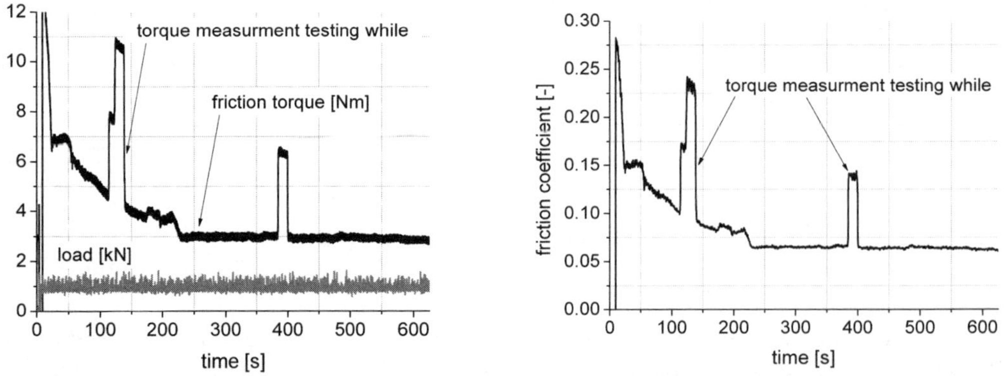

Figure 4. Initial test results for Prototype II illustrating the wear-in process.

The friction coefficient started at about 0.15 and then gradually dropped to a steady value of about 0.06. Note that the large spikes in the frictional torque and friction coefficient (at approximately 120 and 400 seconds) are not due to bearing malfunction, but rather the intentional calibration signals of the torque measurement system. Similar wear-in process was observed each time a new top foil was installed in a bearing. Each time, the process lasted for a few minutes, after which the frictional torque stabilized.

Prototype II bearing was tested at the load values of 1kN, 2kN and 3 kN and two rotational speeds of 800 rpm and 1,600 rpm (specific loads of approximately 100 kPa, 200 kPa and 300 kPa, at sliding speeds of approximately 4 m/s and 8 m/s, respectively). Typical experimental results are presented in Figure 5, which shows the measured load, speed, frictional torque and friction coefficient at constant radial load of 3 kN and for both rotational speeds used.

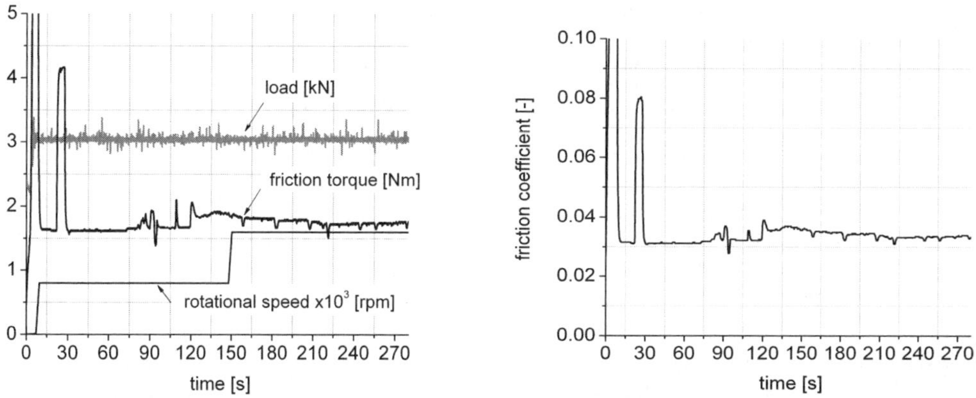

Figure 5. Typical test results for Prototype II: Constant radial load of 3 kN at two rotational speeds of 800 rpm and 1,600 rpm.

The measured friction coefficient of about 0.03 indicates that the lubricating film was not fully developed and that the journal was to some extent rubbing against the bearing. This observation was further confirmed by the elevated water temperature, which increased from the initial value of 25°C to the steady state temperature of 55°C. Post-test inspection of the top foil showed localized wear areas near the bump foil mounting blocks. This finding suggested that the stiffness distribution in bump foil segments (along the angular direction) was not optimal, with the bumps adjacent to the fixed ends being too stiff. Subsequent numerical simulations of the bump foil stiffness (using numerical tool described in [11]) were in agreement with this hypothesis.

Prototype III. Based on the experimental observations made while testing Prototype II, it was decided to design and build third prototype, which would have closer to optimal stiffness distribution in the bump foil structure (Figure 6).

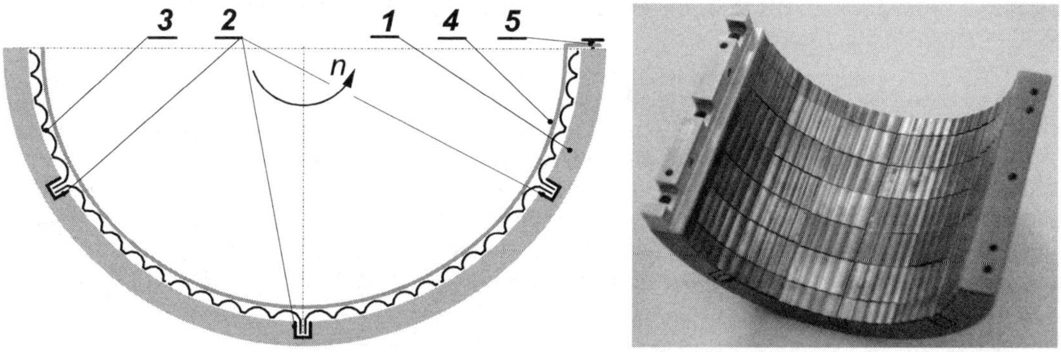

Figure 6. Schematic and photograph of Prototype III with top foil removed (see text for symbol definitions).

The design process utilized numerical tools developed earlier in the project [8], [11]. A significantly more uniform stiffness distribution in the bump foils was achieved by varying the radius of curvature of each bump in bump foil segments, with the largest radius at the fixed end and the smallest radius at the free end (Table 2). The number of bumps in a bump strip was also reduced from 8 to 5, because numerical simulations for larger number of bumps predicted that the first bump (near the fixed end) was always significantly stiffer than other bumps due to bump "pinning" caused by frictional forces. This mechanism was described in detail in [11] and [12]. Prototype II consisted of 42 bump foil segments in 7 rows of 6 segments (Figure 6).

Table 2. Basic dimensions of the bump foil segments and the top foil in Prototype III (bumps numbered sequentially from the fixed end).

bump number	R [mm]	s [mm]	h [mm]	t1 [mm]	t2 [mm]
1	3.00	3.79	0.4	0.150	0.200
2	3.50	4.05			
3	4.00	4.29			
4	4.50	4.51			
5	5.00	4.72			

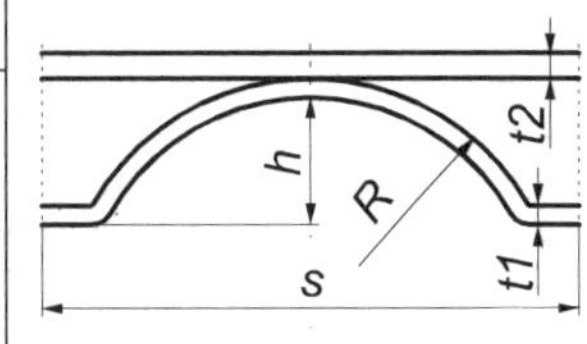

Experimental results of Prototype III. Figure 7 presents typical experimental results of Prototype III collected at a constant radial load of 3 kN (about 300 kPa) and the two rotational speeds of 800 rpm and 1600 rpm.

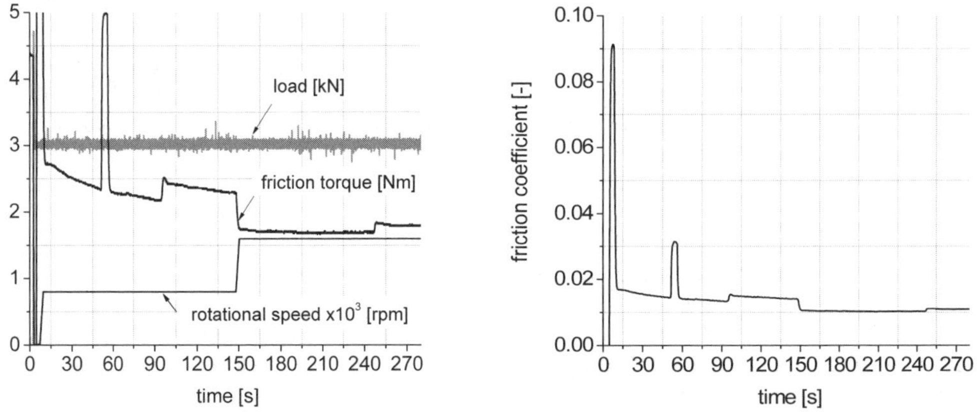

Figure 7. Typical test results for Prototype III: Constant radial load of 3 kN at two rotational speeds of 800 rpm and 1,600 rpm.

It can be seen that soon after startup the frictional torque dropped gradually to about 2.2 Nm. Similar reduction in frictional torque was observed in most experimental runs, and although it was not as significant as during the initial run with new top foils (Figure 4), it suggests further wear-in of the top foil during testing. It is likely that every time the bearing is shut down and then started up the top foil changes its position on the supporting bumps slightly and more wear-in takes place. Stepwise increase in rotational speed resulted in an immediate drop of the torque to about 1.6 Nm, which suggests significant improvement of the developed hydrodynamic film. During the test, the water temperature increased only by 7°C (from 27°C to about 34 °C).

Post-test wear analysis of the top foil. Better operation of Prototype III (over Prototype II) is also clearly visible from the post-test wear analysis of the top foils (Figure 8). The top foil of Prototype II (Figure 8a) shows clear wear spots distributed throughout its surface. High temperatures developed in the bearing caused deterioration of the bearing surface (clear evidence of discoloration, or even burn). Wear spots are most visible at the areas adjacent to the bump foil fixed ends (which have predicted highest stiffness). Surface profilometry (Figure 8c) showed significant buildup of the wear debris immediately before the worn spots, manifested by raised surface and dark residue (magnified image in Figure 8c).

Post-test condition of the top foil of Prototype III (Figure 8b) revealed much fewer wear spots, which confirms closer to optimal stiffness of the bump foil support. The areas adjacent to the bump foil fixed ends were worn to a similar extent as other areas (Figure 8c). However, one row of bump strips (second row from top) shows more wear than other areas, suggesting manufacturing or assembly errors.

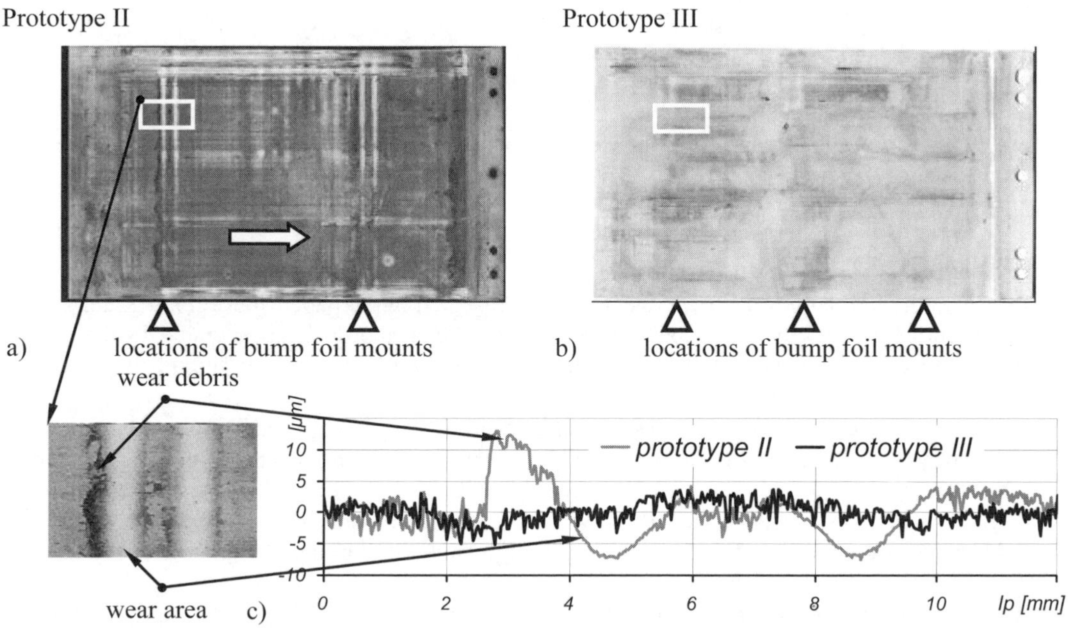

Figure 8. Post-test wear analysis of top foils: a) Prototype II, b) Prototype III, c) surface profilometry of the top foils in the vicinity of the bump foil mounts (profile locations indicated with boxes in Figures 8a and 8b).

Comparison of experimental results for Prototypes II and III. Improved, more uniform stiffness distribution of the bump foil strips in Prototype III resulted in greatly improved bearing characteristics, as compared to Prototype II. Reduced friction coefficient was observed (Figure 9). For Prototype II the lowest friction coefficient was measured at about 0.027 for 3 kN radial load and was independent of the rotational speed. Prototype III demonstrated the lowest friction coefficient of about 0.012 even at 3 kN (triple the load). This significant reduction of the friction coefficient was the result of a closer to optimal film thickness, which allowed for higher developed hydrodynamic pressure and higher load capacity. Increased importance of thick film lubrication in the bearing resulted also in less heat generated, manifested by smaller increase of the water temperature (temperature increase was only about 7 °C for Prototype III, as compared with about 30 °C for Prototype II). Improved stiffness distribution of Prototype III was also evident in the post-test wear analysis of the top foils.

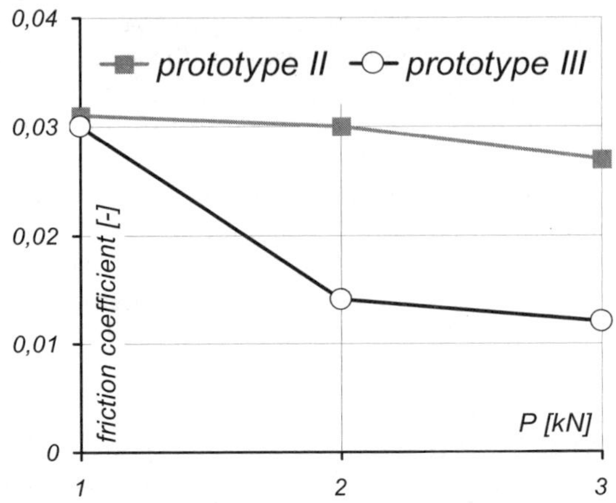

Figure 9. Variation of measured friction coefficient for Prototypes II and III with radial load at constant speed of 800 rpm.

Conclusions

According to the best author's knowledge, the idea presented in the paper of developing and testing of water lubricated compliant foil bearing, is the first such approach documented in the world. Based on the experience gained during design and testing of the three prototype bearings the following conclusions can be drawn:

- The idea of attaching bump foils to the rigid half-sleeve by gluing proved unreliable.
- Geometry and stiffness of the elastic support has strong influence on bearing operating characteristics.
- Uneven stiffness distribution of Prototype II resulted in poor bearing operation. Full film lubrication was never developed, resulting in high frictional torque, heavy wear of the top foil and excessive generated heat.
- Improved, more even stiffness distribution of Prototype III resulted in greatly improved bearing operation. The lubricating film was better developed, leading to reduced frictional torque, wear of the top foil and generated heat.
- Measured friction coefficients for Prototype III were not only lower than the corresponding values for Prototype II, but also showed a decreasing trend with radial load. This trend was likely a result of manufacturing errors, which affected bearing operation stronger at lower loads.
- Experience gained during development and testing of the first two prototypes, as well as utilization of the available numerical tools helped design greatly improved Prototype III.
- Numerical models over-predicted the load capacity of the designed bearings, most likely because the models did not consider such effects as manufacturing errors, geometric tolerances, or surface roughness. Nevertheless, numerical simulations provided useful qualitative input to the design process.
- Second generation of water-lubricated CFBs, with higher load capacity and better tolerance of bearing misalignment, is currently being developed.

Acknowledgments

It is gratefully acknowledged, that this work was funded as a part of research grants from the National Committee of Scientific Research (KBN) No. N502040 32/3013 and NR03003610.

References

[1] H. Heshmat, A Feasibility Study on the Use of Foil Bearings in Cryogenic Turbopumps, (1991) Paper No. AIAA-91-2103, AIAA/SAE/ASME/ASEE 27th Joint Propulsion Conference, Sacramento, California.

[2] G. L. Agrawal, Foil Air/Gas Bearing Technology—An Overview, International Gas Turbine & Aeroengine Congress & Exhibition, Orlando, (1997) ASME Paper No. 97-GT-347.

[3] H. Heshmat, W. Shapiro, S. Gray, Development of Foil Journal Bearings for High Load Capacity and High-Speed Whirl Stability, ASME Journal of Lubrication Technology 104 (1982) 149-156.

[4] H. Heshmat, Advancements in the Performance of Aerodynamic Foil Journal Bearings: High Speed and Load Capability, ASME Journal of Tribology 116 (1994) 287-295.

[5] L. San Andres, T.H. Kim, Thermohydrodynamic Analysis of Bump Type Foil Bearings: A Model Anchored to Test Data, ASME Journal of Engineering For Gas Turbines And Power 132, 4 (2010).

[6] S. A. Howard, Rotordynamic and Design Methods of an Oil-Free Turbocharger, NASA/CR – 208689 (1999).

[7] S.A. Howard, Misalignment in Gas Foil Journal Bearings: An Experimental Study, ASME Journal of Engineering For Gas Turbines And Power 131, 2 (2009).

[8] M. Wodtke, A. Olszewski, P. Hryniewicz, Water Lubricated Foil Bearing – Experimental And Numerical Examination, Proceedings of 13th NordTrib, Finland (2008), Paper No. NT2008-135-30.

[9] C. DellaCorte, K.C. Radil, R.J. Bruckner, S.A. Howard, Design, Fabrication and Performance of Open Source Generation I and II Compliant Hydrodynamic Gas Foil Bearings, NASA technical paper TM—2007-214691, (2007)

[10] A. Olszewski, M. Wodtke, P. Hryniewicz, Fluoropolymer and anti-friction coatings for water-lubricated foil bearings (in polish), Tribologia, 211 (2007), 93-104.

[11] P. Hryniewicz, M. Wodtke, A. Olszewski, R. Rządkowski, Structural Properties of Foil Bearings: A Closed-form Solution Validated with Finite Element Analysis, STLE Tribology Transactions, 52 (2009) 435-446.

[12] C. P. R. Ku, H. Heshmat, Compliant Foil Bearing Structural Stiffness Analysis-Part 1: Theoretical Model Including Strip and Variable Bump Foil Geometry, ASME Journal of Tribology, 114 (1992), 394–400.

Key Engineering Materials Vol. 490 (2012) pp 106-118
doi:10.4028/www.scientific.net/KEM.490.106

Influence of design of hydraulic components on their operation in low ambient temperatures

Ryszard Jasiński

Gdansk University of Technology, Faculty of Mechanical Engineering, Gdansk, Poland

rjasinsk@pg.gda.pl

Keywords: hydraulic drive, hydraulic components, thermal shock, computer simulation, design

Abstract

In many machines and devices there are hydraulic drives. They should operate well in various weather conditions, also in low ambient temperatures. Some malfunctions may occur during operation of hydraulic system in a so called "thermal shock condition", which happens when frozen hydraulic component (e.g. hydraulic pump, motor or directional spool valve) is suddenly supplied with hot oil. Transient thermal state emerges in these conditions. Particular elements of component warm up differently. This results in different thermal expansion of components during warm up, which is changing the size of clearance between cooperating elements.

Experimental tests of hydraulic components in low ambient temperatures were conducted in the hydraulic laboratory of the Faculty of Mechanical Engineering of Gdansk University of Technology. They concerned: orbital motors, satellite motors, gear pumps, spool valves (also proportional), piston pumps, and hydraulic cylinders. It was proved, that in thermal shock conditions some malfunctions may occur, especially with large temperature differences between oil and component.

Based on results of the tests of hydraulic components and systems start-up in low ambient temperatures, one can perform a change in components design, change the type of material of components elements, and even prepare a proper procedure to be followed during start-up of cold hydraulic components and systems.

In the article discusses some designs of hydraulic components resistant to thermal shock, and how to prepare components to work in low temperatures, e.g. by providing system with an additional heater to ensure uniform heating of elements in components. Change in design of hydraulic components resistant to thermal shock conditions can be obtained through computer simulation method.

Analytic and computer simulation methods can be used by engineers who design machines and devices that work in low ambient temperatures.

Introduction

In many machines and devices hydraulic drives are used. They should work well in various weather conditions, even in low ambient temperatures. Startup of a hydraulic system in low temperatures may cause: reduced hydraulic-mechanical efficiency, delayed reaction time of valves, faster wear and more frequent damage to parts (elements) of components, changes in clearance between cooperating elements of components, change of material properties of elements of hydraulic components, increase of flow resistance [1, 2, 3, 16, 17]. Hydraulic components working in low ambient temperature were tested by scientists [3, 16, 17]. The recommendations of machine during start in low ambient temperatures are presented in the position [16]. In other works research of cold motors during start in which the oil in the tank was previously heated (thermal shock conditions) were presented [1, 2].

Malfunction may occur during operation of hydraulic systems during sudden supply of hot oil to cooled hydraulic components, such as: pumps, motors and valves. The author submits some of the designs sensitive to thermal shock, also presents methods of protecting against incorrect work in hydraulic components in thermal shock conditions. An example of component that can ensure correct operation by adjusting design is satellite motor.

Clearances between cooperating elements of components

During start-up of hydraulic component in thermal shock conditions uneven heating process of elements is formed. Most of mobile elements of components have larger ratio of area-to-weight compared to fixed elements (housing). This results in different thermal expansion of components during assembly, which is changing the size of clearance between cooperating elements. There are lots of designs of hydraulic components. In each structure there are characteristic places of cooperation of elements, where the lead to loss of clearance may appear. Due to large variety of structures [1, 3, 14, 16, 17, 22, 23, 24, 25] in the rest of study will be given specific examples of such places. In axial piston pumps and motors (e.g. cam driven commutation, Fig. 1) there may be decay of radial clearance between pistons and cylinders, swash plate and slipper and elements of commutation. In gear pumps and motors there may be loss of axial clearance between faces of gears and side plates or radial clearance between peaks of teeth and housing (Fig. 2).

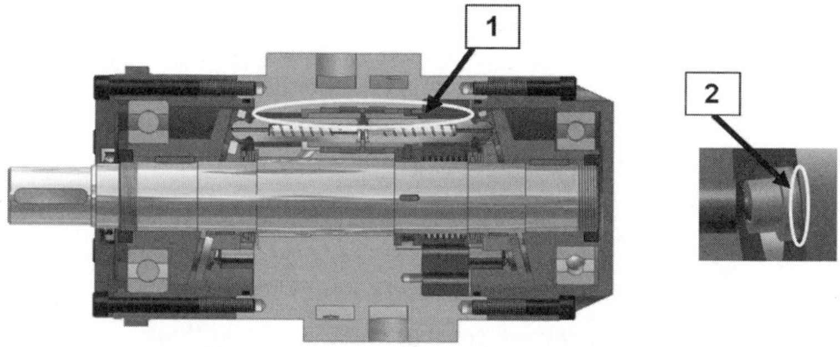

Fig. 1. Axial piston pump PWK-27 [14, 15, 24] with indicated places between cooperating elements where the disappearance of the clearance can occur
1 – radial clearance, 2 – clearance between swash plate and slipper

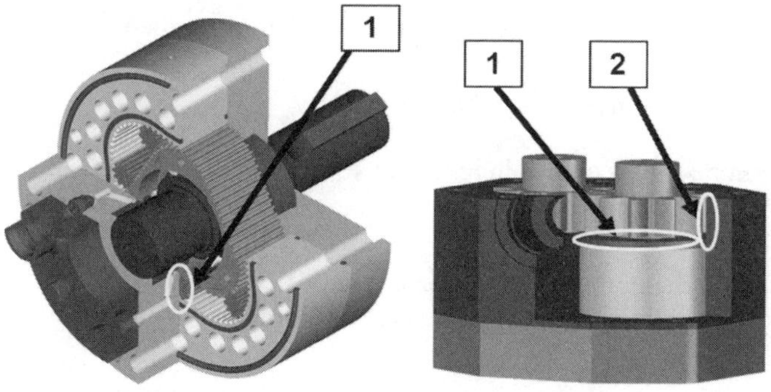

Fig. 2. Hydraulic satellite motor SOK [22] with indicated places where the disappearance of the axial clearance (between satellites and covers) and gear pump [24] with marked locations sensitive to thermal shock conditions
1 – axial clearance, 2 - radial clearance

Some hydraulic components can be sensitive to thermal shock conditions. They should meet requirements of the Polish Standards [19] or in case of marine application the Polish Register of Shipping [20-21]. Power supplies having higher temperature than cooled control components and actuators can be found on ships (winches, mooring arrangements) and in machines. Experimental tests of hydraulic components in low temperatures were carried out in the hydraulic laboratory of the Faculty of Mechanical Engineering of Gdansk University of Technology [1-2, 4-13, 18]. They concerned: satellite motors, orbital motors, gear pumps, spool valves including proportional, servo valve, axial piston pumps and motors, cylinders. It was found that under the thermal shock conditions they may be malfunctioning, especially for high temperature difference between oil and components at the initial moment. Based on research of hydraulic components and systems during startup in low ambient temperatures changes can be performed: in design component, change of material elements of component, and even preparation of proper procedure to start of cold hydraulic components and systems.

Determination of clearance between cooperating elements of component

Changes of clearance between cooperating elements of hydraulic components during start-up of the equipment in these conditions depend on many factors. Figure 3 highlights clearances: geometric l_0, mounting l_m and effective (resulting) l_e. Geometric clearance l_0 is determined by real dimensions of elements.

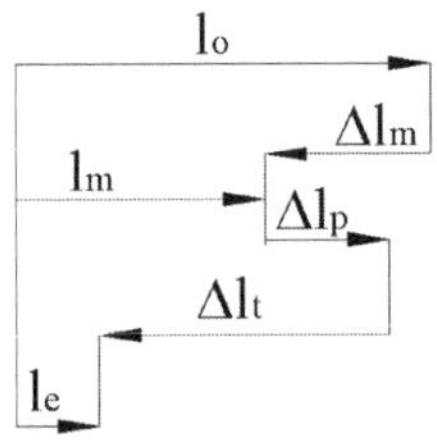

Fig. 3. *Dimensional analysis determining effective clearance between cooperating elements of hydraulic component in thermal shock conditions*

During assembly of hydraulic components geometric clearance l_0 is reduced by elastic deformation of elements Δl_m as a result of mounting clip. Effective clearance l_e depends on mounting clearance l_m, elastic deformation of the elements of the hydraulic components Δl_p, as result of action of oil pressure and difference of linear thermal expansion Δl_t cooperating elements of component. Effective clearance $l_e(\tau)$ at cooled component during start in described conditions is determined by relationship:

$$l_e(\tau) = l_m + \Delta l_p(\tau) - \Delta l_t(\tau) \qquad (1)$$

It is possible to determine the clearance height with several methods [4-11]:

- experimental,
- analytical,
- computer simulation (using the finite element method).

These methods allow to qualify whether a hydraulic component (system) does or doesn't work correctly.

Methods of preventing from the loss of effective clearance between cooperating elements of hydraulic components in thermal shock conditions

Grading of methods preventing from the loss of effective clearance between cooperating elements of hydraulic components in thermal shock conditions is illustrated in Figure 4.

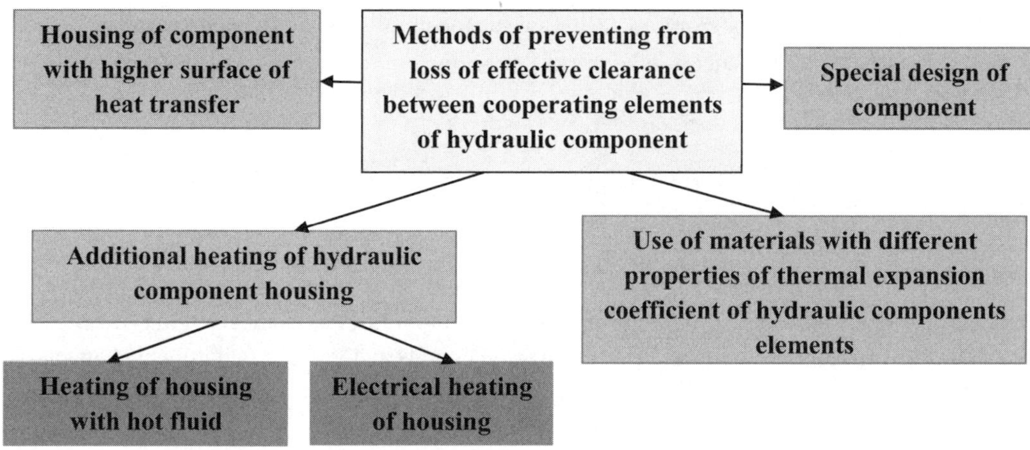

Fig. 4. *Methods of preventing the disappearance of effective clearance in hydraulic components in thermal shock conditions*

Models of heating elements of hydraulic components

Hydraulic pumps and motors consist of two groups of components: mobile and fixed. Figure 5 presents models of heating hydraulic motors. Difference between them lies in fact that second model's (Fig. 5B) fixed elements (housing) have additional channels through which oil flows. During start-up of the component in thermal shock conditions flow of hot oil through an additional channels (cap creates an additional channel) warms up the housing. Additional heat transfer surface in fixed elements (Fig. 5B) will heat up faster than in case shown in Figure 5A.

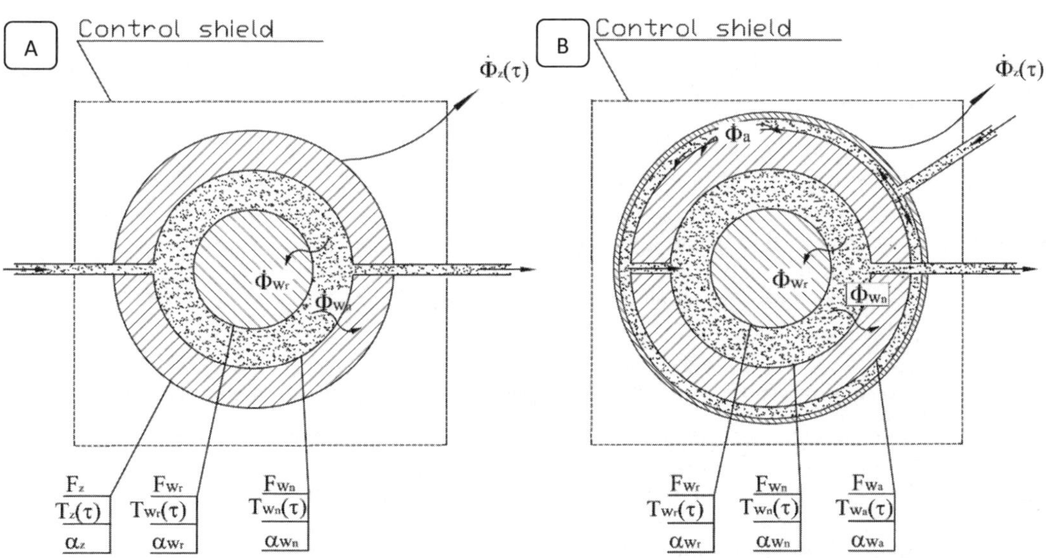

Fig. 5. *Models of heating up cooled motor from flowing hot fluid*
A) no additional channels in fixed elements (housing),
B) with additional channels in fixed elements,
(F - area, T - temperature, α - heat transfer coefficient, $\dot{\Phi}$ - rate of heat,
Lower case: z - outside, r - mobile elements, n - fixed elements, a - additional element creating

Rate of heat $\dot{\Phi}_{\alpha}(\tau)$ flowing from oil to mobile elements $\dot{\Phi}_{Wr}(\tau)$ and fixed elements $\dot{\Phi}_{Wn}(\tau)$ of component without extra channels in housing is equal to:

$$\dot{\Phi}_{\alpha}(\tau) = \dot{\Phi}_{Wr}(\tau) + \dot{\Phi}_{Wn}(\tau) \quad (2)$$

$$\dot{\Phi}_{\alpha}(\tau) = F_{Wr} \cdot \alpha_{Wr} \cdot (T_{ol} - T_{Wr}(\tau)) + F_{Wn} \cdot \alpha_{Wn} \cdot (T_{ol} - T_{Wn}(\tau)) \quad (3)$$

where:
T_{ol} - oil temperature
r - mobile elements, n - fixed elements.

Rate of heat from oil to fixed parts of component with additional channels $\dot{\Phi}_{N}(\tau)$ is equal to the heat flux flowing from oil to inner surface of fixed elements and space for additional channels:

$$\dot{\Phi}_{N}(\tau) = \dot{\Phi}_{Wn}(\tau) + \dot{\Phi}_{a}(\tau) . \quad (4)$$

Assuming that conditions of heat transfer from oil to surface of additional channels will be similar to conditions in channels fixed elements of hydraulic component, the same value of heat transfer coefficients can be assumed:

$$\alpha_{Wn} = \alpha_{Wa} . \quad (5)$$

Equation (4) can be written as:

$$\dot{\Phi}_{a}(\tau) + \dot{\Phi}_{Wn}(\tau) = (F_{Wn} + F_{Wa}) \cdot \alpha_{Wn} \cdot (T_{ol} - T_{Wn}(\tau)) \quad (6)$$

where:
F_{Wa} - area of additional channels of component

Input energy from oil to fixed parts of component is equal to increment of internal energy (temperature rise) and heat flowing through outer surface to environment:

$$(\dot{\Phi}_{Wa} + \dot{\Phi}_{Wn}) d\tau = ((F_{Wn} + F_{Wa}) \cdot \alpha_{Wn} \cdot (T_{ol} - T_{Wn}(\tau))) \cdot d\tau = m_n \cdot c_n \cdot dT_n + \dot{\Phi}_Z \cdot d\tau . \quad (7)$$

Described in equation (7) rate of energy placed into environment is omitted, because heat energy added to environment is small when machine is started. Outside temperature of component is approximately equal to ambient temperature. Increase of heat exchange area from oil to fixed elements cause them heat faster.

Influence of design of orbital motors on their operation in thermal shock conditions

In order to ensure proper operation of hydraulic system in which hydraulic motor is operated in thermal shock conditions, shock-resistant motor in these conditions should be used. These conditions are met by tested orbital motor TF 170 from Parker (Fig. 6).
In housing there are additional channels of oil flow to increase heat exchange surface, resulting in faster heating of housing. It was found (Fig. 7) that in this design solution maximum temperature difference between rotor and housing did not exceed 20 K during start-up of motor in thermal shock conditions for temperature difference oil and ambient temperature $\Delta T_{ol\text{-}ot}$=77 K.

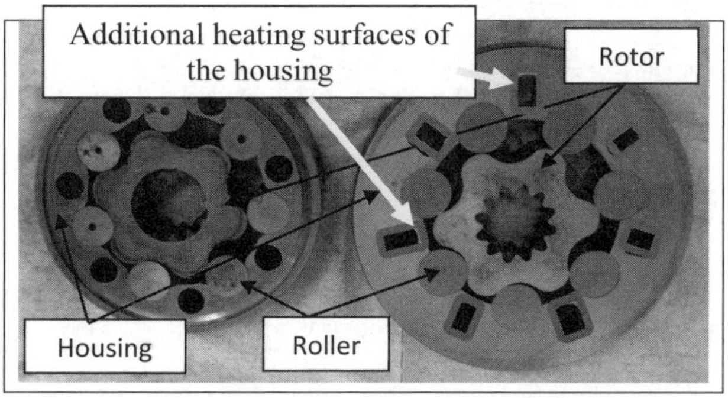

Fig. 6. Elements of orbital motors: GMR 160 (left) and TF 170 (right). In motor TF 170 are additional channels through which oil flows

Orbital motor GMR 160 (Fig. 6) does not have additional heat transfer surface between oil and housing. It was found out (Fig. 7) that maximum temperature difference between rotor and housing did not exceed 46 K for $\Delta T_{ol\text{-}ot}$=78 K [4].

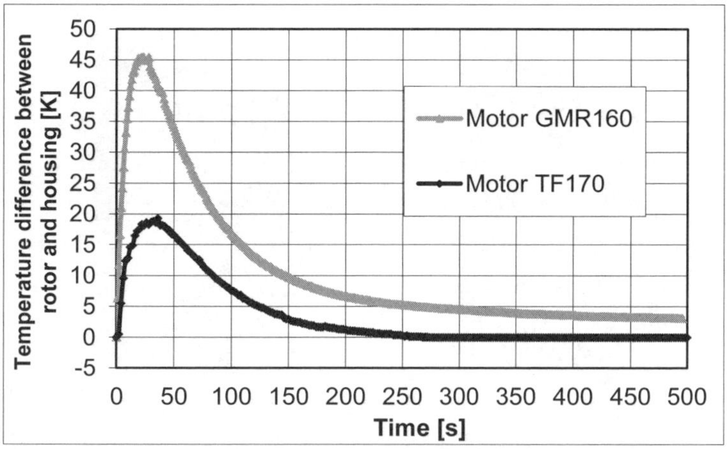

Fig. 7. Temperature difference between rotor and housing of motors TF170 (flow rate 64 dm ³/min, initial temperature -27 °C, oil temperature 50 °C) and GMR160 (flow rate 70 dm ³/min, initial temperature -26 °C, oil temperature 52 °C)

Thanks to substantial initial clearance, motor GMR 160 was resistant to thermal shock conditions and operated correctly when temperature difference between oil and ambient $\Delta T_{ol\text{-}ot}$ = 78 K [4]. In motor TF170 during start in thermal shock conditions small changes of radial clearance between rotor and housing occurred [8, 10, 11]. This signifies high resistance of motors of this design during operation in the specified conditions.

Research of satellite motor SOK in thermal shock conditions

Satellite motors are manufactured without compensation of axial clearance, effective clearances under these conditions should be assured. References [4, 7, 8, 10, 11] show tests results of satellite motor SOK 100 in thermal shock conditions. The motor worked correctly in thermal shock conditions for flow rate to 100 dm^3/min and temperature difference between hot oil and motor to 55 K (Fig. 8).

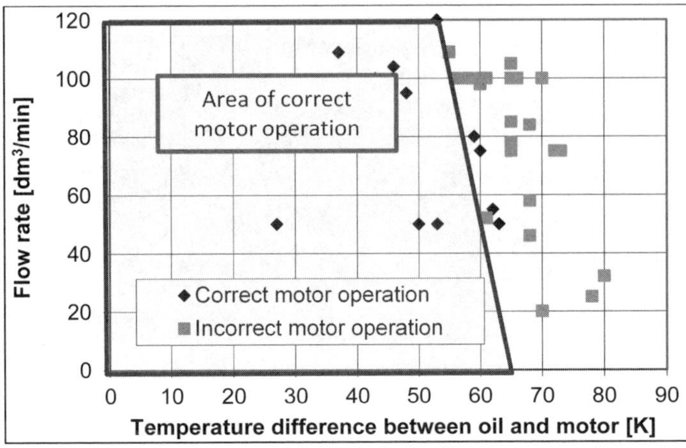

Fig. 8. *Areas of operating parameters of SOK 100 satellite motor in thermal shock conditions*

During experimental test of motor 100 SOK malfunctioned in following conditions: initial temperature T_{ot} = -20 ^{0}C, oil temperature T_{ol} = 50 ^{0}C, flow rate 100 dm^3/min [4].
Computer simulations of warming up of motor SOK 100 (Fig. 9) were performed.

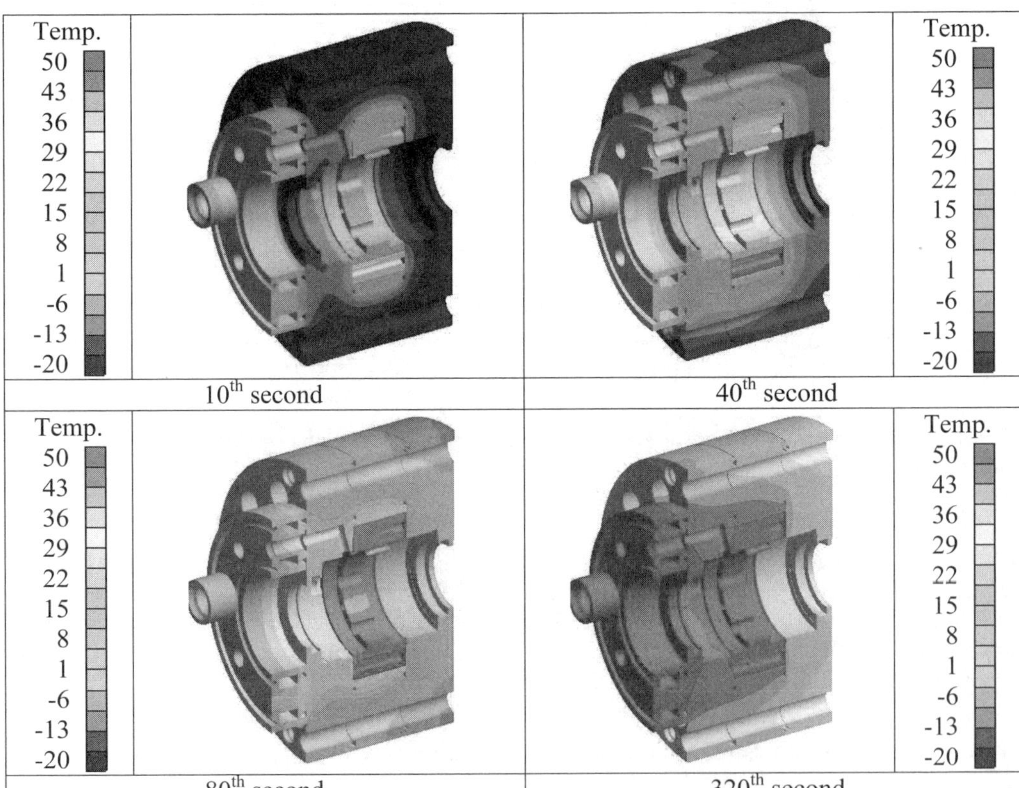

Fig. 9*. Temperature distribution in elements of motor SOK 100*
(initial temperature -20 ^{0}C, flow rate 100 dm^3/min) in 10th, 40th, 80th and 320th second after feeding it with oil at temperature 50 ^{0}C

To perform computer simulation, boundary conditions were adopted: heat transfer coefficient from oil to fixed elements 1460 W/m^2K, heat transfer coefficient from oil to mobile elements 1150 W/m^2K, heat transfer coefficient from outer surface to environment 12 W/m^2K.
On the basis of temperature distribution in motor elements (Fig. 9) an analysis of heating was conducted. The satellites heat up faster than other elements. In 80th second under normal operating conditions, the temperature would be determined.

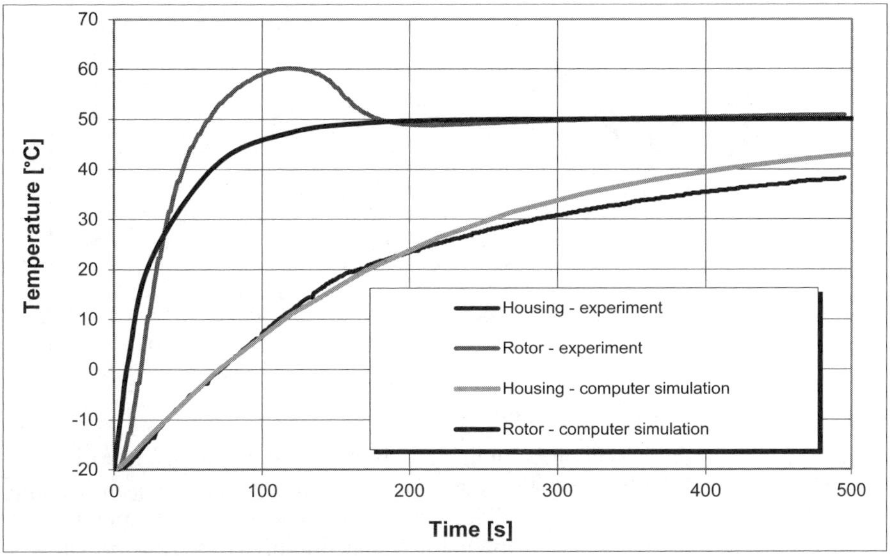

***Fig. 10**. Temperature of housing and rotor in satellite motor SOK 100 obtained from experimental test and computer simulation for initial conditions: motor temperature -20 ^{0}C, oil temperature T_{ol} = 50 ^{0}C, flow rate 100 dm^3/min*

During incorrect motor's operation mobile elements (rotor (Fig. 10), satellites) heat up to temperatures higher than oil temperature. Obtained charts of temperature for satellite, rotor and housing in computer simulation helped to determine temperature differences between satellites and housing and rotor and housing.
Based on temperature difference between satellites and housing from dependence (1) effective clearance between the satellites and covers was determined. In 15th second it comes to effectively closure of the clearance between satellites and covers (Fig. 11). Loss of clearance lasted to 75th second. Malfunction of motor lasted longer, because numerical modeling does not include heat of dry friction between satellites and covers, and later between rotor and covers.

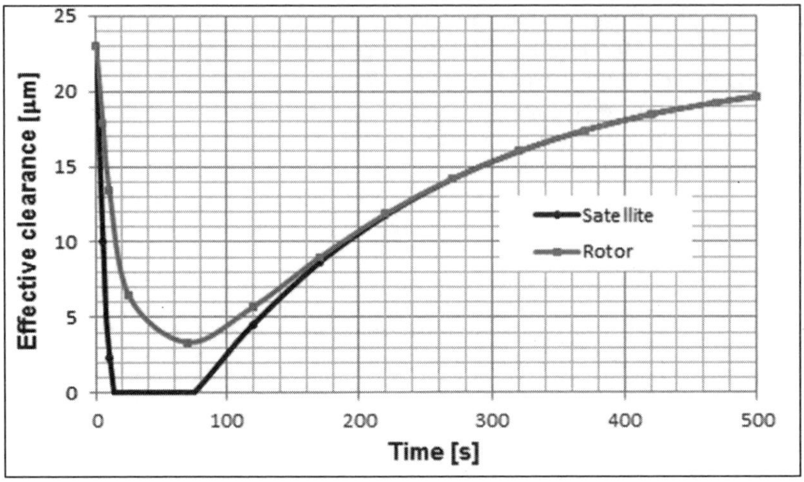

Fig. 11. *Effective clearance between satellites/covers and rotor/covers of motor SOK 100 for parameters: initial temperature T_{ot} = -20 ^{0}C oil temperature T_{ol} = 50 ^{0}C, flow rate 100 dm^3/min*

Additional heating of housing of component

Additional heating of hydraulic component housing to prevent disappearance of efficient clearance can be obtained by flushing the housing with hot oil or other means of heating of housing, such as electrically. Both methods rely on getting additional heat flux to housings of hydraulic components. Satellite motors SOK do not have extra channels for connection of scavenging, which would allow increase in temperature before starting of motor elements in low ambient temperatures. The best method for security reasons is the use of protection system on the motor, ensuring flow of hot oil on outer surface of housing (Fig. 12). Thanks to this fixed parts (housing) heat exchange surface increases.

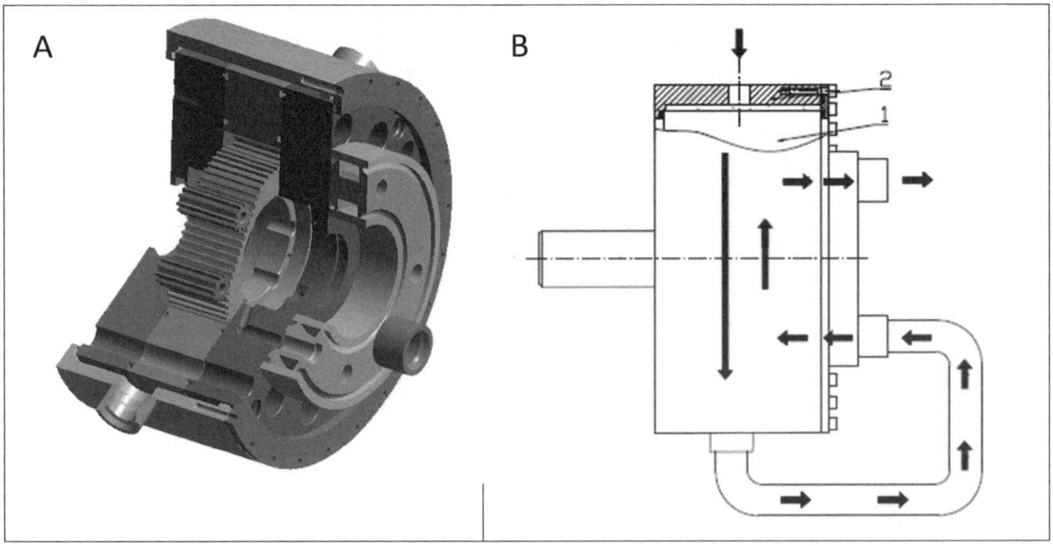

Fig. 12. Construction (A) and method of transferring oil (B) in motor SOK with protection system [26, 27; 1 – motor, 2 - clamping ring

Satellite motors SOK have an advantage over other structures of low-speed motors in form of small size in relation to absorption and easy mounting protection system [26, 27] for motor. In article motor with protection system of several structures is illustrated.
Hot oil flows to the protection system [26] (Fig. 12B). Around outer surface of motor flows oil, heating housing. On the other side of motor oil flows and special pipeline is routed to motor.

Research of heating process of motor SOK with protection system in thermal shock conditions

Computer simulation of elements of motor SOK 100 with protection system heating during start-up was carried out for the same initial conditions for which there was malfunction of motor: ambient temperature $T_{ot} = -20\ ^0C$, oil temperature $T_{ol} = 50\ ^0C$, flow rate 100 dm^3/min.
Protection system was designed to obtain similar average flow velocity in the gap between the clamping ring and outer surface of housing. Under this assumption, calculation assumed the same value of heat transfer coefficient (1460 W/m^2K) as for internal channels of motor. On the surfaces of mobile elements heat transfer coefficient 1150 W/m^2K from oil to those elements was assumed. On outer surface of motor and protection system heat transfer coefficient 12 W/m^2K was adopted.
Use of additional channel in motor, through which flows hot oil, caused rapid heating of housing. After 320 seconds from start-up of motor the conditions of heat flow are steady (Fig. 13), that does not happen in motor with no protection system (Fig. 9).

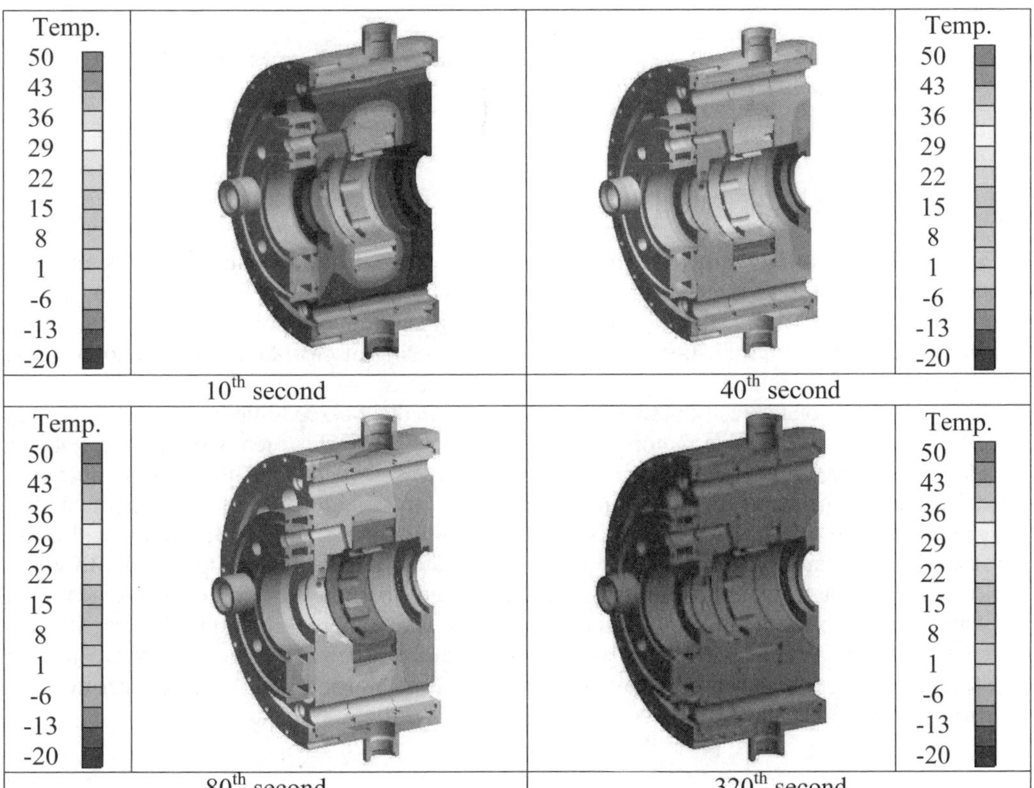

***Fig. 13**. Temperature distribution in elements of motor SOK 100 with protection system (initial temperature -20 ^0C, flow rate 100 dm^3/min) in 10th, 40th, 80th and 320th second after feeding it with oil at temperature 50 ^0C)*

The largest temperature difference between satellite and housing found in 18th second is 38 ^{0}C. For such large temperature difference between cold motor SOK 100 and hot oil is 70 K, clearance between satellites and covers was not erased. Clearance decreased by 18 μm to value 5 μm (Fig. 14).

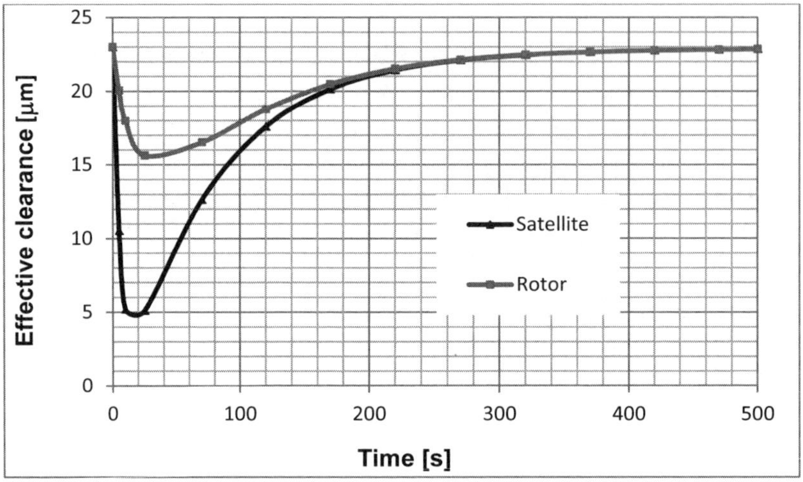

Fig. 14. *Effective clearance between satellites / covers and rotor / covers of motor SOK 100 with protection system for parameters: initial temperature* T_{ot} *= -20* ^{0}*C oil temperature* T_{ol} *= 50* ^{0}*C, flow rate 100* dm^3*/min*

Summary

1. Design of hydraulic components affects boundary changes of clearances during start-up in thermal shock conditions and thus their behavior in these conditions.
2. Using additional channels in fixed elements of hydraulic components can protect them during start-up in thermal shock conditions, from closing of clearance between cooperating elements (malfunctioning).
3. Hydraulic components can operate correctly in thermal shock conditions on specific supply parameters. Tested satellite motor 100 SOK for supply parameters: temperature difference to 55 K at the maximum flow rate of 100 dm^3/min work correctly. For the large temperature difference application of protection systems or warming [26, 27] to the satellite motor should be used. Application of protection or warming system for satellite motor will cause faster heating housing. Computer simulations of motor elements warm-up in order to verify the results obtained from experimental tests were performed. Broad convergence of waveforms of housing and rotor temperature of the computer simulation and experimental tests were obtained. Model motor with the application of protection system, and computer simulations were carried out. It was concluded that used housing heating system assured that effective clearance between the cooperating elements will not be cleared.

References

[1] Balawender A.: *Analiza energetyczna i metodyka badań silników hydraulicznych wolnoobrotowych (Energy analysis and methodology of testing of low-speed hydraulic motors),* Zeszyty Naukowe Politechniki Gdańskiej (Scientific Bulletins of Gdansk University of Technology), Gdansk (1988)

[2] Balawender A., Bieńkowski A.: *Badania silników SOK 160 w warunkach szoku termicznego (Tests of SOK 160 motor in thermal shock conditions),* Report, Gdansk (1981)

[3] Biały J.: *Badania pomp zębatych w obniżonej temperaturze (Research of gear pumps at reduced temperature),* Doctoral thesis, Warsaw (1981)

[4] Jasiński R.: *Działanie wybranych wolnoobrotowych silników hydraulicznych w warunkach szoku termicznego (Operation of some low-speed hydraulic motors in thermal shock conditions),* Doctoral thesis, Gdansk University of Technology, Gdansk (2002)

[5] Jasiński R.: *Badanie eksperymentalne pompy wielotłoczkowej osiowej PWK 27 firmy Hydrotor w niskich temperaturach otoczenia (Experimental research of axial piston pump PWK-27 from Hydrotor company in low ambient temperatures),* Napędy i Sterowanie, 4/2008

[6] Jasiński R.: *Methods of determination of correct operation area for hydraulic component in low ambient temperatures,* Developments in Mechanical Engineering, Gdansk (2008)

[7] Jasiński R.: *Problems of the starting and operating of hydraulic components and systems in low ambient temperature (Part I).* Polish Maritime Research, 4/2008

[8] Jasiński R.: *Problems of the starting and operating of hydraulic components and systems in low ambient temperature (Part II).* Polish Maritime Research, 1/2009

[9] Jasiński R.: *Problems of the starting and operating of hydraulic components and systems in low ambient temperature (Part III).* Polish Maritime Research, 4/2009

[10] Jasiński R.: *Eksploatacja napędów hydraulicznych w niskich temperaturach otoczenia (część 1) (Operation of hydraulic drives in low ambient temperatures),* Hydraulika i Pneumatyka, 1/2011

[11] Jasiński R.: *Wpływ konstrukcji silników hydraulicznych na ich działanie w niskich temperaturach otoczenia (Influence of design of hydraulic motors on their operation in low ambient temperatures),* International Scientific-Technical Conference „Hydraulics and Pneumatics'2007", Wrocław (2007)

[12] Kamrowski Ł.: *Badania eksperymentalne działania gerotorowego silnika hydraulicznego podczas rozruchu w warunkach niskiej temperatury otoczenia (Experimental research of orbital hydraulic motor during start-up in low ambient temperature)*, Graduation paper, Thesis supervisor: R. Jasiński, Gdansk University of Technology, Faculty of Mechanical Engineering (2005)

[13] Lewandowski P.: *Badania eksperymentalne działania wielotłoczkowej pompy PWK-27 podczas rozruchu w warunkach niskiej temperatury otoczenia (Experimental research of piston pump PWK-27 during start-up in low ambient temperature),* Graduation paper, Thesis supervisor: R. Jasiński, Gdansk University of Technology, Faculty of Mechanical Engineering (2006)

[14] Osiecki A.: *Hydrostatyczny napęd maszyn (Hydrostatic drive of machines),* WNT, Warsaw (1998)

[15] Osiecki L.: *Pompy wielotłoczkowe z rozrządem krzywkowym (Piston pump with the cam commutation)*, IX Seminarium: Napędy i Sterowanie 2003, Gdansk (2003)

[16] Przychodzień T. i inni: *Eksploatacja maszyn roboczych w warunkach zimowych (Operation of machines in winter conditions),* WNT, Warsaw (1990)

[17] Szewczyk K.: *Wpływ obniżonej temperatury na charakterystyki hydraulicznych układów maszyn roboczych ciężkich (Effect of reduced temperature on characteristics of hydraulic systems of heavy machines),* Cracow University of Technology (1991)

[18] Zelewski J.: *Projekt modernizacji silnika satelitowego bez kompensacji luzów osiowych odpornego na warunki szoku termicznego w zastosowaniu do hydraulicznej wciągarki pokładowej statku (Modernization project of satellite motor without axial clarence compensating resistant to thermal shock conditions applied to hydraulic winches of ship),* Graduation paper, Thesis supervisor: R. Jasiński, Gdansk University of Technology, Faculty of Mechanical Engineering (2004)

[19] Polish Standardization Office: PN–86/M-73079 - *General conditions of using the uniform hydraulic system*

[20] Polish Register of Shipping: *Publication No. 11P* (1994)

[21] Polish Register of Shipping: *Environmental tests of ship equipment*, Gdansk (1975)

[22] Catalogues from HYDROSTER company

[23] Catalogues from REXROTH company

[24] Catalogues from HYDROTOR company

[25] Catalogues from PARKER company

[26] *Układ zabezpieczający hydrauliczny silnik satelitowy uruchamiany w warunkach szoku termicznego (Protection system to hydraulic satellite motor start-up in thermal shock conditions),* Patent application, P393922, Poland, Inventors: Jasiński R., Zelewski J. (2011)

[27] *Układ podgrzewający oziębiony hydrauliczny silnik satelitowy (System warming the cooled hydraulic satellite motor),* Patent application, P393921, Poland, Inventor: Jasiński R. (2011)

Key Engineering Materials Vol. 490 (2012) pp 119-127
© (2012) Trans Tech Publications, Switzerland
doi:10.4028/www.scientific.net/KEM.490.119

Influence of magnetorheological fluid volume onto obtained critical pressures on rotary shaft seals

Marcin Potoczny[1,a], Bolesław Zachara[1,b]

[1]AGH University of Science and Technology,

Faculty of Mechanical Engineering and Robotics, Cracow, Poland

[a]potoczny@agh.edu.pl, [b]bzachara@agh.edu.pl,

Keywords: magnetic fluids, magnetorheological fluids, MR fluids, magnetic fluids seals, magnetorheological fluids seals, rotary shaft seals

Abstract.

Structure and basal application of magnetorheological fluids as well as their application in rotary shaft seals, have been presented in this study. As the application of magnetorheological fluid in seals is relatively purely known, the conducted studies were aimed at defining their operational parameters. The most essential was manner of the MR fluid delivery into oil gap, mostly the applied fluid volume. Not sufficient volume of the MR fluid often results in non-uniform oil gap filling what in turn decreases the seal efficiency. Results of studies on MR fluid-based seal burst pressure, conducted on specially designed testing stand, have been presented. The results comprise three oil gap sizes of δ=0,15mm, δ=0,3mm and δ=0,5mm, for tests conducted with various volume of MR fluid applied on single sealing stage. Comparative values of the seal burst pressures obtained for three oil gap sizes of multi-stage seal, with the same apparent volume of MR fluid (the same fluid and oil gap volume ratio), were also presented.

Introduction

Magnetic fluids comprise area of materials called as smart materials, which are more and more commonly applied. One of the applications comprises use of magnetic fluids for rotary shaft seals. In such case, explaining the influence of various magnetic fluid physical properties and seal constructional features onto operational features, i.e. tightness, motion resistance and durability, is very important.

Tightness keeping ability via magnetic fluid seal is determined by critical pressure (burst pressure). It is maximal pressure within the seal, over which magnetic fluid is ejected from oil seal and partial or total tightness loss is observed. That depends on the degree of the gap filling with magnetic fluid, as well as on pressure difference at both seal sides.

Actually, ferrofluids are commonly used in sealing techniques [1,2]. They are characterized with small values of frictional resistance and perfect tightness in high vacuum conditions and suitable type of the ferrofluid applied. Insufficient efficiency at higher pressures of the sealing agent and relatively high cost of the sealing, what is related with high cost of the ferrofluid, are major disadvantages of these seals.

Similar character of the magnetorheological fluids (MR fluids) and ferrofluid behavior within magnetic field stimulated an idea of their application in sealing techniques. The first studies have been executed in scientific centers in Japan at the Akida University and Belorussian Academy of Sciences [3,4,5]. The studies proved that magnetorheological fluid-based seals as compared with ferrofluid-based seals have the highest burst pressure values. It is related with a fact that magnetorheological fluids have definitely higher values of saturation magnetization. Higher resistance to motion as compared with ferrofluid-based seals, what is related with increase of magnetorheological fluid viscosity under influence of magnetic field, is major disadvantage of the MR-based seals.

Magnetorheological (MR) fluids

Magnetic fluids can be defined as colloidal suspensions of ferromagnetic particles within non-magnetic carrier fluid (base fluid). Fluids having low vapor pressures satisfying required standards depending on the fluid destination and operational conditions, are usually used as carrier fluids. Synthetic oil or light mineral oil are commonly applied, but hydrocarbons, water, fluorine derivatives of hydrocarbons, esters, diesters, poliphenyloesters etc., are also used [6]. Ferromagnetic particles are covered with mono-molecular layer of the surface active agent (for example oleinic acid), which protects against coagulation and sedimentation of particles [7].
Magnetic fluids are classifird as non-Newtonian fluids described by Bingham model. Boundary shear depends on magnetic induction occurring within the fluid. The model describes changes of shear stress in function of deformation rate during shearing.

$$\tau = \tau_{0B}(B) + \eta \cdot \dot{\gamma} \tag{1}$$

τ_{0B} - shear stress induced by magnetic field,
B - magnetic induction,
η - dynamic viscosity,
$\dot{\gamma}$ - shear rate

For given value of magnetic field, when linear relation between magnetic induction (B) and magnetic field intensity (H) is observed, the shear stress can be expressed as a function of magnetic field intensity *(H)*:

$$\tau = \tau_0(H) + \eta \cdot \dot{\gamma} \tag{2}$$

The equations are satisfied when the fluid shear stress is bigger than the boundary stress. Below the yield stress, for small deformations ($\gamma < 10^{-3}$) , the material behaves as elastic body (quasi-constant) and the following relation is obligatory (3),[8]:

$$\tau = G(H) \cdot \gamma, \qquad \tau < \tau_0 \tag{3}$$

Change of viscosity within magnetic field is one of the basic features of ferrofluids: ferrofluid particles suspended within carrier fluid, under influence of external magnetic field, form chain-shaped particle clusters being oriented and concentrated along magnetic field lines (Fig.1).
With respect to the ferrofluid structure, two basal types are distinguished:
1) Ferromagnetic fluids (ferrofluids – FF), which are also determined as nanomagnetic fluids.
2) Magnetorheological fluid (MR fluids), called also as micromagnetic fluids.

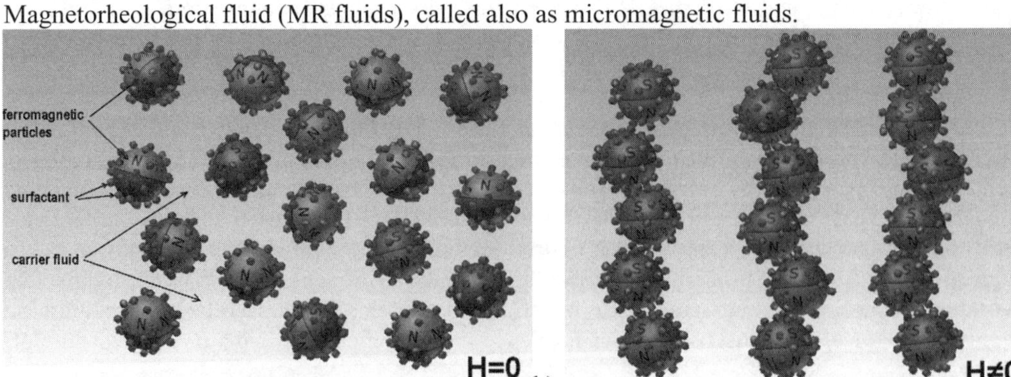

a) b)

Fig. 1. Magnetic fluid: a) without magnetic field (H=0), b) within magnetic field (H≠0)

Magnetorheological fluids (MR fluids) as compared with ferrofluids are characterized with considerably higher diameter of magnetic particles amounting from 0,5μm to 10μm. Their percentage content in the fluid amounts from 20% to 85%, what influences high values of saturation induction, reaching up to 2T. Depending on applied base it preserves its properties within temperature range from –50° to 150°C. High viscosity changes already at relatively low magnetic field intensity are one of the basic features of MR fluid. At external magnetic field intensities amounting from 150 to 250 kA/m, shear stresses on the level from 50 to 250 kPa, occur within the MR fluid. Susceptibility to gravitational sedimentation is one of disadvantageous features of the MR fluid. It results in deposition of particles on the container bottom, if the fluid is left stand-still for a longer time.

As the time of MR fluid reaction for magnetic field change is relatively short, such fluids can be applied in systems, which require high operation dynamics. Velocity change both in result of increase and decrease of the magnetic field intensity is possible during several microseconds. It is also essential that manufacturing of magnetorheological fluids is very simple, what considerably influences lower price of ferrofluids. MR fluids are used in various mechanical devices for controlled energy dissipation or transfer. The fluids are applied in controlled mechanical systems: breaks, clutches, vibration dumpers, valve without movable elements, laudspeakers, step motors, CD and DVD drive units and also in seals [7].

Testing stand MRU

The examinations were conducted on laboratory testing stand MRU (Magneto-Rheological Unit) designed for defining boundary operational conditions of magnetic fluid-based seals (ferrofluid or magnetorheological fluid), particularly defining the seal burst pressures and resistance to motion in dependence on their constructional and operational parameters (Fig. 2,3), [10]. The testing stand is composed of testing head, drive unit, supplying system together with air system and computer-aided measuring and control systems.

Fig. 2. Testing stand MRU – general view

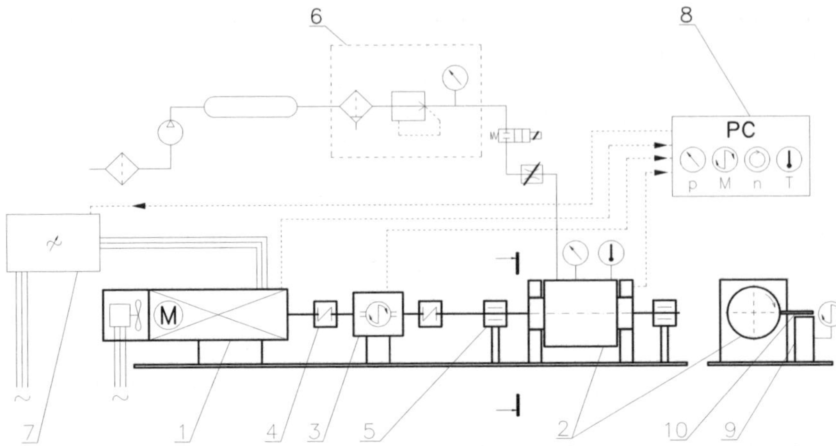

Fig. 3. Testing stand scheme: 1 – motor, 2 – testing head, 3 – torque meter, 4 – clutch, 5 – rolling bearing, 6 – air supply system, 7 – frequency converter, 8 – computer-aided measuring system, 9 – pressure sensor, 10 – additional pressure lever for torque measurements

Geometry of tested MR-based seals

Magnetorheological fluid MRF-122EG manufactured by Lord (USA), was used for the examinations [11]. The tests were conducted for MR fluid-based seal of nominal diameter of D=50mm in static (n=0min^{-1}) and dynamic (n=1000min^{-1}) conditions. Geometrical and magnetic parameters of the tested seal are shown in Fig. 4 and Table 1. The tests were conducted in ambient temperature of about 23°C, humidity of about 50% and atmospheric pressure of about 730÷750mmHg (970÷1000hPa). Single-stage seal made on measuring bush connected with internal seal bush and set on rotary drum, has been tested. The bushes were placed inside pole pieces in testing head. Seals with oil gap of δ=0,15mm, δ=0,3mm and δ=0,5mm filled with various volumes V [mm^3] of the MR fluid, have been tested.

Neodymium magnets Ø8x5 of coercive force 1050 KA/m, placed between pole pieces, uniformly on the seal circumference were used as a source of magnetic field. Magnetorheological fluid was applied into ring-shaped gaps δ between sealing stages of testing bushes and internal surfaces of pole pieces. The fluid was kept in position by magnetic field forces. Air pressure was generated in the space between pole pieces and sealing stages of the measuring bush.

Table 1. Geometrical and magnetic parameters of the MR fluid-based seal.

Number of sealing stages Z_w-Z_z	Geometrical parameters					Magnetic parameters	
	D [mm]	δ [mm]	t [mm]	b [mm]	α [°]	B_{max} [T]	H_{max} [kA/m]
6-1	50	0,15mm	0,6	5	45	1,35	750
		0,3mm	1,2			0,92	325
		0,5mm	2			0,67	180

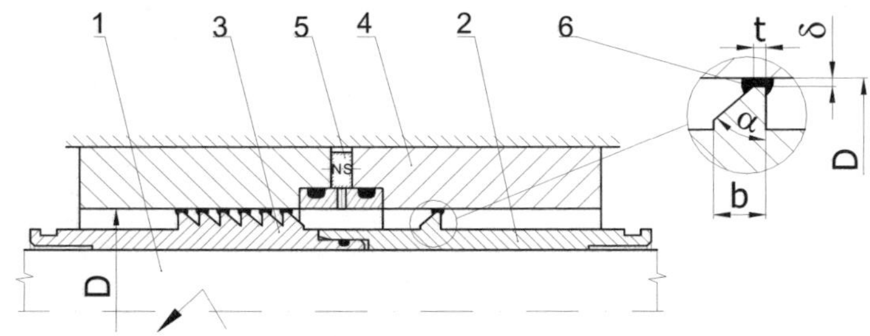

Fig. 4. Geometry of MR fluid based seal: 1-shaft, 2-measuring bush, 3-basic bush, 4- pole piece, 5-magnet, 6- MR fluid.

Testing results

The executed tests allowed defining the relation between seal burst pressure and MP fluid volume. Volume change in the seal was determined in form of relative volume factor V/V_n, defined as a ratio of the volume applied onto sealing stage and volume of nominal working gap. The seal burst pressures were measured via relative volume factors $\frac{V}{V_n}=1;3;5;10$ for each gap height δ. The obtained results are gathered in Table 2 and in Fig. 5, 6, 7. Static p_{krs} and dynamic p_{krd} values of single stage seal burst pressure, with gap δ=0,15mm; δ=0,3mm and δ=0,5mm high, at rotary speed $n=0min^{-1}$ and $n=1000min^{-1}$, have been presented.

Table 2. Static and dynamic seal burst pressure for various $\frac{V}{V_n}$

Number of sealing stages Z_w	Number of sealing stages Z_z	Gap height δ[mm]	Gap nominal volume V_n [μl]	Relative volume factor of MR fluid V/V_n	Volume of fluid applied into the gap V [μl]	Static seal burst pressure p_{krs} [kPa]	Dynamic seal burst pressure p_{krd} [kPa]
6	1	0,15	15	1	15	0	0
				3	45	55	18
				5	75	67	36
				10	150	81	44
6	1	0,3	60	1	60	42	12
				3	180	54	34
				5	300	74	44
				10	600	79	53
6	1	0,5	160	1	160	34	14
				3	500	81	38
				5	800	87	42
				10	1600	98	42

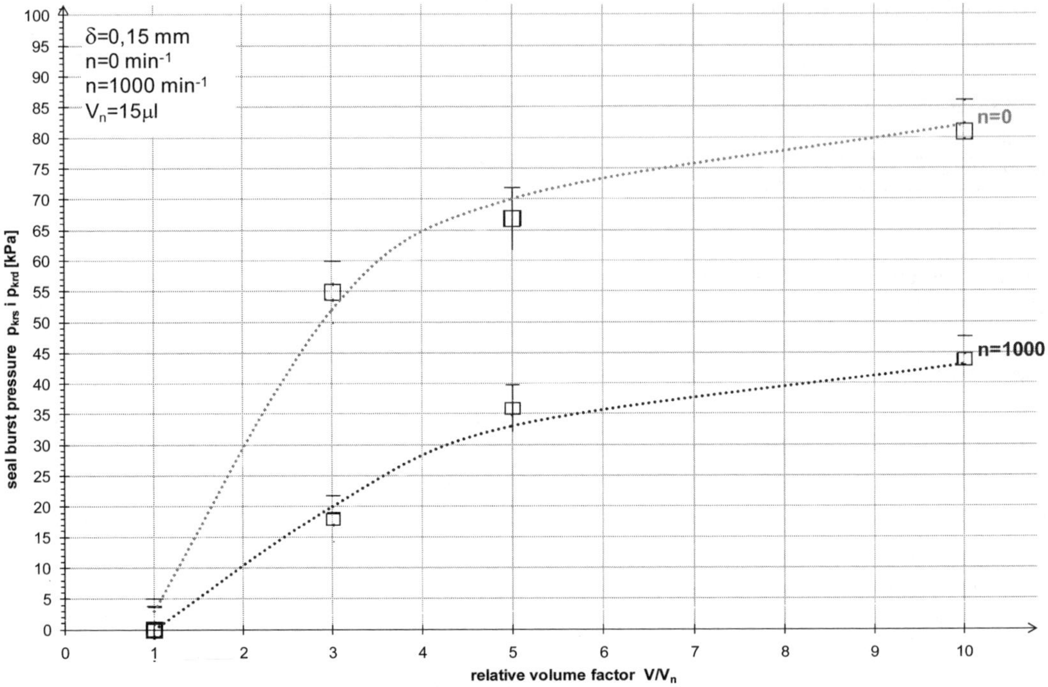

Fig. 5. Seal burst pressure p_{krs} (for n=0min^{-1}) and p_{krd} (for n=1000min^{-1}) - number of sealing stages Z_z=1, with gap δ=0,15mm, in function of relative MR fluid volume $\frac{V}{V_n}$.

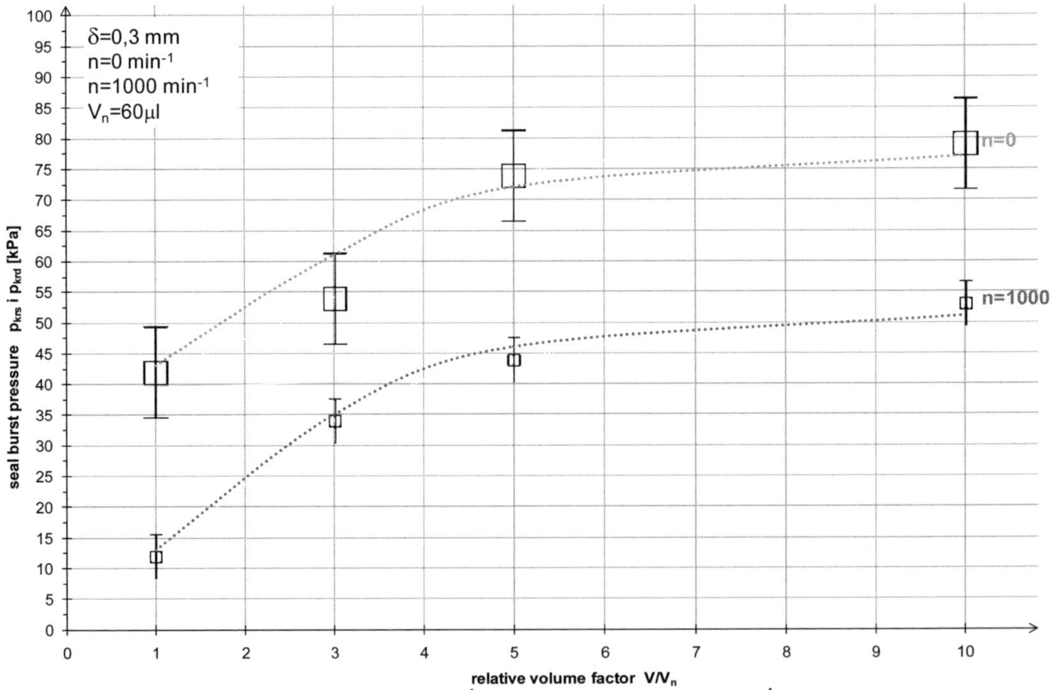

Fig. 6. Seal burst pressure p_{krs} (for n=0min^{-1}) and p_{krd} (for n=1000min^{-1}) - with number of sealing stages Z_z=1, with gap δ=0,3mm, in function of relative MR fluid volume $\frac{V}{V_n}$.

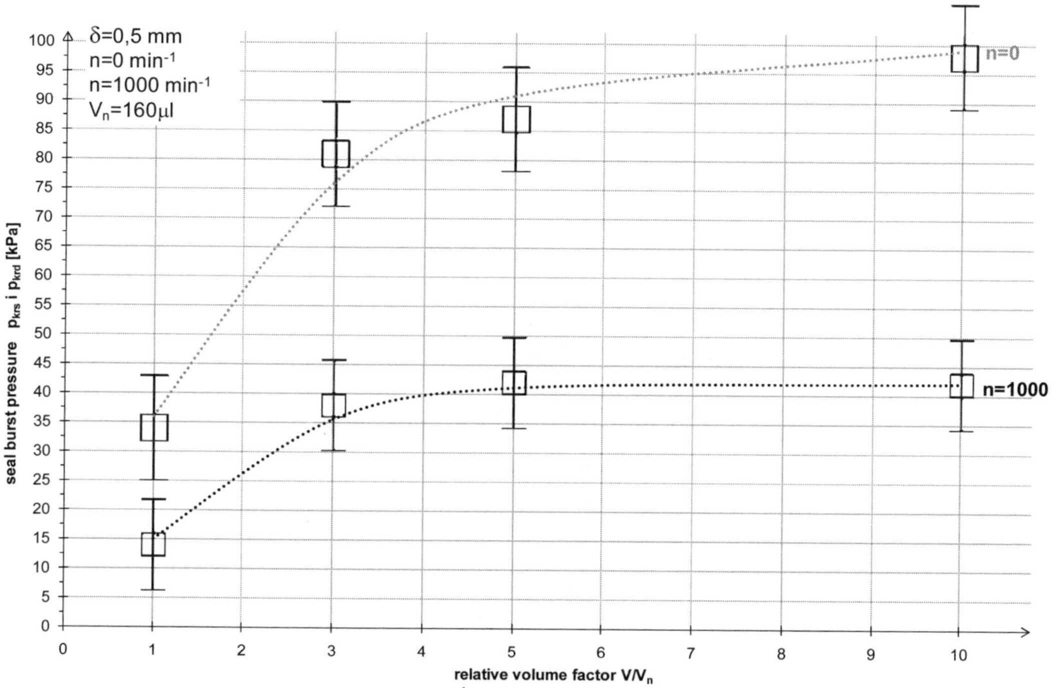

Fig. 7. Seal burst pressure p_{krs} (for n=0min^{-1}) and p_{krd} (for n=1000min^{-1}) - number of sealing stages Z_z=1, with gap δ=0,5mm, in function of relative MR fluid volume $\frac{V}{V_n}$.

Additionally, examinations of multi-seals at constant value of the factor $\frac{V}{V_n}$ =3,5 have been executed in dynamic conditions for n=1000min^{-1}. Results of these examinations are presented in Table 3 and in Fig.8 for three gap heights δ=0,15mm, δ=0,3mm and δ=0,5mm, with number of sealing stages Z_z=1÷5.

Table 3. Dynamic seal burst pressure p_{krd} for factor $\frac{V}{V_n}$ =3,5

Gap height δ[mm]	Nominal gap volume V_n [μl]	Relative volume factor V/V_n	Fluid volume in the gap V [μl]	Number of seal burst sealing stages		Dynamic seal burst pressure p_{krd} [kPa]
				Z_w	Z_z	
0,15	15	3,5	50	6	1	63
				6	2	105
				6	3	146
				6	4	153
				6	5	183
0,3	60	3,5	200	6	1	61
				6	2	80
				6	3	91
				6	4	93
				6	5	126
0,5	160	3,5	550	6	1	43
				6	2	47
				6	3	57
				6	4	61
				6	5	58

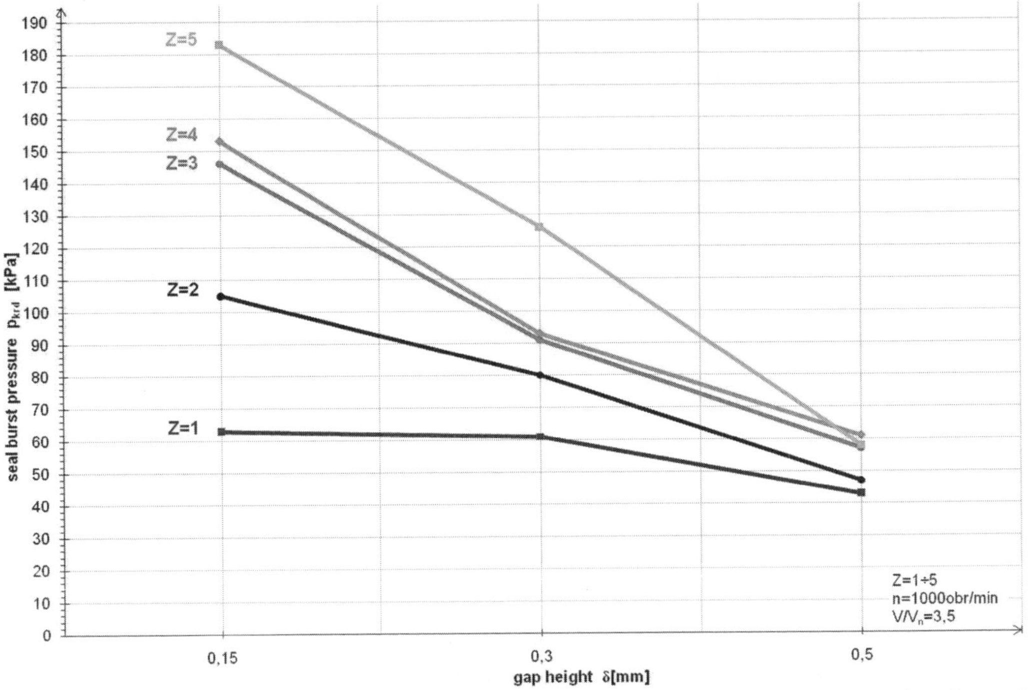

Fig. 8. Seal burst pressure of the sealing with number of sealing stages Z_z=1, 2, 3, 4 i 5, for various gap heights δ, at rotary speed n=1000min^{-1} and constant factor $\frac{V}{V_n}$ =3,5

Summary

Examinations of seal burst pressure at various values of MR fluid volume applied into operational gap proved high influence of the MR fluid volume V onto value of seal burst pressure.

On the basis of obtained results we can conclude that for each gap δ, at definite value of a factor V/V_n, further increase of the seal burst pressure is relatively small. Thus we can assume for each gap height δ, such relative volume factor V/V_n, which can be considered as optimal, and determine optimal MR fluid volumes for individual gap heights. The following values of factors V/V_n were assumed as optimal:

For a gap of the height δ = 0,15mm, V/V_n =10;

For a gap of the height δ = 0,3mm, V/V_n =5;

For a gap of the height δ = 0,5mm, V/V_n =3.

Based on the obtained results of examinations, which were conducted with constant relative volume factor we can conclude that a tendency of obtained seal burst pressure drop in result of increasing operational gap height and reduction of number of sealing stage, has been observed. Different drop of seal burst pressure values for a gap with δ=0,5mm was observed in case of 5-stage seal. It can indicate for magnetic field dissipation on numerous sealing stages and, in a case of such large gap, it can indicate for the magnetic field concentration in points below value of the MR fluid saturation. However, explanation of this phenomenon calls for further examinations.

References

[1] B.M. Berkovsky, V.G. Bashtovoy: Magnetic fluids and application handbook, Begell House Inc.,New York-Wallingford, 1996

[2] B.M. Berkovsky, V.F. Medvedev, M.S. Krakov: Magnetic fluids; an engineering applications, Oxford Science Publications, 1993

[3] T. Fujita, K. Yoshimura, Y. Seki, G. Dodbiba, T. Miyazaki: Characterization of magnetorheological suspension of seal, Journal of Intelligent Materials Systems and Structures, vol.10, 1999

[4] S.R. Gorodkin, W.I. Kordonski, N.A. Protasewich: Ocenka wozmozhnosti ispol'zowanija magnitoreologicheskikh syspenzijj w uplotnitel'nykh ustrojjstwakh. Magnitnaja gidrodinamika, 1988, № 4, 128-131.

[5] W.I. Kordonski, S.R. Gorodkin: Magnetorheological fluid-based seal, Journal of Intelligent Materials systems and Structures, vol.7,1996

[6] B. Wiślicki: Magnetic Fluids – properties and applications. Liquids supplies. Fuels, oils and lubricants in the exploitation. 2002, R.11, No. 93, 4-13.

[7] A. Milecki: Electro- and magnetorheological fluids and their applications in technology. Publisher of Poznan University of Technology, Poznań 2010

[8] M.R. Jolly, J.W. Bender, J.D. Carlson: Properties of application of magnetorheological fluids, Journal of Intelligent Materials Systems and Structures, vol.10, 1999

[9] T. Socaciu, M. Bucur , S. Mihai : Experimental research regarding magnetorheologic fluids. Interdisciplinarity in engineering. Scientific international conference, Tg. Mures – Romania, 15-16 November 2007. Inter-Ing 2007

[10] Patent application P-390575. Measuring head to determine the boundary operational conditions of magnetic fluid seals.

[11] MRF-122EG, Magneto-Rheological Fluid, Lord Technical Data (USA)

Key Engineering Materials Vol. 490 (2012) pp 128-134
© (2012) Trans Tech Publications, Switzerland
doi:10.4028/www.scientific.net/KEM.490.128

Influence of Shaft Misalignment on Water Lubricated Turbine Sliding Bearings with Various Bush Modules of Elasticity

Wojciech Litwin[1, a], Artur Olszewski[2,b] Michal Wodtke[3,c]

[123]Gdansk University of Technology; ul. Narutowicza 11/12; 80-233 Gdansk, Poland

[a]wlitwin@pg.gda.pl, [b]aolszews@pg.gda.pl, [c]mwodtke@pg.gda.pl

Keywords: sliding bearings, water lubricated bearings, angular misalignment, hydropower unit

Abstract. Hydropower industry increasingly frequently uses water lubricated bearings in turbines. This stems from the lack of negative impact of such solution on the environment. The paper presents an analysis of the influence of shaft misalignment in a turbine with water lubricated main shaft bearing on its hydrodynamic capacity. The theoretical analysis was based on the results of calculations for experimentally verified elastohydrodynamic lubrication (EHL) model. The obtained results showed impact of the stiffness of bush material and the degree of misalignment on bearing's hydrodynamic capacity.

Introduction

The constantly rising degree of environmental awareness results in an ever increasing number of machines and devices which have to meet restrictive environmental protection requirements. The search for environmentally friendly solution also applies to bearing systems in which oil leakage may cause significant degree of pollution to the environment. Bearing systems applied in hydropower plants are one of the examples of uses in which starting a few decades ago the traditional solutions of oil lubricated bearings were progressively replaced with environmentally friendly water lubricated ones. The use of water lubricated bearings in river based water turbines finds a particularly strong justification, as any leakage of oil to the natural body of water is absolutely unacceptable [1].

Since the viscosity of water is substantially lower than that of oil, water lubricated bearings have a significantly lower load-carrying ability and are characterized by a smaller minimum water film thickness [2]. The most common method of increasing the carrying ability of a water lubricated bearing is increasing its bush length. On the other hand, the greater bush length allows for a significantly smaller angle of shaft misalignment. Two basic solutions are employed in order to decrease the impact of shaft misalignment - either self-alignment of bush housing or use of an elastic polymer bush material. Another way is the self-alignment of housing during assembly, following which the bearing may be locked in a desired position. There also exist solutions where the self-alignment is automatic while the machine is actively working, however in case of water lubricated bearings these are much less common [1]. It should be emphasized that in an actual water turbine there always appear situations where the shaft placement is not perfectly parallel. This is caused by assembly and workmanship faults in the bearing, housing and shaft, as well as by elastic deformations of components, shifting of turbine foundations etc

Rigid bushes of water lubricated bearings are usually made of sintered bronze with graphite and require the use of hardened, stainless steel journals of the minimum hardness of 30HRC. In these types of solutions rather small bearing clearances are used with relative clearance usually below 0.001. The use of such small clearances is possible since the bush and journal material is rigid, easily machined and maintains a very stable shape in water (no water soaking). In addition, due to the low viscosity of water the problem of excessive increase in temperature due to heat generation in bearing interspace generally does not appear in water lubricated bearings, therefore the clearance does not have to be increased for that reason.

Elastic bearings are made of polymer materials of good tribological properties and relatively low degree of water soaking. The use of polymer materials of relatively low hardness allows operation of these types of bearings with a non-hardened journal which constitutes a significant cost advantage. The relatively low rigidity of bush materials allows for tolerating a certain degree of shaft misalignment.

Description of the problem

Both manufacturers and users of water turbines are frequently faced with the problem of choosing between a rigid and an elastic bearing. On one hand, a rigid bearing due to its inflexible material and smaller clearance has greater load-carrying capacity; on the other an elastic bearing is inherently less sensitive to shaft misalignment. The aim of the authors was to carry out a comparative analysis of the load-carrying capacity of typical rigid and polymer bearings working with an ideally parallel journal, as well as under non-optimum conditions of shift-misalignment.

Calculations

In calculations the elastohydrodynamic lubrication (EHL) model was used. The employed program consisted of three modules. The first module was used for calculating bearing interspace geometry and allowed for taking into account shaft misalignment. The second module was utilized in calculating the hydrodynamic pressure in water film and was based on the hydrodynamic lubrication theory and the Reynold's equation. The third module allowed for calculating elastic bush deformation using the finite element method (FEM) [4,5,6,7].

In the analysed case, the lubrication agent consisted of fresh water at the temperature of 5°C with viscosity of approximately 1.55 mPa·s. The calculations were conducted for the isothermal model, which was satisfactorily accurate for the analysed case [3].

The correctness of the used model was positively verified through an experiment. Presented below are examples of pressure field distribution obtained through measurements and calculations (Fig. 1 and Fig. 2.). There are at least two reasons for the appearance of certain differences in the results of the calculations and the experiment. It may be well supposed that certain inaccuracies in measurements may stem from the imperfect shape of the bush. The calculation results may carry certain errors stemming from the accepted constant value of the elasticity module which in the analyzed case was only 800MPa.

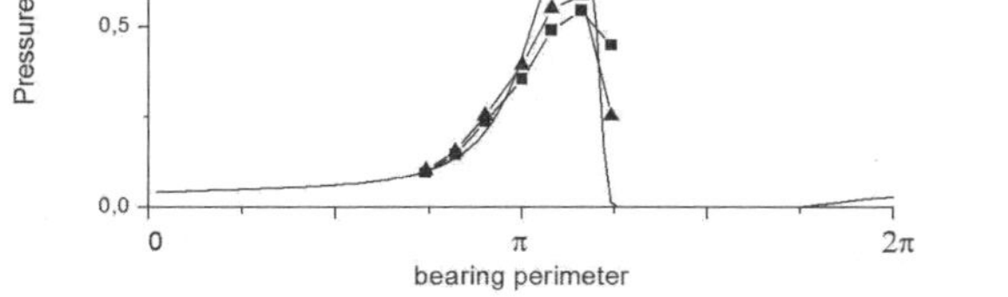

Figure1. Comparison of calculation (C) and measurement results (A, B) at the test-rig for elastic bush bearing, shaft diameter 100mm, shaft speed 11 [rev/s], load 0.1MPa

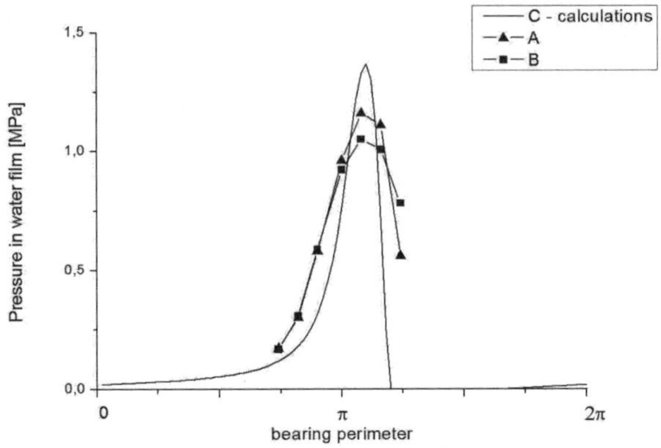

Figure 2. Comparison of calculation (C) and measurement results (A, B) at the test-rig for elastic bush bearing, shaft diameter 100mm, shaft speed 11 [rev/s], load 0.2MPa

Data accepted for calculations

The calculations were conducted for two types of bush material with significantly differing elasticity modules. Subjected to the analysis was elastic polymer having the elasticity module of approximately 800MPa, as well as rigid sintered bronze with graphite with the elasticity module of approximately 110GPa. The above values are representative of typical properties of bearing materials employed in elastic polymer bushes and rigid bushes of sintered bronze with graphite. It was accepted that in comparing the properties of individual bearings the calculations would be conducted with the same, assumed minimum lubricating film thickness of 5μm. Calculations were carried out in order to compare the maximum theoretical hydrodynamic capacity of bearings with differing properties of bush material. As it was previously mentioned, in actual working conditions shaft misalignment regularly occurs, therefore it was decided to conduct calculations for two angles of shaft misalignment, as well as for the ideal case of no shaft misalignment.

The research was carried out on the bearings of 100mm shaft diameter and the bush length of 200mm (L/D=2). It was assumed that the shaft speed was 600 rpm which is illustrative of average speeds of shafts of this diameter used in water turbines. Detailed data employed in the calculation is presented below (Table 1).

Accepted for the elastic bush bearing (cases E1, E2, E3) was the manufacture recommended clearance of 0.25mm. Calculations for the rigid bush were carried out for two clearance values. The calculations for the clearance of 0.25mm (cases SN1, SN2 and SN3) were only meant to illustrate the impact of drastic increase in bush material on its maximum theoretical hydrodynamic capacity. However, in a practical case of making a bronze bearing, the clearance would be much smaller amounting only to 0.1mm (cases SO1, SO2 and SO3).

Table 1. Data accepted for calculations

number	bearing code	bush material type	module of elasticity	bearing clearance / shaft diameter / bush thickness [mm]	misalignment - angle [°]	minimum film thickness
1	E1	elastic polymer	800 MPa	0.25/100/12	0	5μm
2	E2				0.00057	5 ÷ 7μm
3	E3				0.00143	5 ÷ 10μm
4	SN1	sintered bronze, oversized clearance	110GPa	0.25/100/12	0	5μm
5	SN2				0.00057	5 ÷ 7μm
6	SN3				0.00143	5 ÷ 10μm
7	SO1	sintered bronze, reduced clearance	110GPa	0.1/100/12	0	5μm
8	SO2				0.00057	5 ÷ 7μm
9	SO3				0.00143	5 ÷ 10μm

Calculation results

The calculation results were presented in the form of graphs illustrating the pressure field as a function of bearing length (0 ÷ 200) and circumferential angle (0 ÷ 2π).

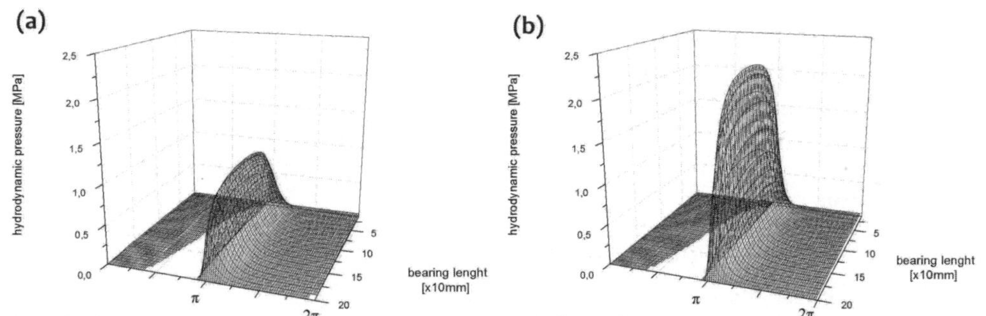

Figure 3. Calculated pressure field distribution in lubricating film for elastic bush bearing E1 (a) and rigid bush bearing SN1 (b); no shaft misalignment

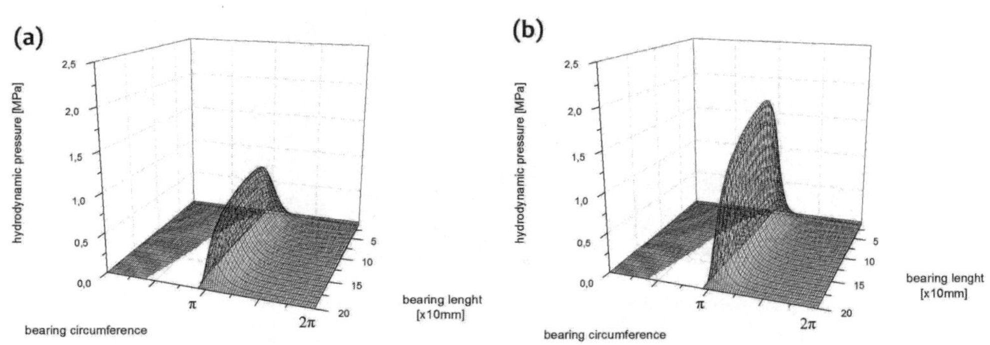

Figure 4. Calculated pressure field distribution in lubricating film for elastic bush bearing E2 (a) and rigid bush bearing SN2 (b); low value of shaft misalignment

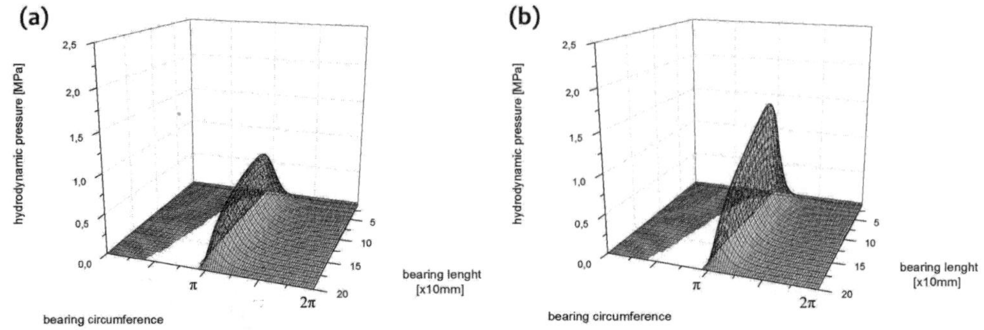

Figure 5. Calculated pressure field distribution in lubricating film for elastic bush bearing E3 (a) and rigid bush bearing SN3 (b); maximum value of shaft misalignment

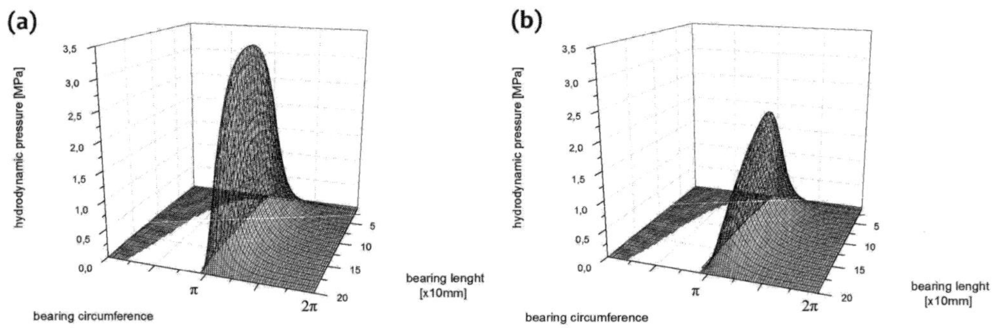

Figure 6. Calculated pressure field distribution in lubricating film for rigid bush bearing with recommended limited diameter clearance for ideal working conditions of SO1 (a) and with shaft misalignment SO3 (b)

Results comparison

Presented in Figure 7 is a group comparison of calculated hydrodynamic capacity of three types of bearings (E, SN, SO) taking into account shaft misalignment with the assumed minimum film thickness of 5μm.

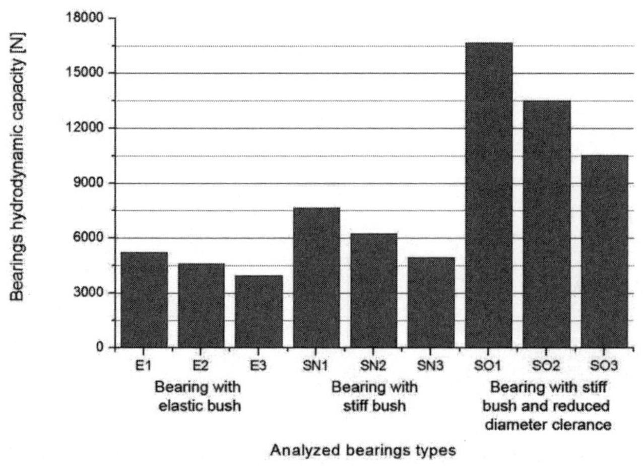

Figure. 7. Calculated maximum theoretical load carrying capacity of three bearings with optimum shaft placement and with shaft misalignment

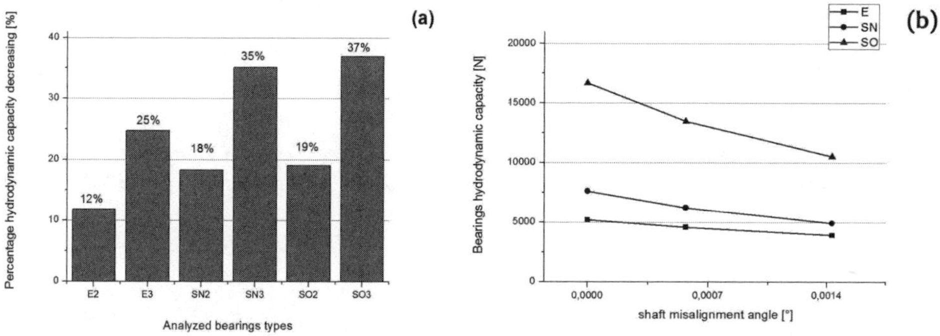

Fig. 8. Percentage decrease of hydrodynamic load carrying capacity of analyzed bearings (a) and load capacity of analyzed bearings as a function of shaft misalignment angle (b)

Conclusions

- Under the conditions of the journal being parallel to the bearing shell (no journal tilting) the rigid shell offers greater hydrodynamic load capacity as compared to a shell made of a more compliant material (less rigid). The assumed comparison condition is, that in both cases the minimum thickness of water film in the bearing is the same.
- In all the analyzed cases decrease in hydrodynamic capacity is of proportional character (Fig. 8a)
- Decrease of hydrodynamic capacity is nearly linear to angular shaft misalignment (Fig. 8b).
- Maximum pressure in the lubricating film of the rigid bush bearing is over twice as large as in the elastic bearing and in the analyzed case exceeds the value of 2MPa. The difference in the maximum value stems from differing shape of the pressure field.
- Based on the conducted analysis it can not be unequivocally determined which of the bearings (rigid or elastic) is the better solution for use in water turbines. Rigid bearings have greater hydrodynamic capacity, however they are also more expensive. Furthermore, in their case there is a serious risk of rapid wear which might lead to significant shaft and bush damage if loss of lubricating film occurs.
- Elastic bearings have lower hydrodynamic capacity (compared with rigid bearings). In case of substantial shaft misalignment mixed friction takes place which in time leads to running-in.
- In designing water lubricated bearings of turbines, work conditions of both rigid and elastic bush bearing should be individually examined. The calculations should be conducted taking into account the maximum expected value of shaft misalignment, as well as for the actual, accepted bearing modules of elasticity.

References

[1] Neyman A., Olszewski A., Wasilczuk M.: *Ecological bearing systems for water turbines two research programs aimed at making water turbines more eco-friendly*; Tribology science and application. Eds. F. Franek, C. Kajdas. Vienna: Sci. Centre Pol. Acad. Sci. 2004

[2] Litwin W.: *Influence of main design parameters of ship propeller shaft water-lubricated bearings on their properties*; POLISH MARITIME RESEARCH 4(67) 2010 Vol 17

[3] Litwin W.: *Water – An Extraordinary, Ordinary Lubrication Liquid. Influence Of Water Salinity, Pressure And Temperature On Water Lubricated Bearings Properties* ASME/STLE International Joint Tribology Conference October 17-20, 2010, San Francisco, California, USA

[4] Nikolakopoulos P. G., Papadopoulos C. A.: *A study of friction in worn misaligned journal bearings under severe hydrodynamic lubrication*; Tribology International 41 (2008)

[5] Jun Sun, Gui Changlin: Hydrodynamic lubrication analysis of journal bearing considering misalignment causedby shaft deformation; Tribology International 37 (2004)

[6] Yansong W., Q. J. Wang, Chih L.: *A mixed-EHL analysis of effects of misalignments and elastic deformations on the performance of a coupled journal-thrust bearing system;* Tribology International 39 (2006)

[7] J. Y. Jang, M. M. Khonsari: *On the Behavior of Misaligned Journal Bearings Based on Mass-Conservative Thermohydrodynamic Analysis*; ASME Journal of Tribology JANUARY 2010, Vol. 132

Key Engineering Materials Vol. 490 (2012) pp 135-142
© (2012) Trans Tech Publications, Switzerland
doi:10.4028/www.scientific.net/KEM.490.135

The Influence of the $K_{H\beta}$ Factor Determination Method on the Evaluation of the Gear Tooth Surface Durability and Bending Strength

Stanislaw Krawiec*

Wroclaw University of Technology, Poland
email: stanislaw.krawiec@pwr.wroc.pl

Keywords: gears, strength, $K_{H\beta}$ factor

Abstract. The references concerning the strength analysis of gears include a few methods of calculating the value of the face load factor (K_{HB}). In the paper, discussed are the C2 ISO method [1, 2, 3] and a method presented in [4, 5]. The two methods differ mainly in the way the bending and torsional deformations in the shafts and gears are calculated. On the example of the calculations of the first stage in a two-stage helical gear presented schematically in Fig.1, the influence of the these two different approaches on the value of the safety factors for pitting (S_H) and tooth breakage (S_F) has been analysed. The influence of four accuracy classes used in the manufacture of gearing (5, 6, 7 and 8) on the value of the safety factors S_H and S_F has also been evaluated.

Introduction

Gear transmissions are the mechanisms most frequently used for transmitting energy in various machines. Their enormous popularity is owed primarily to the stability of the transmission ratio, very high efficiency, higher than the one observed in other mechanical, hydraulic, or electrical transmissions. Another advantage of gear transmissions is their compact structure and the high power and high speed applicability. In designing modern transmissions dedicated to heavy duty applications, the focus is on a robust but economical design. This requirement can be met if, along with the precise manufacture and proper assembly, the correct strength calculations of the designed gears are made. The core of calculations is always the evaluation of the gear tooth surface durability (pitting) and bending strength. While designing transmissions that operate at high rotational speeds, one should also check the teeth against resistance to seizure. The aim of the strength calculations is to determine values of the pitting (S_H) and breakage (S_F) safety factors, as well as their evaluation. According to [1, 2, 3], these factors are calculated from the following formulas:

$$S_H = \frac{\sigma_{H\lim}}{\sigma_{H0}\sqrt{K_A K_v K_{H\beta} K_{H\alpha}}} Z_{NT} Z_L Z_R Z_V Z_w Z_x \tag{1}$$

$$S_F = \frac{\sigma_{F\lim}}{\sigma_{F0} K_A K_v K_{F\beta} K_{F\alpha}} Y_{NT} Y_{ST} Y_{\delta relT} Y_{RrelT} Y_x \tag{2}$$

where:

σ_{Hlim} – allowable stress number (contact),
σ_{Flim} – the nominal stress number (bending),
σ_{H0} – nominal contact stress at the pitch point,
σ_{F0} – nominal tooth root stress,
Z_{NT} – life factor for contact stress for reference test conditions,
Y_{NT} – life factor for tooth root stress for reference test conditions,
Z_L – lubricant factor,
Z_R – roughness factor affecting surface durability,
Z_V – velocity factor,
Z_w – work hardening factor,
Z_x, Y_x – size factors (pitting, tooth root),
$Y_{\delta relT}$ – relative notch sensitivity factor, the quotient of the gear notch sensitivity factor of interest divided by the standard test gear factor,

Y_{RrelT} – relative surface factor, the quotient of the gear tooth root surface factor of interest divided by the tooth root surface factor of the reference test gear,
K_A – application factor,
K_v – dynamic factor,
$K_{H\alpha}$, $K_{F\alpha}$ – transverse load factor (contact stress, root stress),
$K_{F\beta}$ – face load factor (root stress)
$K_{H\beta}$ – face load factor (contact stress).

In the case of the coefficients listed above, the face load factor $K_{H\beta}$ is the most difficult for calculations. The factor is defined as the maximum load intensity (local load per unit length increment of the face width) compared to the mean load (F_m/b):

$$K_{H\beta} = \frac{(F/b)_{\max}}{F_m/b}. \tag{3}$$

The value of this factor can be experimentally determined by measuring the pattern of the loaded teeth contact in an already manufactured transmission or by the analytical calculation of the effective distortion of the tooth contact line. Calculating the distortion is very difficult because it is influenced by numerous parameters, the values of which depend on bearing slackness, manufacture imprecision, elastic deformations (of the shafts primarily). The magnitude of the distortion, as well as the load and the deformation related to it, changes randomly. Therefore, one can talk only about the approximate and statistical values, defining them as a probability of the occurrence of a specific, effective distortion of the tooth line. Its exact calculation is time consuming and monotonous, hence the need to employ approximation methods. In the paper, the influence of two methods for the calculation of the face load factor $K_{H\beta}$ on the value of the pitting and tooth breakage safety factors has been analysed. These are the C2 method recommended by ISO [1, 2, 3] and a method described in [4, 5]. The two methods differ mainly in the way the bending and torsional deformations in the shafts and gears are calculated. The subject analysis was carried out by means of a computer program developed by the author, which enhances the design process of gear transmissions. The program is currently used by the students of the Faculty of Mechanical Engineering of the Wroclaw University of Technology who registered on the design course of the Fundamentals of Machine Design.

A Short Description of the Analysed Methods

In the C2 method described in ISO [1], the $K_{H\beta}$ factor is calculated from the expression, the form of which depends on whether the gearing transmits the load throughout the total face width (formula a), or only on a part of it (formula b).

$$\text{a)} \quad K_{H\beta} = \sqrt{\frac{2 \cdot b \cdot F_{\beta y} \cdot c_\gamma}{F_m}}, \quad \text{b)} \quad K_{H\beta} = 1 + \frac{b \ F_{\beta y} \ c_\gamma}{2F_m} \tag{4}$$

where:

$F_{\beta y}$ – effective equivalent misalignment (after running-in),
c_γ – mean value of mesh stiffness per unit face width,
F_m – mean transverse tangential load at the reference circle relevant to mesh calculations, $F_m = F_t K_A K_v$.

This effective equivalent misalignment $F_{\beta y}$ resulting from manufacture deviations, as well as elastic deformations of the elements transmitting the load, such as shafts, gear bodies, the case, and thermal deformations can be derived from the following formula (the initial running-in process is accounted for):

$$F_{\beta y} = F_{\beta x} - y_\beta = \chi_\beta \cdot F_{\beta x} \tag{5}$$

where:

$F_{\beta x}$ – initial equivalent misalignment (before running-in).

Distortion $F_{\beta x}$ is calculated from the formula:

$$F_{\beta x} = f_{ma} + f_{Sh1} + f_{Sh2} + f_{be} + f_{ca} + f_{te} \tag{6}$$

where particular components of this formula account for:

f_{ma} – mesh misalignment due to manufacturing deviations,

$f_{Sh1,2}$ – equivalent misalignment due to deformations of pinion and wheel shafts,

f_{be} – equivalent misalignment due to bearing deformation,

f_{ca} – equivalent misalignment due to deformations of the case,

f_{te} – thermal deformations.

The C2 ISO method considers only the f_{ma} and f_{Sh1} distortions. The other distortions ($f_{Sh2}, f_{be}, f_{ca}, f_{te}$) are not taken into account. It is also assumed that when calculating the deflection line, the load is applied in the middle of the rim width and the line distortion can be approximated by a straight line, namely:

$$F_{\beta x} = 1{,}33 f_{Sh1} + f_{ma}\,. \tag{7}$$

The f_{Sh1} distortion is calculated from the formula:

$$f_{Sh1} = \frac{F_m}{b} \cdot f_{Sh0}\,. \tag{8}$$

The f_{Sh0} distortion is a one-sided deformation generated by the application of a unilateral force. Its value for spur gears featuring no crowning or modifications of the tooth tips is determined by the formula:

$$f_{Sh0} = 0{,}023\gamma\,, \tag{9}$$

$$\gamma = \left[\left(1 + K^{I}\frac{l_S}{d_1^2}\cdot\left(\frac{d_1}{d_{Sh1}}\right)^4 - 0{,}3\right) + 0{,}3\right]\cdot\left(\frac{b}{d_1}\right)^2 \tag{10}$$

where:

K^{I} – constant for calculating the dimensional proportion ratio (according to Table 1),

d_{Sh1} – equivalent diameter of the constant diameter shaft that deforms to the same magnitude as a real shaft,

d_1 – pitch diameter of the pinion,

l – bearing span,

s – asymmetrical position of the gears in relation to bearings.

According to references [4, 5], the face load factor $K_{H\beta}$ can be calculated from the following formula:

$$K_{H\beta} = 1 + \frac{Cf_{ky}}{K_A K_v f_{z0}} \tag{11}$$

where:

C – constant (0.4 is for case hardened teeth; 0.3 is for normalized or quenched and tempered teeth),

f_{ky} – effective distortion of the tooth contact line from its ideal location,
f_{z0} – deformation of the manufacture error-free mesh

Table 1 Values of the K^{I} constant [1]

Diagram number	K' factor		Arrangement of gears, shafts and supports - diagrams
	with stiffening $\frac{d_1}{d_{sh}} \geq 1{,}15$	without stiffening $\frac{d_1}{d_{sh}} \leq 1{,}15$	
1	0,48	0,8	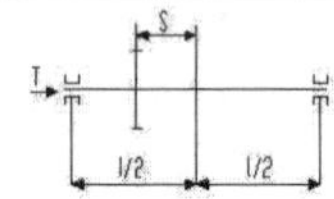
2	-0,48	-0,8	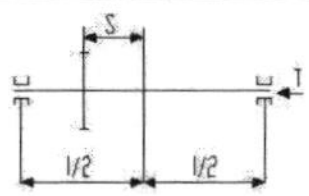
3	1,33	1,33	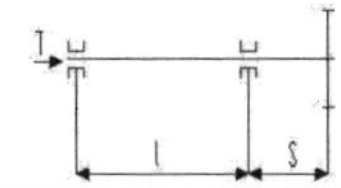
4	-0,36	-0,6	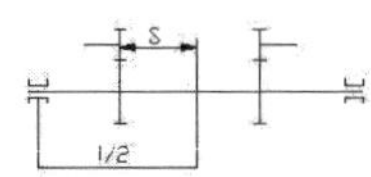
5	-0,6	-1	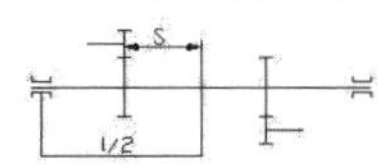

$$f_{z0} = \frac{F_t Z_\varepsilon^{\,2}}{bc^{''} \cos\alpha_t} \tag{12}$$

where:

F_t – (nominal) transverse tangential load at reference cylinder per mesh,
$Z\varepsilon$ – contact ratio factor (pitting),
b – tooth width,
c' – maximum tooth stiffness per unit face width (single stiffness) of a tooth pair,
α_t – transverse pressure angle.

$$f_{ky} = |f_{kE}| + f_{kZ} - f_{Kv} \tag{13}$$

where:

f_{kE} – tooth misalignment caused by elastic deformations (of the shafts mainly) and by the bearings slackness,
f_{kZ} – misalignment resulting from the allowable manufacture deviations,
f_{Kv} – misalignment (alleviation of the misalignment) as a result of running-in.

The misalignment f_{kE} is influenced by the elastic distortion of shafts, bearings, bodies, as well as the torsional deformations of the pinion and the wheel. It can be expressed as:

$$f_{kE} = f_w = |f_{w1} + f_{w2}| \tag{14}$$

For a shaft arrangement of the designed transmission that is consistent with one of those presented in Table 2 and for values of the auxiliary parameters included in the range shown in the table, the shafts distortions are described by the formula:

$$f_{W1,2} = A_{1,2} \quad \Phi^2{}_{bd1,2}\left[\left|0{,}7 + B_{1,2}\right| + 0{,}3\right] \frac{F_t}{b} \quad K_A K_v, \tag{15}$$

where:

$$B_{1,2} = \frac{K'_{1,2}\Psi^2{}_{ld1,2}\Psi_{sl1,2}}{\Psi^4{}_{dWd1,2}}, \quad \Psi_{bd1,2} = \frac{b_w}{d_{1,2}}, \quad \Psi_{dWd1,2} = \frac{d_{W1,2}}{d_{1,2}},$$

$$\Psi_{ld1,2} = \frac{l_{W1,2}}{d_{1,2}}, \quad \Psi_{SL1,2} = \frac{S_{1,2}}{l_{W1,2}}, \quad \Psi_{dd1,2} = \frac{d_{1,2}}{d_{II}}.$$

The misalignment resulting from the acceptable manufacture deviations f_{kz} is defined by:

$$f_{kz} = \sqrt{f^2_{\beta q_1} + f^2_{\beta q_2} + f_y^2 + \left(f_x \quad \tan\alpha_t\right)^2} \quad \cos\alpha_t \cos\beta_b \tag{16}$$

where:

$f_{\beta q_1}, f_{\beta q_2}$ – nominal tooth position deviations,

f_x – nominal deviation of the gear axes parallelism in the common plane,

f_y – nominal gear axes misalignment.

Table 2 Constants $K'_{1,2}$ and $A_{1,2}$ for calculating the shafts distortion $f_{W1,2}$ [4]

Arrangement of gears, shafts and supports – diagrams	Shaft	$K'_{1,2}$	$A_{1,2}$ for $(B_{1,2}+0{,}7)\le 0$	$A_{1,2}$ for $(B_{1,2}+0{,}7)>0$
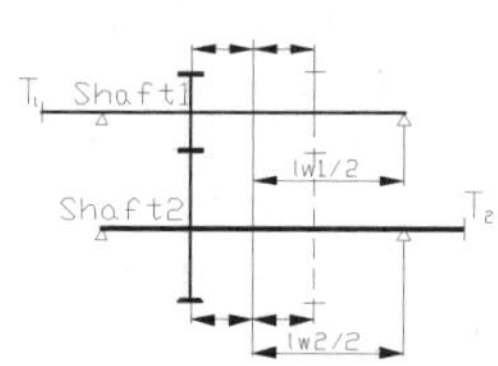	1	0,5	-0,0233	0,0233
	2	-0,5	0,0233	-0,0233
	1*	-0,5	-0,0233	0,0233
	2*	0,5	0,0233	-0,0233
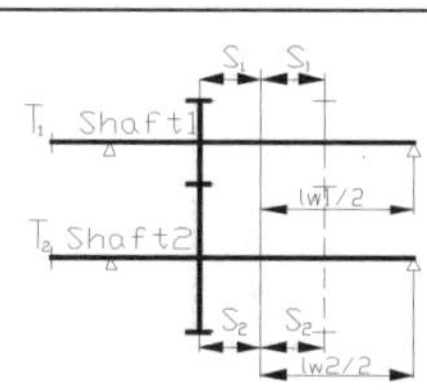	1*	0,5	-0,0233	0,0233
	2*	0,5	-0,0233	0,0233
	1	-0,5	-0,0233	0,0233
	2	-0,5	-0,0233	0,0233
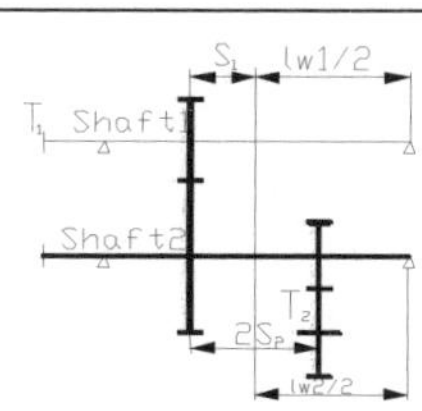	1	0,5	-	0,0233
	2	-0,5(1+1,8ψdd2- -3,18ψdd2ψS12)	0,0233	-0,0233
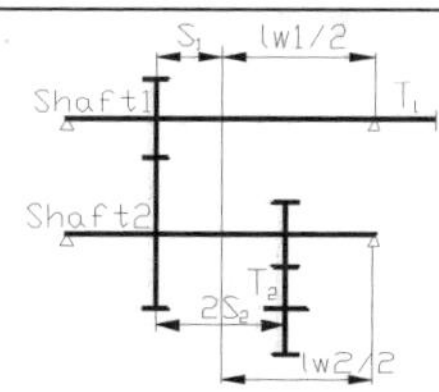	1	-0,5	0,0233	-0,0233
	2	-0,5(1+1,8ψdd2- -3,18ψdd2ψS12)	0,0233	-0,0233

The running-in process of the gears always alleviates the negative misalignment phenomenon of the mating teeth, and it is accounted for by the f_{kv} factor. Its value is assumed as approximately equal to the value of wear-in y_{β}, which is taken from [4].
The face load factor (root stress) $K_{F\beta}$ in both analysed methods is derived from the formula:

$$K_{F\beta} = K_{H\beta}^{\ N_F}. \tag{17}$$

$$N_F = \frac{1}{1 + h/b + (h/b)^2} \tag{18}$$

where:
b – rim width,
h – tooth height.

Results of Calculations and conclusions

The analysis of the influence of the two above presented methods of determining the $K_{H\beta}$ factor on the value of the safety factors S_H and S_F was conducted on the example of the design calculations of the first stage in a two-stage gear transmission,. The design scheme of this transmission together with dimensions necessary for calculations has been shown in Fig. 1. The transmission is designed to transmit a power of P=10.0 kW at a rotational speed of the pinion n=1400 rpm. The analysed stage of the transmission has a ratio of i = 3.15. The geometrical features of the toothing are: z_1 = 22, z_2=79, the helix angle β = 14°, toothing width b= 32 mm, correction coefficients x_1=0.62, x_2=0.77, normal module m_n=3, normal pressure angle α_n=20°. The material used for both gears is 41Cr4 steel, the teeth quenched and tempered to 43 HRC for the pinion and 40 HRC for the driven wheel. Both gears are full-disc gears, and the pinion is mounted on the shaft. For the shafts of the calculated toothing, assumed was the equivalent diameter d_{Sh1} =32 mm and d_{Sh2} =42 mm. The toothed wheel rim in the pinion is Φ = 46 mm in diameter, and in the driven wheel, Φ = 221 mm. It was decided that the gears would be made without the crowning and modification of the tooth ends. The design documentation contains information on the accuracy classes of the gears. It was also assumed that the gears' position would be adjusted, and the initial running-in of the gears would be conducted. The transmission should work flawlessly at the application factor K_A =1.25 for a time of 25000 h.

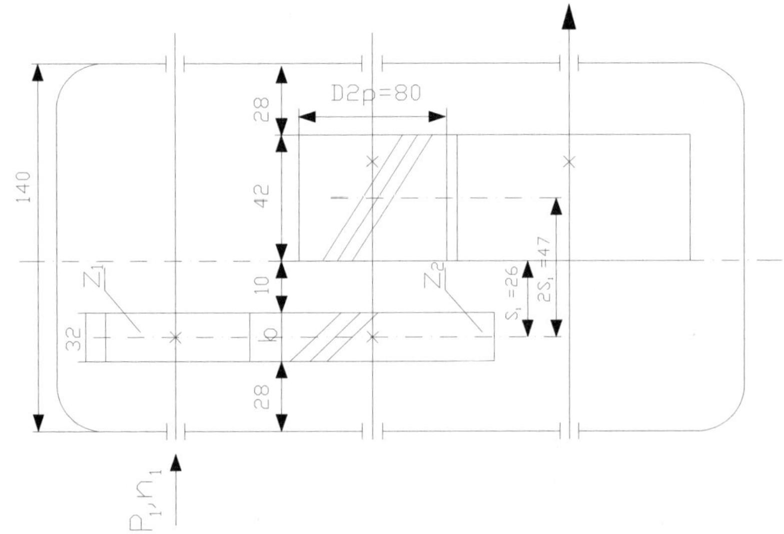

Fig. 1. Design scheme of the analysed transmission

As the accuracy class of gearing is one of the essential determinants when discussing $K_{H\beta}$, analysed were also four most popular gear accuracy classes, namely 5th, 6th, 7th, and 8th. Values of manufacture errors of the gears for the calculated accuracy classes were determined according to [6]. The calculations, as mentioned earlier, were carried out by means of a dedicated, author's computer program that works in accordance with the presented above analytical formulas. The results of the analysis of the influence of the compared methods on the value of the face load factor $K_{H\beta}$, as well as values of the safety factors s_H and s_F, for four accuracy classes of gears, have been presented in Fig. 2 (for the pinion) and in Fig.3 (for the gear).

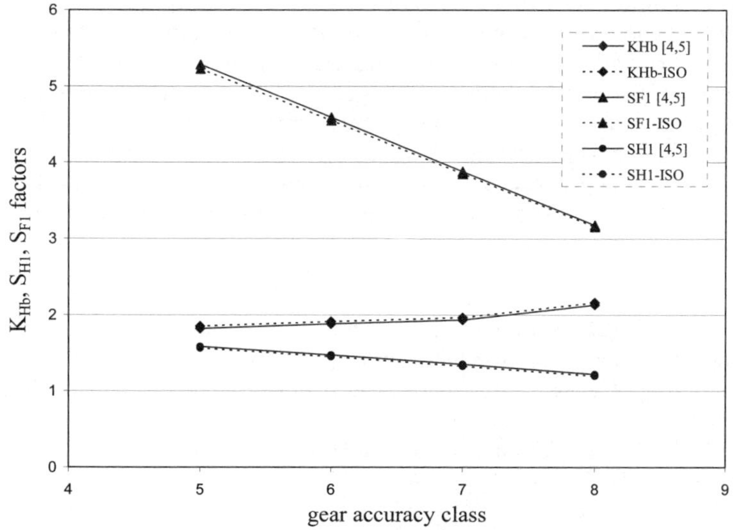

Fig. 2. $K_{H\beta}$, S_{H1}, S_{F1} factors for the pinion calculated according to ISO and [4,5]

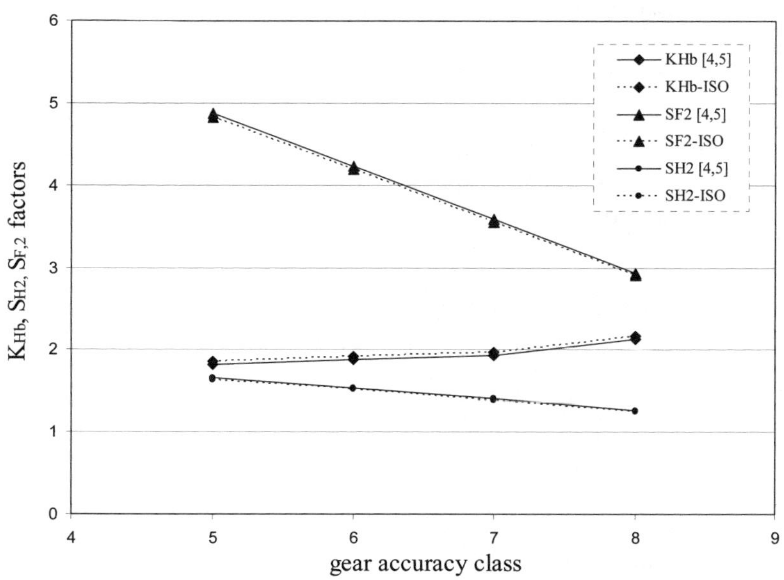

Fig. 3. $K_{H\beta}$, S_{H2}, S_{F2} factors for the gear calculated according to ISO and [4,5]

The broken line in the figures indicates values of the factors calculated by means of the C2 ISO method, whereas the solid line, the method described in [4, 5]. Similar analyses have also been carried out for different gear widths (b=24, 28 and 32 mm) and for more rigid shafts (d_{Sh1} =36 mm and d_{Sh2} =46 mm). No essential differences in the results (when compared to those in Fig. 1 and 2) were observed. The analysis of values of the face load factor $K_{H\beta}$ calculated by means of the two above-mentioned methods, as well of their influence on the gear strength enable to draw the following conclusions:

1. The choice of one of the two analyzed methods of determining the value of the face load factor $K_{H\beta}$ is unimportant when evaluating the strength of cylindrical gears. The values of the S_H (pitting) and S_F (tooth root) safety coefficients, regardless of what method of determining $K_{H\beta}$ factor had been applied, were different by not more than 2%.
2. The influence of the accuracy class of the gears on the gear tooth breakage strength is considerably greater than on the gear tooth surface durability. For instance (Fig.2), in the case of the pinion, a change of its accuracy class from 5 to 8 results (according to ISO C2) in a reduction of the breakage safety factor S_{F1} value from 5.22 up to 3.16, namely by 40%, and the gear tooth surface durability factor S_{H1} value from 1.57 up to 1.2, namely by 24%.

References

[1] PN-ISO 6336-1: 2000, Calculation of load capacity of spur and helical gears Part 1: Basic principles, introduction and general influence factors [in Polish].
[2] PN-ISO 6336-2: 2000, Calculation of load capacity of spur and helical gears Part 2: Calculation of surface durability (pitting) [in Polish].
[3] PN-ISO 6336-3: 2001, Calculation of load capacity of spur and helical gears Part 3: Calculation of tooth bending strength [in Polish].
[4] A. Dziama, M. Michniewicz, A. Niedźwiecki, Gears, PWN, Warszawa, 1989 [in Polish).
[5] M. Maziarz, S. Kuliński, Durability Calculations of gears according to ISO standards, AGH Kraków, 2005 [in Polish].
[6] PN-ISO 1328-1:2000, PN-ISO 1328-2:2000, ISO System of Accuracy - Definitions and Allowable Values of Deviations Relevant to Corresponding Flanks of Gear [in Polish].

Key Engineering Materials Vol. 490 (2012) pp 143-155
© (2012) Trans Tech Publications, Switzerland
doi:10.4028/www.scientific.net/KEM.490.143

Investigation of Dynamic Magnetic Fluid Seal Wear Process in Utility Water Environment

Zbigniew Szydlo[1,a], Marcin Szczech[1,b]

[1] AGH-University of Science and Technology, 30-059 Krakow, Poland

[a]zbszydlo@agh.edu.pl, [b]marcin.szczech@gmail.com

Keywords: magnetic fluid seal, water environment, rotating shaft, working life of seal, critical pressure, critical motion velocity.

Abstract. Serious problems arise when the magnetic fluid seal technology is needed to use in assemblies working in water or other liquid environment. Some efforts have been up to time made to design hybrid, two stage sealing structures with shields or mechanical seals used as protection measures to the magnetic fluid seal. Anyway, earlier or later there is always problem with direct contact between magnetic fluid and the environmental liquid. In the paper are presented results of experiments carried out with one stage rotating magnetic fluid seal operating in an direct contact with utility water. The special test procedures have been elaborated and practically used, with the aim to define main characteristics of magnetic fluid seals and to simplify experiments. These characteristics were: *critical pressure*, *critical motion velocity* and *working life* of the seal. Four grades of commercial, silicon based magnetic fluids of hydrophobic properties were tested in a single stage sealing system. The results of tests are presented in the form of tables and diagrams. It is concluded that hydrophobic commercial magnetic fluids could be efficiently used in rotating shaft seals however within limited and rather low range of motion velocity.

Introduction

Magnetic fluids are successfully applied in rotating shafts seals and can be also used to resolve tightness problems in other technical appliances operating in gaseous or vacuum condition. The actual state-of-art of this technology has been achieved as a result of many researchers' experimental and analytical work focused on recognition of magnetic fluid properties and seal characteristics. The principles of magnetic fluid seal design including magnetic material and magnetic fluid selection are formed in fundamental works of R.E. Rosensweig [1], B.M. Berkovsky [2], and E. Blums [3].

However, development of theoretical and practical principles of the magnetic fluid seals adopted to operation in water environment (generally in liquids) is a far more complex task when compared to gaseous or vacuum conditions. This is mainly due to the fact that two liquid phases; i.e., sealing magnetic fluid and the environmental liquid (water) are in a direct contact in this case.

The indispensible requirement for successful operation of magnetic fluid seal contacting with other liquid is possibly high (expectable absolute) immiscibility, and high surface tension at the interface between magnetic fluid and the environment liquid.

In static conditions (covers, pipe connections) and/or without any movement of the environmental liquid in relation to the magnetic seal fluid surface, the mutual immiscibility of liquids is practically sufficient to secure long life of the seal.

Anyway, the high interface tension and immiscibility of liquids may occur insufficient in dynamic condition of the seal operation and/or the flow of the environmental liquid in the vicinity of the seal. Relative movement of two liquids causes surface instabilities over the liquids interface that grow up with increased relative velocity [1,4]. Intensity of instabilities attains the critical value when the critical velocity is approached, resulting in rapid wear out of magnetic fluid and the loss of the seal structural tightness.

Magnetic fluid seals in liquid environment were investigated in some research centres and there are some information on their operation in such condition. In papers [5,6,7] are described experiments on the workability of magnetic fluid seals against water environment, with the seal operating without protective measures. Different water pressure and the seal velocity conditions were investigated.

A. F. Chorney and W. Mraz [8], propose the conception of magnetic fluid seal protected against immediate attack of an external liquid by the use of a hybrid, two stage seal, consisting of the master mechanical with secondary magnetic fluid seal. The authors recommend that design as an efficient, environmentally friendly solution to process pumps leakages.

There is also known a medical application of the magnetic fluid seal in a rotary blood pump, clinically used to solve problems of artificial hearts. The rotor of this precise blood pump sealed with a magnetic fluid seal operates a long time without wear. This pump and the seal have been developed by the team under the supervision of Y. Mitamura [9].

The Author of this paper has been since a time engaged in laboratory tests of magnetic fluid seals operating in utility water environment [10,11]. The experiments were carried out with a special stands adapted to operation in water environment condition [12,13]. Multistage magnetic seals with commercial magnetic fluids were used in these tests. Owing to the project tasks, which were solution of some technological problems, the seals were designed as models of the real structures. Despite difficulties in the analysis of results, mainly due to complexity of the seal design, the experiments were promising and confirmed the opinion that some commercial magnetic fluids, manufactured on the base of hydrophobic liquids are usable to the seals operating in water condition.

Problem Presentation

Physical phenomena at the water/magnetic fluid interface result in surface instabilities due to velocity difference across the interface and are the main reason of the magnetic fluid seal wear and subsequent leakage [9]. Certainly, there are many additional factors both geometrical and dynamic that affect the ability of the seal efficiency in water environment. The complexity of the problem has been increased by the magnetic field whose direction and intensity may offer stabilization or deepen the surface instabilities [1,4].

In actual state of knowledge the experiment seems to be more promising than theoretical consideration, in identification of the interface phenomena and estimation of the basic characteristics of a magnetic fluid seal operating in water conditions. Anyway, the experiments should be possibly simple, both as concerns selection of dynamic parameters and the geometrical forms of the tested seals.

Previously, in experimental works [6,7,10,11] the symmetrical multistage sealing systems were tested. The results of such experiments [10,11] were troublesome in interpretation, mainly due to uncertainties concerning proper estimation of the leakage moment. Hence, it was decided to simplify the seal geometry by the use of a single stage magnetic fluid seal as the basic experimental structure.

Single stage magnetic fluid seal is rather unique solution in sealing systems of technological devices. In practical applications the multistage seals are used, with an even number of sealing stages in symmetrical arrangement [2]. The symmetrical design is a simple consequence of the magnetic circuit physical shape where two poles are required to close magnetic loop with gaps filled with magnetic fluid.

This situation is shown in Fig.1, that presents the scheme of operation of two stage and a single stage seal in water. Up to the certain pressure, the water is blocked by the first stage of the seal, Fig.1a. Once the critical pressure of the first stage is attained the water penetrates between two stages area, Fig.1b, and initially is hold in this position. With the progressing wear of the seal (both the first and the second stage) the actual critical pressure is exceeded and finally the water leakage is observed . Anyway, the observations in this arrangement are disturbed by two step leakage process as we cannot define the moment of the first stage leakage.

When a single stage seal is applied as presented in Fig.1c, the identification of the seal failure (leakage) is easier because the water penetrating the seal stage is immediately directed to the leakage tank. The single stage magnetic circuit of the seal has no symmetric magnetic fluid gap but a cylindrical gap without magnetic fluid. Cylindrical gap has a sufficient width to minimize magnetic reluctance of the magnetic circuit and the sufficient thickness to facilitate outflow of leakages.

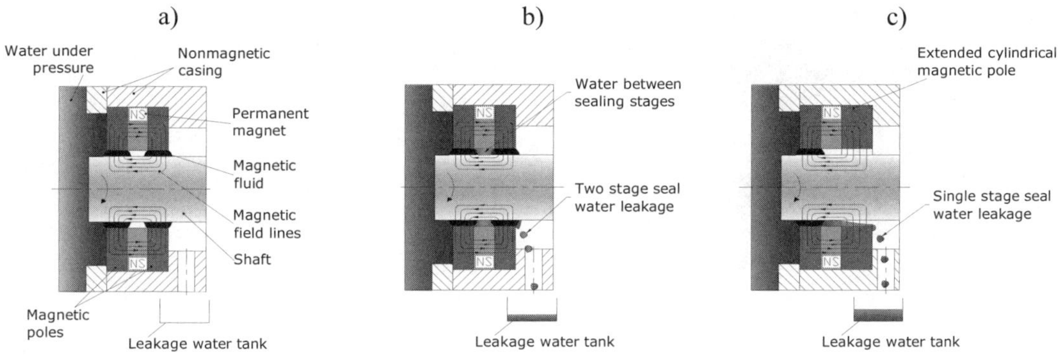

Fig.1. Schematic presentation of two stage and a single stage magnetic fluid seal of a rotating shaft operating in water: a) operation of the seal below critical pressure of water, b) penetration of water between seal stages and leakage trough the second stage, c) immediate leakage after failure of a single stage seal.

The experimental test stand – denoted "MASTL1-V2" enables application of the presented and other forms of magnetic fluid seals. However all tests described in the paper were performed with seals of the same, single stage geometry. Four grades of magnetic fluids of high hydrophobic properties were investigated. All magnetic fluids are commercial products delivered by Ferrolabs Inc., Sterling, VA, USA [14].

The test stand, the seal geometry and the magnetic fluids characteristics are described in more detailed way in the following chapters.

Test stand

Test stand has been designed to meet different requirements expected in magnetic fluid seals research. The structural sectional view of the test stand head is shown in Fig.2. The head has been adapted to realize experiments with water in direct contact with magnetic fluid seal surface. Details, including main dimensions of the magnetic fluid single stage seal unit are shown in Fig.3.

Compressed air is delivered through the channel 11 (yellow), to the membrane holder 9, and by the elastic membrane 8, pressurizes water 12 (blue), which directly contacts with the seal unit elements. The seal unit (detailed view in Fig.3) consists of the single-stage sealing sleeve 1, of O.D.: $d_s = 49.80\ mm$, cylindrical magnetic poles 2, of I.D.: $d_p = 50\ mm$, permanent magnets assembly 3, (14 magnetic pieces of diameter: d_m=15mm and thickness: h_m=5mm), and the magnetic fluid volume 4, hold by magnetic field in the gap of thickness: δ=0.1mm, formed between the sealing stage and the cylindrical magnetic pole. The sleeve has a cylindrical part of the outer diameter the same as the sealing stage with no magnetic fluid, closing the magnetic circuit by air. The water, after penetration the single MF ring is freely evacuated, through the cylindrical gap to the outflow 17.

The magnetic field circuit is closed sufficiently well through cylindrical air gap, despite lack of magnetic fluid in this area due to extended length of the sleeve cylindrical part. This problem has been positively examined in numerical simulations of the magnetic field distribution that are not presented in the paper.

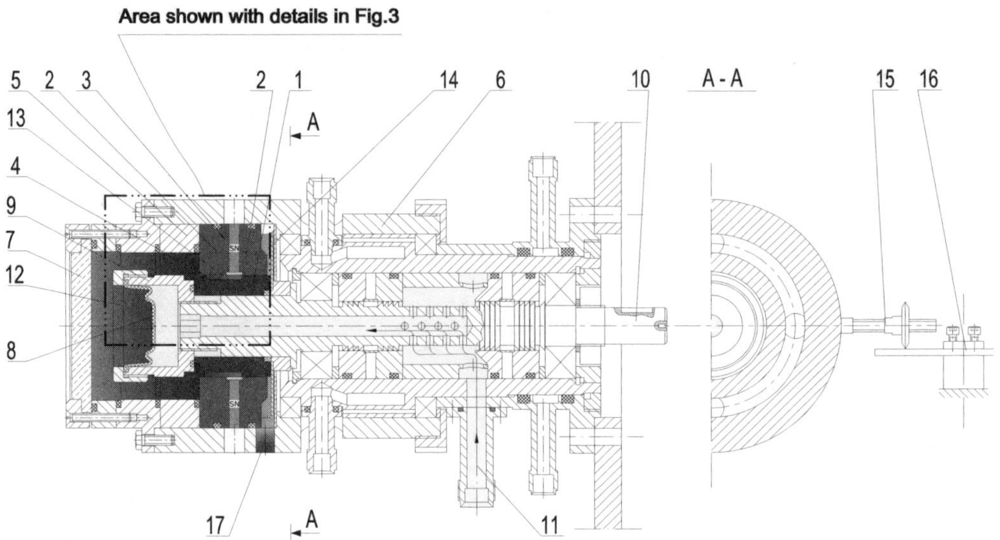

Fig. 2. The structure of the test head for experiments of magnetic fluids seals in water environment. 1 – single-stage sealing sleeve, 2 – magnetic poles, 3 – permanent magnets, 4 – magnetic fluid, 5, 6 – casing of the rotational part of the head, 7 – transparent cover of water chamber, 8 – elastic rubber membrane, 9 – membrane holder, 10 – rotating shaft, 11 – compressed air channel, 12 – pressurized water-filled chamber, 13 – area of liquid penetration in the sealing sleeve area, 14 – area of liquid outflow due to leakage, 15 – the torque-meter arm, 16 – torque measurement strain gauge beam, 17 – leakages outlet.

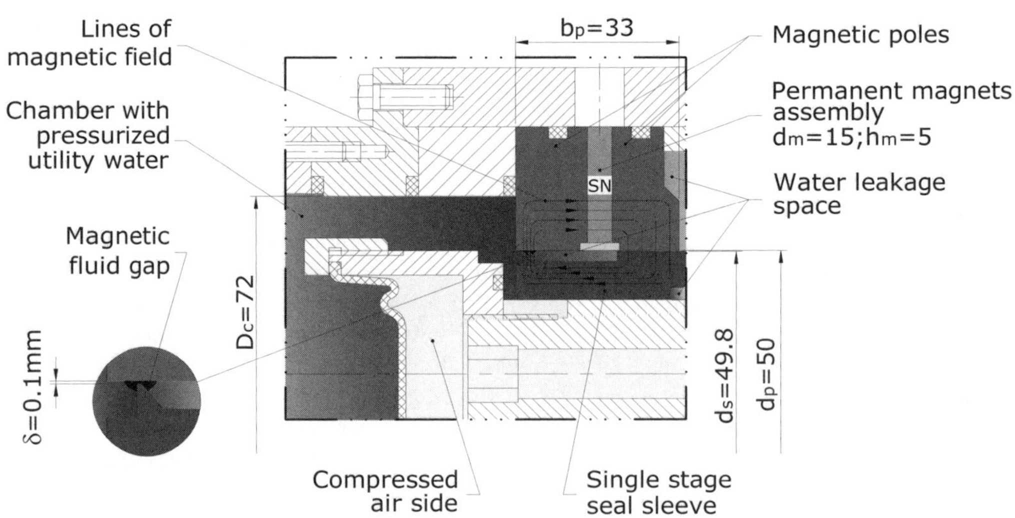

Fig.3. Design details of the seal single stage seal unit applied in experiments.

The general view of the laboratory test stand and the measurement-recording system is shown in photograph, fig.4.

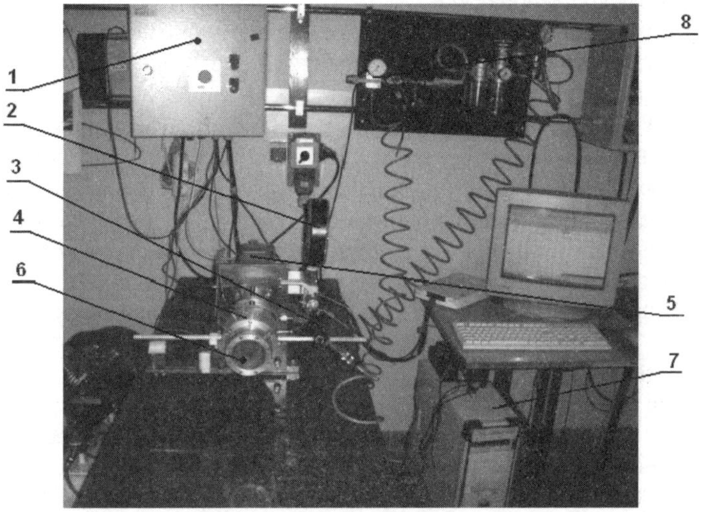

Fig.4. Laboratory test stand for magnetic fluid seals experiments in water environment and the measurement – recording system.

1 – control box, 2 – pressure gauge, 3 – compressed air supply line, 4 – test head, 5 – driving motor, 6 – transparent observation cover of water chamber, 7 – PC with measurement equipment, 8 – compressed air control block.

The water-filled chamber 6, is pressurized by the compressed air delivered from the air control block 8. The actual pressure, is indicated by the pressure gauge 2, and measured by MBS 32 Danfoss pressure transducer (not shown on photograph) of the measurement range of 0...1 MPa and the accuracy 0.2% of the maximum range.

Rotational speed of the driving motor 6, is controlled by the frequency converter located in control box 1, which also indicates rotational speed of the driving shaft and the actual power of driving motor.

Signals of pressure, torque, temperature, rotational speed and power are transferred to the PC measurement-recording equipment 7. The multi-channel measurement chart, type NI USB-6221 M Series was used in experiments, in cooperation with LabVIEW 8.5 data processing programme, both the products of National Instruments.

Experiment

Conception of Experiment. Research conception was based on the assumption that the motion velocity is the factor deciding on the magnetic fluid seal durability.

The scope of experiments has been projected to make possible the quantitative examination of that assumption within a series of short-time and long-time tests.

Short-time tests were aimed at determination of the *critical operational characteristics* of the magnetic fluid seal: *critical pressure* and *critical motion velocity.*

Long-time tests were aimed at determination of the *working life* of the magnetic fluid seal depending on the motion velocity of the seal.

Critical Pressure. Critical pressure has been determined in the static tests (non rotating shaft) with water pressure in the test chamber continuously increased until the water leakage was observed. The value of leakage pressure was denoted as the *critical pressure* of the seal.

Relative pressure. The water pressure level is an important factor that affects the magnetic fluid seal operation in water conditions, but that problem was not the subject of that study. Anyway, due to significant differences between values of *critical pressure* of seals obtained with different magnetic fluids it was not possible to establish common value of water pressure applied in all tests. Hence, the unified value of the factor: *water pressure/critical pressure*, was used in all experiments, called: *relative pressure,* and denoted: p/p_{cr}.

Further experiments, both *critical motion velocity* measurements and *working life* tests (with different motion velocities of the seal) were realized with the values of water pressure in the test chamber set on the basis of *relative pressure*: $p/p_{cr} = 0.5$.

Critical Motion Velocity. Critical motion velocity of the seal was measured in tests with the rotational velocity of the seal shaft being continuously increased, starting from zero till the velocity of the first leakage observed. The rate of velocity increase (rotational acceleration), unified for all tests was $1s^{-2}$, since the maximal rotational velocity 120 s^{-1} was achieved after 120s. The water pressure maintained in the test chamber was determined from the *critical pressure* value, specified for the particular magnetic fluid to obtain *relative pressure:* $p/p_{cr} = 0.5$.

The value of rotational velocity the first leakage was observed has been denoted as *critical motion velocity*.

Working Life of the Seal. The *working life* of the magnetic fluid seal vs. motion velocity of the seal was measured in the series of long-time tests carried with the steady motion velocity, steady water pressure in the test chamber (*relative pressure:* $p/p_{cr} = 0.5$) and the magnetic seal temperature maintained, all the test time, within the range (25°C - 30°C). The test has been continued till first leakage was observed. The total number of the shaft revolutions until first leakage was observed was called the magnetic fluid seal *working life.*

Leakage was the main criterion of the seal operation however the friction moment, magnetic fluid seal temperature, rotational velocity and the water pressure were unceasingly measured and recorded. These values have been observed to control the test process.

The series of the magnetic fluid seal *working life* tests started with the *critical motion velocity* and have been continued with lower and lower velocities till the *working life* higher, than 1 million revolutions has been obtained. The *working life graphs* have been constructed on the base of the results obtained in subsequent tests of a series.

Procedure. The structure of the test head and specially the simple geometrical form of the single stage seal were designed to avoid research uncertainties connected with complex phenomena, discussed in chapter **"Problem Presentation".** Moreover, all tests have been conducted on the base of unified procedures rigorously kept in the course of the whole experiment with the objective to eliminate incidental errors and mistakes. Experiments were performed in the steady sequence of short-time and long-time tests with some operational data (e.g. magnetic liquid quantity) selected prior to experiment, and other (pressure, velocity) set on the base of current results of tests. Real time acquisition of current test parameters (pressure, resistance torque, temperature, shaft velocity and the drive power) enabled correct determination of *critical pressure*, *critical motion velocity* and the seal *working life* based on recorded values . Additionally, analysis of acquired data facilitated identification of the source of experimental problems in the case of tests, which produced considerably different results.

Sequence of Tests. Experiments were performed in the following sequence:

1. Measurement of the magnetic fluid seal *critical pressure* in static condition test.
2. Measurement of the magnetic fluid seal *critical motion velocity* in tests with water pressure determined from *critical pressure* value, on the basis of *relative pressure*, $p/p_{kr}=0.5$.
3. The series of *working life* tests with water pressure determined from *critical pressure* value, on basis of *relative pressure*, $p/p_{kr}=0.5$. *Working life* tests were performed with different motion velocities subsequently reduced in the test series to obtain *working life* above 1 million revolutions.

Repeatability of Test Results. Most tests were long time, *working life* experiments, which were extremely time consuming. It was generally accepted that each test was repeated with the same obligatory procedures. If the test results difference was kept within ca.10% of the higher value, the test has been classified as positive and the final result was the arithmetic mean of two values. In the case of greater differences between subsequent tests results, the acquired data of tests have been analyzed to identify the source of possible errors and third repetition was decided. Practically there were a few cases that the third repetition was necessary and the main reason of faults were problems with stabilization of operational pressure in the water chamber.

Materials. Materials, which are of direct effect on the experiment results are: magnetic fluids, soft magnetic materials (magnetic poles and sealing sleeve) and hard magnetic materials (permanent magnets).

Magnetic Fluids. Four grades of magnetic fluids specified in Table1 were tested in experiments. All fluids are commercial products of Ferrolabs Inc. [14].
The selection was made within two kinds of fluids of different destination and with regard to their values of saturation magnetization, M_s, and the plastic viscosity, η_{pl}. These two kinds were:

- FLS - FerroLiquid for Seals,
- FLA - FerroLiquid for Audio.

Within each kind there were selected one fluid of higher magnetization and lower viscosity and the other of higher viscosity and lower magnetization.

FLS magnetic fluids are typically used for sealing purposes, with broad range of applications including seals in biopharmaceutical industry and high-speed drivers for CD and DVD.

FLA type fluids are recommended as audio-liquids used in higher brand of full-range speakers, woofers, and sirens. However they proved quite well behaviour in experiments with sealing application in water condition.
The specification of selected fluids is given in Table1.

Table 1. Specification of magnetic fluids used in tests. Products of FerroLabs, Inc.[14].

No	Magnetic fluid	Carrier liquid	Magnetic particles	Saturation magnetization M_s, [kA/m]	Plastic viscosity at 20°C, η_{pl}, [Pa·s]	Critical temperature, [°C]	Operational temperature, [°C]
1	FLS 040.040	Silicon fluid	Magnetite Fe_3O_4	45.00	0.550	+200	-70...+150
2	FLS 250.020	Silicon fluid	Magnetite Fe_3O_4	25.00	1.120	+200	-60...+100
3	FLA 003.45	Siloxane	Magnetite Fe_3O_4	45.00	0.450	+200	-70...+150
4	FLA 002.25	Siloxane	Magnetite Fe_3O_4	25.00	0.725	+250	-70...+150

Magnetic Poles and Sealing Sleeve. Magnetic poles and sealing sleeves were made of high magnetic permeability (relative permeability μ_r>2000) carbon steel.

Permanent Magnets. Permanent magnets used in test seal were Neodymium 38 of coercivity H_c=912 kA/m and energy density $(BH)_{max}$=294 kJ/m^3 formed as discs Ø15x5 mm [15]. Discs (15pieces) were arranged evenly on the circumference in the recess between magnetic poles.

Characteristics of tested seals. Below are listed the main geometrical and operational data of tested magnetic fluid seals.

Geometrical data (Fig.2 and 3):

• Magnetic poles I.D. (The seal rated diameter):	$d_p = 50$ mm
• Sealing sleeve O.D.:	$d_s = 49.8$ mm
• Gap thickness:	$\delta = 0.1$mm
• Water chamber I.D.:	$D_c = 72.0$ mm
• Number of sealing stages:	$z = 1$
• Total width of magnetic assembly (magnetic poles+permanent magnets):	$b_n = 33$mm
• Diameter of permanent magnet element:	$d_m = 15$mm
• Thickness of permanent magnet element:	$h_m = 5$mm
• Number of permanent magnet elements:	$L = 14$

Operational data:

• Quantity of magnetic fluid applied to the gap:	100 ±2 μl
• Operational medium in test chamber:	utility water
• Pressure in test chamber:	$p = (0{,}02 \div 0{,}1)$ MPa
• Relative pressure in test chamber:	$p/p_{cr} = 0.5$
• Range of rotational velocity:	$n = (0 - 120)\text{s}^{-1}$

Results. Results of all tests are listed in Table 2. Table contains values of operational parameters and the values obtained in tests carried with particular fluids for different shaft rotational velocities. The values of test results listed in table are the arithmetic mean of two values obtained in measurements as described in subchapter "**Repeatability of Test Results**".

The *working life diagrams* have been prepared on the base of listed data. Basically the diagrams show the relation between number of revolutions a magnetic fluid seal could withstand to the first leakage vs. the seal motion velocity *(rps and m/s)*. The time intervals since start (in hours) are added at some points representing tests made with longer operational periods.

Working life diagrams are shown separately in Figs.5÷8. Graphs of the trend lines best fitted with test results are also shown in the diagrams.

The *working life diagram* of the magnetic fluid FLS 040.040 is shown in Fig.5. First measurement point on diagram presents working life obtained in the test with *critical velocity* of the seal: $73s^{-1}$.

Trend line equation has the form: $y = -7.14\ln(x) + 124.83$.

The *working life diagram* of the magnetic fluid FLS 250.020 is shown in Fig.6. First measurement point on diagram presents working life obtained in test with critical velocity of the seal: 22[s^{-1}].

Trend line equation has the form: $y = -1.54\ln(x) + 31.08$.

The *working life diagram* of the magnetic fluid FLA 003.45 is shown in Fig.7. First measurement point on diagram presents working life obtained in test with critical velocity of the seal: 82[s^{-1}].

Trend line equation has the form: $y = -7{,}03\ln(x) + 127{,}3$

The *working life diagram* of the magnetic fluid FLA 002.25 is shown in Fig.8. The time interval to leakage at the critical velocity of the seal (55 s^{-1}) was not measured. First measurement of working life was made with slightly lower motion velocity: 50 s^{-1}.

Trend line equation has the form: $y = -3{,}85\ln(x) + 81{,}075$.

Tabl.2. Results of tests of four grades of magnetic fluids used in seals operating in utility water. Sealing gap thickness: δ= 0.1mm; quantity of magnetic fluid applied to the gap: 100μl; relative pressure: $p/p_{cr} = 0.5$.

No	Magnetic fluid	Critical pressure, p_{cr} [MPa]	Test pressure, p [MPa]	Critical velocity, [s^{-1}]	Rotational velocity in test, [s^{-1}]	Number of revolutions to leakage	Operational time to leakage [s]
1	FLS 040.040	0.0776	0.039	73	73	730	20
2					60	30000	500
3					40	200000	5000
4					20	980000	49000
5	FLS 250.020	0.0616	0.031	22	22	250	12
6					20	500	25
7					19	20900	1100
8					18	32400	1800
9					10	530000	53000
10					7	980000	140000
11	FLA 003.45	0.0868	0.043	82	82	850	22
12					75	1375	28
13					60	31200	520
14					50	32000	640
15					40	82000	2050
16					30	1800000	60000
17	FLA 002.25	0.0443	0.022	55	55	Not measured	
18					50	2000	40
19					45	9000	200
20					40	384000	9600
21					20	1428000	71400

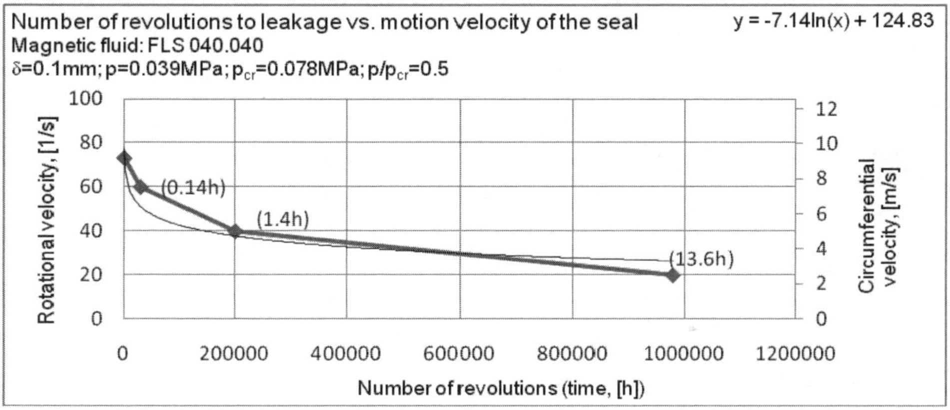

Fig.5. Working life vs. motion velocity of the seal with magnetic fluid FLS 040.040 operating in utility water. Pressure of the water in test chamber: *p= 0.039MPa*

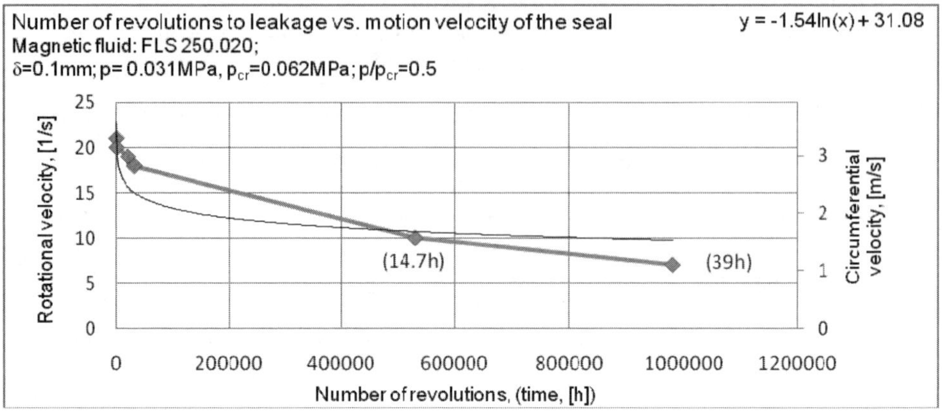

Fig.6. Working life vs. motion velocity of the seal with magnetic fluid FLS 250.020 operating in utility water. Pressure of the water in test chamber: *p= 0.031MPa*

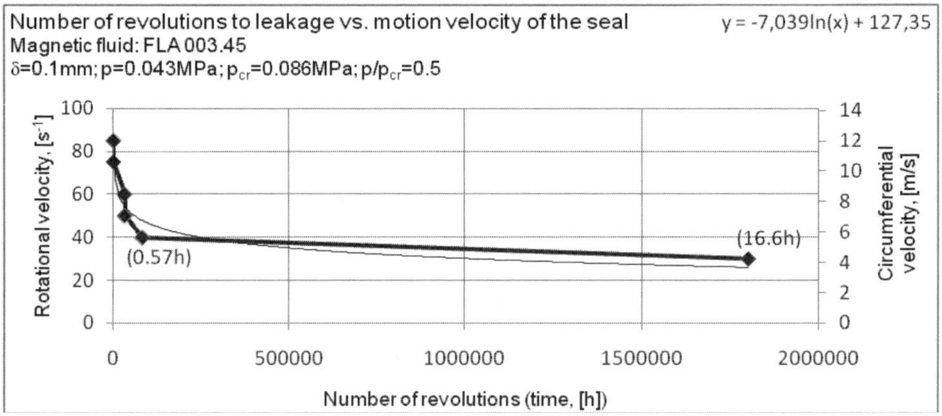

Fig.7. Working life vs. motion velocity of the seal with magnetic fluid FLA 003.45 operating in utility water. Pressure of the water in test chamber: *p= 0.043MPa*

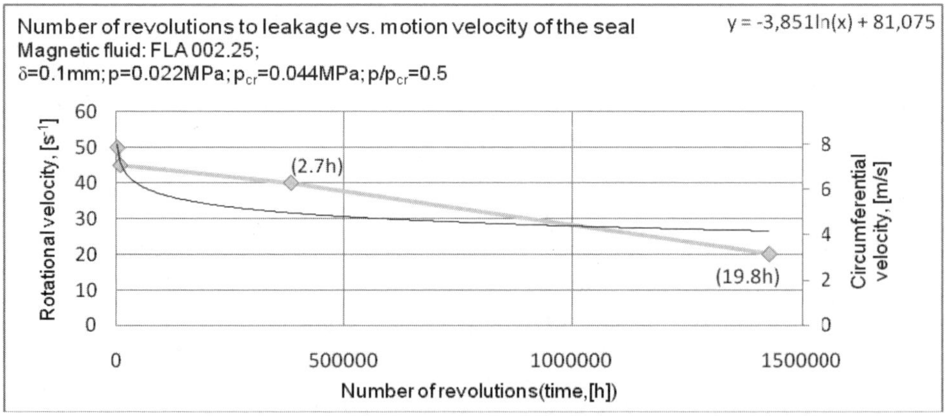

Fig.8. Working life vs. motion velocity of the seal with magnetic fluid FLA 002.25 operating in utility water. Pressure of the water in test chamber: *p= 0.022MPa*

Comparison of Results and Discussion

Magnetic saturation and *plastic viscosity* of magnetic fluids, compared with *critical motion velocity* and *critical pressure* data obtained in tests are summarized in Table 3.

Maximal *critical motion velocity*, 12.8822 m/s was obtained in tests with magnetic fluid FLA 003.45, manufactured on the base of *siloxane* as the carrier liquid. That type of fluids is destined to acoustic devices.

Minimal *critical motion velocity*, 3.4562 m/s was obtained in tests with magnetic fluid FLS 250.020, manufactured on the base of silicon fluid as the carrier liquid. That fluid is typical for sealing applications.

Further comparison shows that other acoustic type magnetic fluid, FLA 002.25 has also proved higher *critical motion velocity* then the other sealing type fluid, FLS 250.020, despite similar magnetic saturation values of both fluids.

Tabl.3. Magnetic and viscous properties of magnetic fluids compared with *critical pressure* and *critical motion velocity* values obtained in experiments with magnetic fluid seals operating in water.

Magnetic fluid	Saturation magnetization M_s, [kA/m]	Plastic viscosity at 20°C,	Critical pressure, p_{cr}, [Pa]	Test pressure, p, [MPa]	Critical motion velocity	
					[s^{-1}]	[m/s]
FLS 040.040	45.00	0.550	0.0776	0.038	73	11.4683
FLS 250.020	25.00	1.120	0.0616	0.031	22	3.4562
FLA 003.45	45.00	0.450	0.0868	0.043	82	12.8822
FLA 002.25	25.00	0.725	0.0443	0.022	55	8.6405

Anyway the conclusions should be careful, as all tests were performed on the principle of the unified value of *relative pressure:* $p/p_{cr} = 0.5$. Acceptance of that principle resulted in significant difference in actual values of the *water pressure* in test of different types of magnetic fluids.

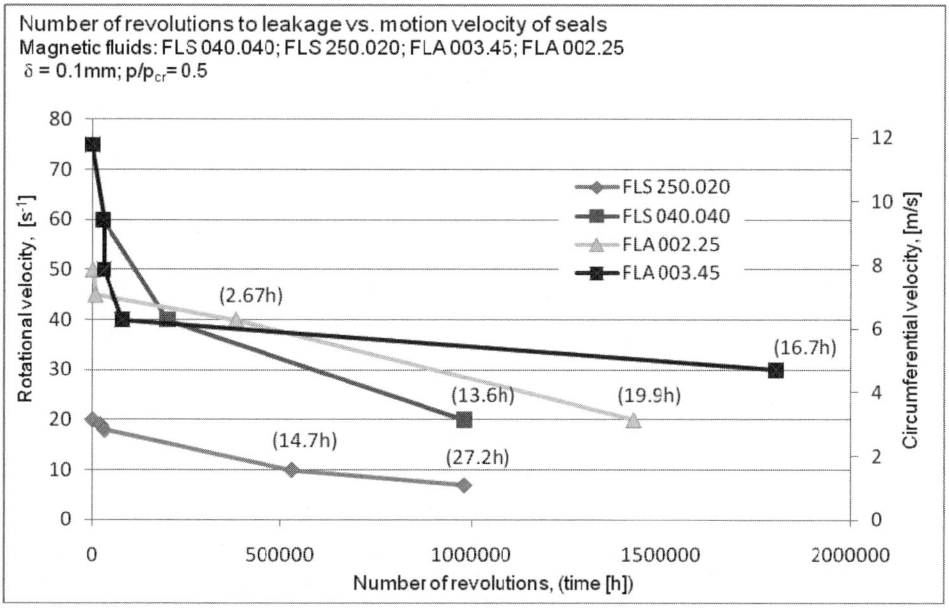

Fig.9. Working life vs. motion velocity of seals with four magnetic fluids grades used in experiments of magnetic fluid seal operation in utility water. Values of *water pressure* in the test chamber: FLS 040.040, $p=0.038MPa$; FLS 250.020, $p=0.031MPa$; FLA 003.45 $p=0.043MPa$; FLA 002.25, $p=0.022MPa$.

As an example, let's consider supposedly prospective test of the fluid FLA 002.25 with its *critical motion velocity,* 55 s^{-1} considerably higher than the *critical motion velocity* of the fluid FLS 250.020, 22s^{-1}, despite much higher *critical pressure* of the last fluid. Further analysis of data, explains that tests with FLA 002.25 have been made with pressure 0.022MPa, which is considerably lower than the pressure of tests with the fluid FLS 250.020, which was 0.031MPa.

Diagrams of *working life* of all magnetic fluids tested in experiments of magnetic fluid seals operating in water are set together in Fig.9.

Comparison of diagrams in the region of motion velocities below ca. 40s¯1, indicates that acoustic type magnetic fluids FLA 003.45 and FLA 002.25, show higher *working life* values than sealing type magnetic fluids FLS 250.020 and FLS 040.040. This observation, similar as above discussed problem of *critical motion velocities* has to be considered carefully due to different values of water pressures in tests with different grades of magnetic fluids.

Unexpectedly, the magnetic fluid FLS 250.020 shows *working life* values running significantly below other diagrams, despite relatively high *critical pressure* value, as shown in Table 3. This characteristics may be an effect of relatively high value of the *water pressure* applied in test with that magnetic fluid, resulting from the high value of its *critical pressure*.

Analysis of *working life* diagrams (probably with an except for the FLS 250.020 seal), suggests that graphs could be separated into higher velocity range and the lower velocity range, the two area that differ in the rate of working life increase with the seal motion velocity decrement.

The wear process of the magnetic fluid seal is very intensive in the high velocity range, and results in shorter *working life*. The lower limit of high intensity wear region is in the vicinity of the rotational velocity: ~ 40s^{-1} (linear velocity: ~ 6.283m/s).

The region of low wear intensity starts below the velocity value 40s^{-1}. As the result of reduced velocity we can see significantly longer and longer *working life*. This behaviour is particularly observed in experiments with magnetic fluids of higher saturation magnetization: FLA 003.45 and FLS 040.040.

Conclusions

Experiments with the magnetic fluid seals operating in water environment confirmed satisfactory correctness of the research conception and experimental strategy realized with the test stand MAST L1-V2 .

The simple single stage geometrical form of the seal has been tested due to the main objective of research, which was to recognize the fundamental relation: time to failure vs. motion velocity of the seal. Most experiments were time consuming *working life* tests hence the range and diversity of parameters were limited.

Despite these limitations the results allow to formulate the following remarks and conclusions:

1. Magnetic fluids based on silicone type hydrophobic carrier liquids offer the possibility of successful use in seals operating in water but in the range of motion velocities highly reduced in comparison with gaseous application.
2. The intensity of wear of magnetic fluid seal depends to the highest degree on the motion velocity of the seal. In the lower range of the seal motion velocities we can expect significant increase of working life values with subsequent reduction of motion velocity.
3. Results of some experiments suggest that the pressure exerted on the magnetic fluid has a serious effect on values of critical velocity and working life of the seal operating in water. Experiments with different pressure values should be carried out to explain the influence of that factor on characteristics of magnetic fluid seals operating in water.

4. The structure and control-measurement characteristics of the test stand MAST L1-V2, designed specifically for experiments with magnetic fluid seals, enable easy and correct operation. In the course of long-time tests emerged a problem with automatic leakage detection. This problem has to be resolved in view of long-time tests expected in future investigations of magnetic fluid seals in water environment.
5. Further research are necessary to explain the possibility and dynamic characteristics of ultra long-time or unlimited operational periods of magnetic fluid seals in water environment.

Acknowledgements:

This study has been developed on the base of Research Project no N N502 4568 33 granted by Polish Republic Government.

References

[1] R.E. Rosensweig, Ferrohydrodynamics, Dover Publications, Inc. Mineola, New York, 1997.

[2] B.M. Berkovsky, V.F. Medvedev, M.S. Krakov , Magnetic fluids engineering applications, Oxford Science Publications, 1993

[3] E. Blums, A. Cebers, M.M. Maiorov, Magnetic fluids, New York : Walter de Gruyter, 1997.

[4] C. Rinaldi, A. Chaves, S. Elborai, X. He, M. Zahn, Magnetic fluid rheology and flows, Current Opinion in Colloid & Interface Science Vol: 10 Issue: 3-4, October, 2005.

[5] D. B. Orlov, at all., Eksperimentalnoe issledovanie resursa magnitozidkostnych uplotnenij pri germetazicii zidkich sred, Magnitnaya Gidrodinamika, No 4, c.127-130, 1989. (in Russian).

[6] J. Kurfess, H.K. Muller, Sealing liquids with magnetic fluids, Journal of Magnetism and Magnetic Materials ,85,(1990)

[7] J. Vihersalo, at all., Sealing of Liquids with Magnetic Fluid Seals, Proc. of 6th Nordic Symposium on Tribology, NORDTRIB'94, Uppsala, Sweden, 12-15 June 1994.

[8] A. F. Chorney, W. Mraz, Hermetic Sealing with Magnetic Fluids, Machine Design; May 7, 1992; 64, 9; ProQuest Science Journals pg. 79.

[9] Y. Mitamura, S. Arioka, K. Sakota, M . Azegami, Application of a magnetic fluid seal to rotary blood pumps, J. Phys.: Condens. Matter 20 (2008) 204145 (5pp).

[10] Z. Szydlo, L. Matuszewski, Ring thruster – a preliminary optimisation study of ferrofluid seal and propeller, Polish Maritime Research, 2007 spec. iss. S1 s. 71–74.

[11] Z. Szydlo, L. Matuszewski, The application of magnetic fluids in sealing nodes designed for operation in difficult conditions and in machines used in sea environment, Polish Maritime Research, 2008 vol. 15 no. 3 s. 49–58.

[12] Test Head for Investigation of Magnetic Fluid Seal Parameters. AGH – University of Science and Technology in Krakow; Inventor: SZYDŁO Zbigniew, OCHOŃSKI Włodzimierz, ZACHARA Bolesław. Polish Patent PL 377598 A1. – Appl. 2005-10-12 ; Publ. 2007-04-16., Polish Patent Office Bulletin, 2007 no 8 p. 22. (in Polish).

[13] Z. Szydlo, Experiments with magnetic fluid seals (water – air) determination of working life and permissible linear velocity, Research Project No 6.6.130.261, AGH 2008. Not published. (in Polish).

[14] Ferrolabs, Inc. · Sterling, VA United States, http://www.ferrolabs.com/, 2011.

[15] P.P.H.U. "ENES" Pawel Zientek, Warszawa, http://www.magnesy.pl/, 2011.

Key Engineering Materials Vol. 490 (2012) pp 156-165
doi:10.4028/www.scientific.net/KEM.490.156

Methods of Increasing Loadability for the Plastic Cycloidal Gears

Krzysztof Biernacki [1,a]

[1] Ul. Lukasiewicza 7/9; 50 – 371 Wroclaw.
Wroclaw University of Technology, Poland.

[a] krzysztof.biernacki@pwr.wroc.pl

Keywords: cycloidal gears, plastic, stress, deformation, finite elements method.

Abstract. The working unit of the hydraulic gerotor machine consists in cycloidal gears. In the paper, ways of increasing the load limit for the plastic cycloidal gears have been presented . The objective has been achieved by means of structural and material methods. The modifications have been introduced on the basis of the theoretical analysis of stress and deformation that are generated in the cycloidals gears while in operation. The analysis has been carried out with the use of the finite elements method. The outcome of the analysis is determination of the maximum working pressure value at which the hydraulic gerotor machines featuring the plastic cycloidal gears can operate.

Introduction

The use of plastics in the structure of hydraulic machines (cylinder, valve, pump) creates new possibilities of their development and broadens the scope of their applications. Using plastics is advantageous for a number of design, technological, and economic reasons. The structure of a hydraulic system featuring plastic elements is lighter than the steel structure, and can work with ecological fluids such as emulsions or water. It is worthwhile noticing that such a system will generate a lower level of noise. Using plastics instead of steel is also beneficial both for technological and economic reasons because the elements of the system can be manufactured by means of a simple and cheap injection mould method.

The focus should be on introducing plastics in the design of the elements most exposed to loading, as well as those manufacturing of which is the most complicated. Such elements are cycloidal gears. The gears stand for the main working unit of the hydraulic gerotor machines.

In the previously published papers, considered was the possibility of applying a plastic as a construction material for making the cycloidal gears in the gerotor pump [1,2,3]. The authors of the papers managed to specify the mechanism of generating stresses and deformations in the gears, as well as to estimate their values. On the grounds of the findings, it can be observed that it is possible to use plastics instead of steel in making the cycloidal gears [2, 3]. However, in comparison to steel, plastics show lower strength and higher deformability, which results in the fact that the working pressure for the pump with the plastic cycloidal gears will be lower than the working pressure for the steel gear pump. In spite of the lower strength of the plastic, the pump performed up to p=5MPa of the working pressure range. At that load, the plastic gear pump achieved the volumetric efficiency level of approx. h_V=72% for the rotational velocity n=1500rev/min [1,3].

Achieving the value of p=5MPa of the working pressure is a good result and it indicates that plastics can and ought to be used in hydraulic systems. However, it is also necessary to look for solutions that will make it possible to apply plastics for operation in higher working pressure conditions. Hence, a necessity of introducing design changes in the structure of the cycloidal gears, which will cause the increase of the loadability for the working unit of the machine.

Such solutions have been suggested in this article. At the next stage, the analysis of the solutions by means of FEM (Finite Elements Method) was carried. The analysis was aimed at:

- determining the mechanism of generating stress and deformations of the gears, and consequently defining the internal clearances,
- specifying the theoretical range of the gear system loadability in the context of the allowable values of stress and deformations.

Operation Principle of the Cycloidal Gears in the Hydraulic Gerotor Machine.

The operation principle of the cycloidal gears has been depicted in Fig. 1.

The cycloidal gear system (1) (2) is driven into the rotational motion by means of torque M. Between the gears featuring a special internal cycloidal toothing of the tooth difference z_2-z_1=1, made are the intertooth displacement chambers [6,7].

The working fluid goes into the pump through the inlet channel (Inlet) and fills in the intertooth displacement chambers. Next, as a result of the gear unit's revolution, the working fluid is charged towards the outlet channel. While the process is in progress, pressure of the fluid increases.

As a consequence of the cycloidal gear unit's revolution, the unit remains under the mechanical load resulting from the activity of torque M, as well as under the hydraulic load resulting from the pressure of the displaced fluid.

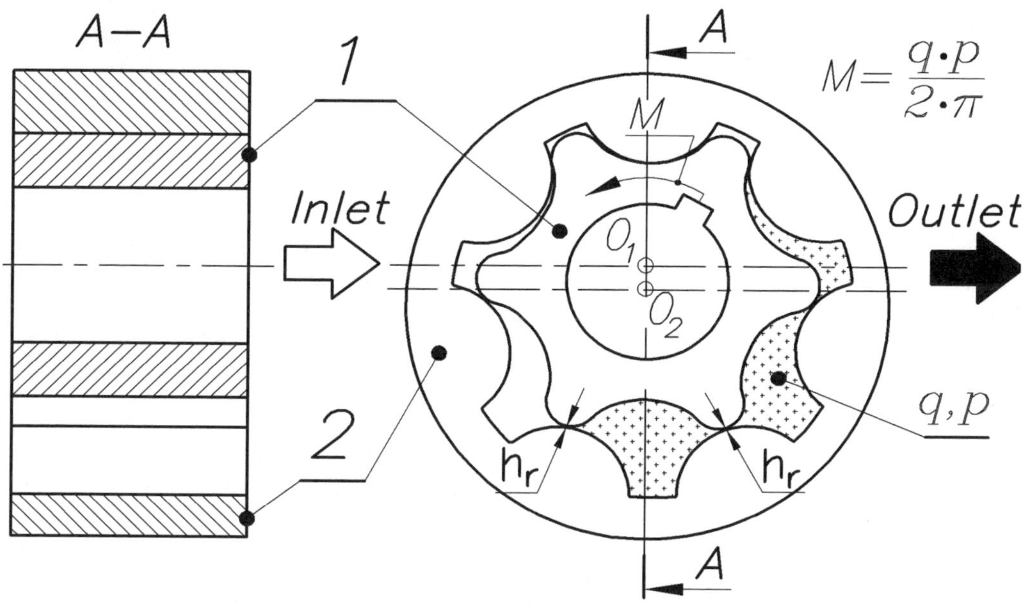

Fig. 1. Design and operation principle of the cycloidal gear system in the hydraulic gerotor machine: 1 – driving wheel, 2 – driven wheel.

Such a load distribution makes the cycloidal gear unit remain in balance, and the pressure in the displaced fluid is generated by torque M.

In the analysed cycloidal gear system, the mechanical energy is transmitted from the drive shaft onto the externally toothed driving wheel (1), through the splineway, and then the energy is carried from the driving wheel, through the intertooth contact points, onto the internally toothed driven wheel (2). As a result of the gear unit's revolution, the charged fluid pressure gets higher, and the mechanical energy gets transformed into the hydraulic energy.

The summary load of the gears causes their deformation and the occurrence of the so called intertooth radial clearance h_r. The occurrence of the clearance is a negative phenomenon, because it makes the hydraulic machine get untight and consequently, leads to the lowering of its efficiency.

Fig. 2 illustrates the distribution of stress and shifts for the polioxymethylene (POM) cycloidal gear unit while operating in the gerotor pump that was loaded with working pressure p=4MPa and torque M=7,16Nm [2].

The pictures show clearly that the effort of the driving wheel is bigger than the one of the driven wheel, and a considrable concentration of stress occurs in the area of the splineway.

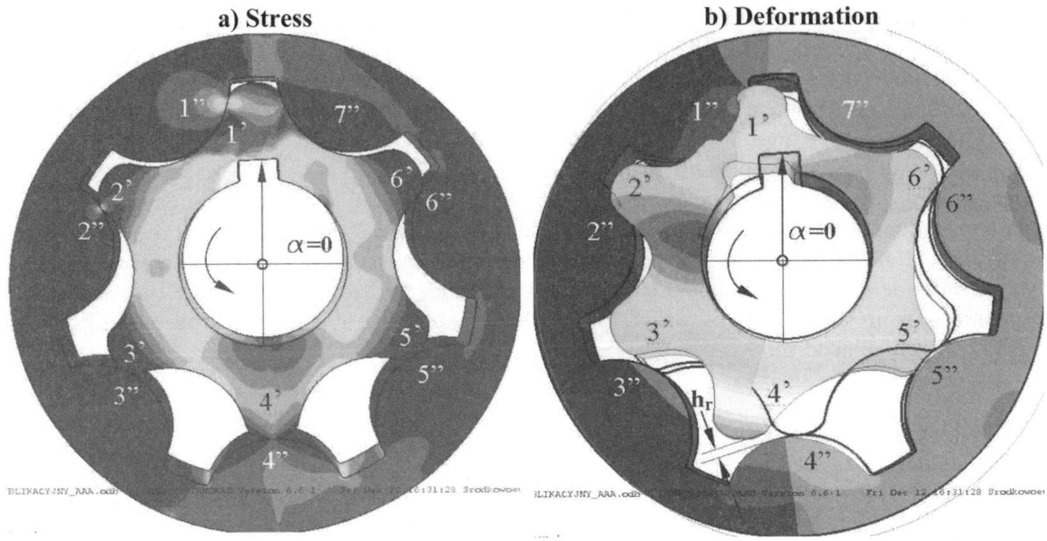

Fig. 2. Distribution of stress and deformations in XY plane in the polioxymethylene cycloidal gears model, the position of the outer wheel α=0 [2]

The maximum value of the stress was recorded on the left side of the splineway in the driving wheel, and it was approximately 26MPa, whereas the stress on the contact of 1'-1" pair was approx. 24 MPa.

However, in the case of the hydraulic machines, more important are the deformations, because they lead to the pump getting untight and to the lowering of its efficiency. Particularly dangerous for the pump is the occurrence of radial clearance h_r at the border of the inlet and the outlet zone. The maximum value of the clearance was defined as hr_{dop}=0,1mm [2].

For the cycloidal gear system depicted in Fig. 2a, the maximum deformation was observed at the top of tooth 4', and radial clearance h_r generated by the deformation was h_r=0.082mm. That value is close to $hr_{dop.}$.

Such a stress and deformation distribution is caused by the assymetrical distribution of loading pressure p, as well as the local application of torque M.

Therefore, it is necessary to introduce modifications in the cycloidal gears' design in order to widen the range of operational parametres for the gears. The modifications should include both the material change, as well as some changes in the gear system's design.

Material Selection, Design Modifications, Research Programme

Material Selection.

A typical plastic used for making cycloidal gears is the pure polioxymethylene (POM) [1]. However, as it has been proved by the analysis of the material, its strength parametres do not make it possible to use elements made of this plastic for high pressure fluid power units [2,3]. Therefore, it was decided that a modified plastic of higher strength parametres than the polioxymethylene (POM) would be used.

The plastic was produced as a result of research and development process carried out in the Institute of the Fundamentals of Machine Design and Tribilogy at Wroclaw University of Technology. The strength limit for the modified plastic was specified on the research basis and is $\sigma_{dop}\cong 160$MPa.

Table.1 Comparison of the modified plastic and the polioxymethylene parametres

No.	Material	Young's modulus E [MPa]	Yield point Re [MPa]
1.	Modified plastic	14000	160
2.	Polioxymethylene (POM)	3000	60

The modified plastic was properly selected as a material for the making of the model of the gear system, and then numerically analysed.

Table 1 shows that the strength parametres of the polymer plastics are considerably worse than the ones of the steel, but there is still a rich selection both in terms of range as well as modification of their qualities. Not unimportant are also such characteristics of the plastics as a lower price, lower density, corrosion resistance, and also, unlike steel, the possibility to form elements of the material.

Design Modifications.

The modified cycloidal gear system has been shown in Fig. 3.

For the analysis it was assumed that the cycloidal gears would be made of the modified plastic presented in table 1, and then subjected to the design changes.

The design modifications in the gear system included the geometry and the material structure of the gears. No modifications in the tooth profile of both gears were introduced. The driving wheel tooth profile is the one of the epicycloid, whereas the driven wheel tooth profile resembles the arches of a circle. The tooth profiles of the gears were drawn based on the equations published in [6,7].

The design modifications concerning the geometry focus primarily on the driving wheel, because it is exposed to greater load. Changes in the wheel include a number of splineways, which has been increased to six, as well as their profile (Fig. 3a). The splineways have bent walls and are rounded in the corners. The modifications are aimed at a more uniform distribution of load resulting from the application of torque M, as well as diminishing the concentration of stress in the area of the splineways.

The geometrical changes in the passive wheel have been limited only to lengthening the fillet radius at the base of the gear teeth.

However, the most crucial modification regarded the structure of the material used for making the analysed gear system. For the manufacture of the cycloidal gear unit, a metal core was introduced. The core's profile has been marked grey in Fig. 3b.

Introduction of the metal core is supposed to strengthen particular gears. As the material for making the core, an aluminium alloy was selected, because it is easier to form elements made of it than the of the ones made of steel.

For the gear system shown in Fig. 3b , the contact surface of the teeth is made of the modified plastic that has been marked yellow.

Introduction of the metal core should reduce the deformations of the gear system, thus limit the possibility of generating radial clearance h_r.

Apart from the reduction of the clearances, the modification was also aimed at the improvement of the gears' effort. The metal core will move the loads connected with transmitting torque M from the drive shaft onto the driving wheel.

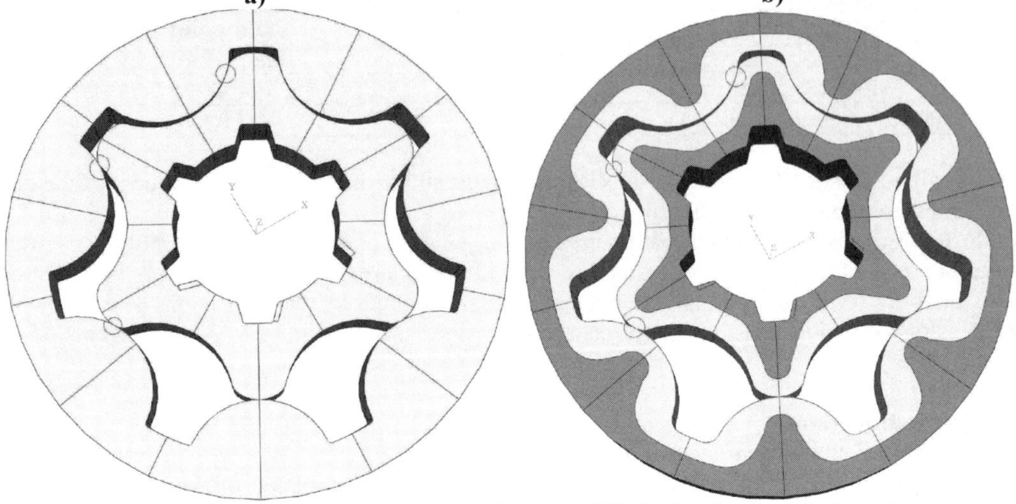

Fig. 3. The cycloidal gear system made of the modified plastic after design changes: a) – the six-splineways-per-a-key system, b) – the system featuring gears strengthened with a metal core

Research Programme.

According to the research programme, a systematic analysis was conducted on the described gear systems in the ranges of loads with pressure *p* of values: 4, 6, 8, 10, 12, 14, 16, 20 and 24MPa, as well as a proportionally higher value of torque M. The found values of reduced stresses σ and of radial clearance h_r have been presented in table 2.

Computational Model of the Gear System

Using the work diagram of the cycloidal gear system shown in Fig. 1, developed has been a geometrical model presented in Fig. 4 together with the main measurements. The toothing profiles of both gears have been defined on the basis of special equations [6, 7].

The mechanical load of the model results from transmitting torque M by the drive shaft of d_1=25mm diameter onto the driving wheel through the keys. Next, from the driving wheel, the drive is transmitted onto the driven wheel of the internal toothing. The hydraulic load of the model results from pressure p of the working fluid in the intertooth displacement chambers of the gear unit. The gear unit remains in balance, because torque M applied onto the shaft will be equal to the torque produced by pressure p.

The model ought to be fixed, and the way of fixing also stems from the operation principle of the cycloidal gear system presented in Fig. 1. Fig. 1 shows the driving wheel fixed on the shaft, which can only revolve on axis O_1. Respectively, in the model presented in Fig. 4, the gear is fixed radially on the d_1=25mm diameter shaft and can revolve on axis Z.

The driven wheel is fixed in the port of the hydraulic machine's body and it can revolve on axis O_2. Respectively, in the model shown in Fig. 4a, the gear is fixed radially on the d_2=75mm diameter, but it can revolve on axis Z.

The gear system is also loaded with pressure p working on the front surfaces of the gears. Hence, in the model presented in Fig 4a, the front surface of the gear system is loaded with pressure p, whereas the opposite surface is fixed.

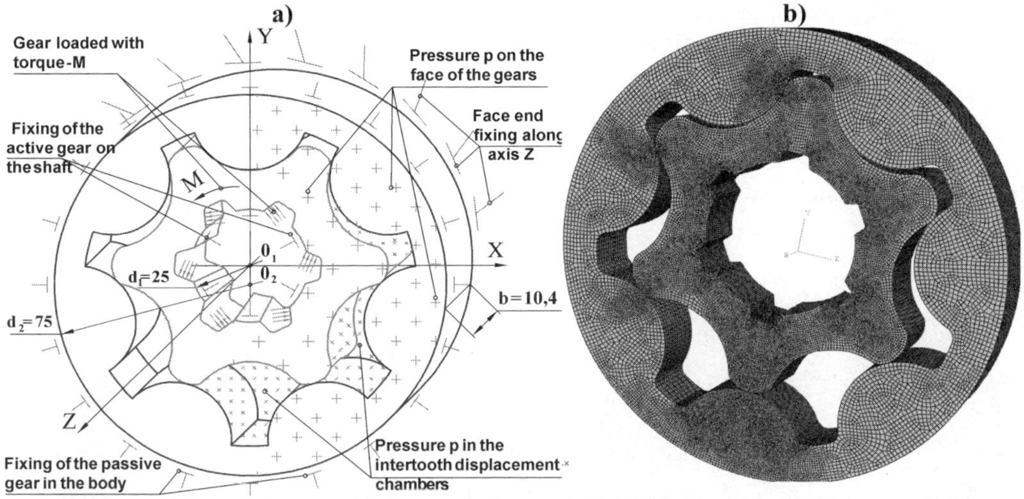

Fig. 4. Model of the cycloidal gears: a) – load distribution, b) – HEXA type finite elements grid

The numerical model of the cycloidal gears was made by means of ABAQUS system version 6.7-5, which was provided by *Dassault Systèmes* Simulia Corp. The licence of the program (the licence number: 05UWROCLAW) was provided by Wrocławskie Centrum Sieciowo-Superkomputerowe at Wroclaw University of Technology.

For the development of the finite elements grid shown in Fig. 4b, a hexagon HEXA type element have been used, which is a typical element for making solid models [5]. The finite elements grid for the gear unit have been made of approx. 500 000 elements. The elements make a profile of the gear system in XY plane. In order to model the thickness of the gear unit, 15 layers of HEXA elements have been created. The layers were arranged approximately every 0.7mm along OZ axis as far as the thickness b=10,4mm of the gear system were achieved (Fig. 4a). A HEXA element is cubic in shape and has 8 knots.

In anticipated points of stress accumulation, the grid was more densely structured, which can be observed in Fig. 4b. The denser places are the contact points of the mating gears on the inlet side (the left side of the model in Fig. 4b) as well as the splineways in the driving wheel.

The ABAQUS system also allowed the modelling of the collaborating surfaces by means of a special algorithm characteristc for the system. The contact was assumed in the contact point of the teeth on the inlet side of the gear system (left side of Fig. 4). Friction coefficient μ for the contact was selected considering the kind of plastic as well as conditions in the areas of the contact (eg. high thrust). The friction coefficient value of μ=0,4 was specified on the basis of research [4].

The numerical model made it possible to precisely determine the stress and deformations for the gear system. The digitization error size has been estimated based on the relations available in literature [5] and does not exceed 0.02% for the deformations and approx.1.3% for the stress.

Numerical Analysis Results

Fig. 5 shows the results of the research carried out on the model of the modified plastic cycloidal gear system that was loaded, as previously, with working pressure p=4MPa as well as torque M=7,16Nm.

The stress distribution presented in Fig. 5a is asymmetrical, whereas the driving wheel (featuring the external toohing) is exposed to greater effort than the driven wheel (featuring the internal toothing). It is to see in Fig. 5a that the increase in the number of splineways caused the effort in the driving wheel to be more uniform than the effort in the driving wheel shown in Fig. 2a. The maximum value of strength was recorded on the contact of gear tooth pair 1'-1and it is 25.6MPa.

It is to see in Fig. 5a that practically the whole thrust is focused on the contact of the gear tooth pair 1'-1". A very little part of the thrust is generated by the contact of tooth pair 2'-2", whereas in pair 3'-3" the intertooth forces (stresses) do not occur.

Fig. 5b shows that the maximum deformation for the modified plastic gears has been recorded for tooth 6'in the driving wheel. However, the value is of not so great importance for the tightness of the hydraulic machine. It is only the deformation in tooth pair 4'-4" that generates the radial clearance of h_r=0,006mm, and through that, the possibility of losing the tightness in the hydraulic machine.

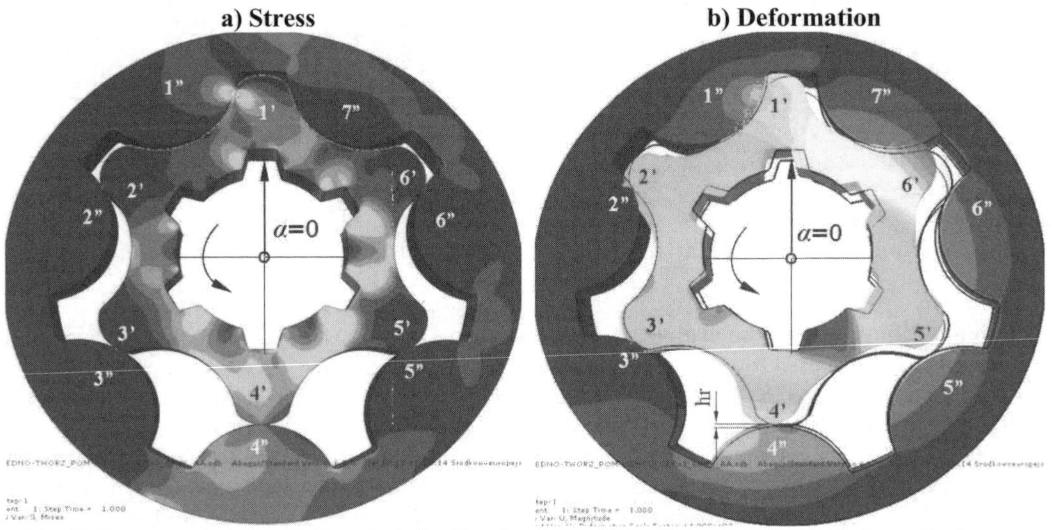

Fig. 5. Stress and deformation distribution in the cycloidal gear system made entirely of the modified plastic; a) – stress distribution, b) – deformation distribution

Fig. 6 depicts the stress and deformation distribution in the modified plastic cycloidal gear system, the gears of which have been strengthened with the metal cores.

The driving wheel (Fig. 6a) is exposed to greater effort, and the core in the wheel is clearly seen on the background of the stress distribution of the whole system. It can therefore be stated that the metal core has taken over the main part of the stress. The maximum stress has been recorded at the base of the splineway under tooth 5' (Fig. 6a). The value of the stress is 38MPa, and it is far lower than the allowable stress Re for the aluminium alloys.

The stress in the part of the gears made of the modified plastic is insignificantly lower than the value recorded in the system (Fig. 5). For the configuration illustrated in Fig. 6A, the maximum value of the stress in the plastic is 25MPa, and it has been recorded on the contact of gear tooth pair 1'-1". The recorded value of stress on the contact of tooth pair 1'-1" is far lower than the value of the stress allowable for the modified plastic (table. 1).

The maximum deformation value has been recorded for the tip of tooth 1', similarly to the configuration in Fig. 6. The deformation in this point is approx. 0.025mm, but it is not really important for the tightness of the hydraulic machine. The value of radial clearance h_r here is much more significant and is now approx. h_r=0.0024mm. It is a really considerable decrease in comparison to h_r=0.082 for the system presented in Fig. 2 [2], and is noticable comparing to the system depicted in Fig. 6.

The results of the numerical analysis conducted by means of the finite elements method at the higher load parametres for the considered configuration of the gear system has been presented in table 2.

In table 2 it is to see, that the cycloidal gears made of the pure polioxymethylene POM and without any design modifications can work from the working pressure of approx. 5MPa. And that is so because for this working pressure p, the value of radial clearance h_r reaches the value of the allowable clearance of hr_{dop}=0.1mm [2].

a) Stress **b) Deformation**

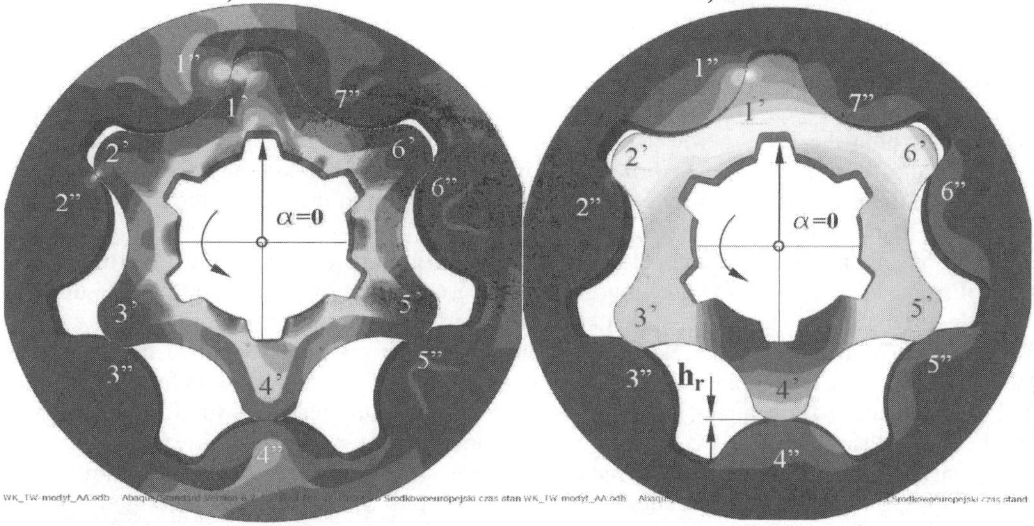

Fig. 6. Stress and deformation in the cycloidal gear system made of the modified plastic, the gears of which have been strengthened with metal cores

Introduction of the design changes has brought a positive effect of increasing the loadability for so modified cycloidal gear systems.

Table 2 shows that the cycloidal gear system made of the modified plastic, where the driving wheel features six splineways, can work at the pressure range **even up to p=24MPa**. At such a value of pressure p, the tooth contact strength gets close to the maximum strength for the modified plastic ($\sigma_{dop}\cong$160MPa, Fig. 3), whereas the allowable radial clearance hr_{dop}=0,1mm is not exceeded.

Table 2. Results of the strength analysis carried out on the cycloidal gear models by means of FEM

No.	Working load p [MPa]		Gear system made of POM		Gear system made of the modified plastic			
					6-splineway gear system		Metal core gear system	
	Pressure p [MPa]	Torque M [Nm]	σ [MPa]	h_r [mm]	σ [MPa]	h_r [mm]	σ [MPa]	h_r [mm]
1	4	7,16	26	0,082	25,6	0,006	25	0,0024
2	6	10,74	39	0,123	38,4	0,009	38	0,0036
3	8	14,32	52	0,164	51,2	0,012	50	0,0048
4	10	17,9	65	0,204	64	0,015	63	0,006
5	12	21,48			76,8	0,018	75	0,0072
6	14	25,06			89,6	0,021	88	0,0084
7	16	28,64			102,4	0,024	100	0,0096
8	20	35,8			127,8	0,03	125	0,012
9	24	42,96			153,6	0,036	150	0,0144

Table 2 shows that the modified plastic cycloidal gear system where both driving wheels have the metal cores also can work to **value p=24MPa** of the working pressure. At that value of pressure p, the stress on the tooth contact gets close to the maximum strength of the modified plastic ($\sigma_{dop}\cong$160MPa, Fig. 3), whereas the allowable radial clearance hr_{dop}=0,1mm is not exceeded.

Conclusions

The stress distribution is closely related to the direction of the mechanical energy flow through the examined cycloidal gear system. In the analysed gear unit, the mechanical energy flows from the drive shaft onto the driving wheel through the splineways. At the next stage, the energy is transmitted from the driving wheel, through the intertooth contact points onto the driven wheel. Such a flow of the mechanical energy causes stress concentration in the area of the splineways and on the contact points of both gears.

Introduction of modifications into the gear systems in a form of metal cores, increasing the number of splineways, as well as changing the material used for making the gears have resulted in the changing of the stress and deformation distribution, and in the strengthening of the gears.

The possibility of applying the plastic gears in the gerotor machines ought to be considered together both from the viewpoint of the allowable stress and of the allowable deformations. It is crucial to make sure that the stress in the gears does not exceed the plasticity limit for a particular material, thus causing damage of the gears. Simultaneously, it must be secured that the gear deformations and the resulting from it radial clearance h_r do not exceed the maximum values assumed for a given class of a hydraulic machine. According to the author, the maximum value of the clearance must be kept below hr_{dop}=0,1mm [2]. The deformation condition is in this case more important than the stress condition because the deformations cause leakage that lowers the efficiency of the hydraulic machine.

The theoretical analysis carried out by means of the finite elements method FEM shows that the cycloidal gear made of the modified plastic and featuring six splineways can work at the working pressure up to p=24MPa and torque M=42,96Nm. For those values of p and M, neither the maximum allowable values for the stress in this material ($\sigma_{dop}\cong$160MPa, table1) nor the allowable value of the radial clearance for the hydraulic machine hr_{dop}=0,1mm are not exceeded.

Although the introduction of the metal cores in the gears makes their manufacture complicated, it increases the loadability of the gear system.

As a result of strengthening the gear system with the metal cores, further increase of its loadability up to the working pressure value of p=24MPa and torque M=42.96Nm was achieved.

In practice, considered must be the assembly inaccuracies, thermal dilatation, and some other factors. In that case, the loadability must be lowered. The author is of an opinion that the following loadability limits can be assumed:

- p=4-5MPa of the working pressure for the polioxymethylene (POM) unmodified cycloidal gear system,
- p=16-18MPa of the working pressure for the cycloidal gear system made of the modified plastic and featuring six splineways,
- p=16-18MPa of the working pressure for the cycloidal gear system made of the modified plastic and featuring metal cores.

References

[1] S. Bednarczyk, J. Stryczek, and Z. Zienowicz. Gerotor pumps with plastic gears. Hydraulika i Pneumatyka, 2006, (5), pages 16-19. (available at www.simphip.not.pl).

[2] K. Biernacki, J. Stryczek. Analysis of stress and deformation in plastic gears used in gerotor pumps. The Journal of Strain Analysis for Engineering Design, October 2010, vo.45, Issue 7, pages 465-479.

[3] K. Biernacki, J. Stryczek. Design of plastic cycloidal gears used in the gerotor pump. HYDRAULIKA I PNEUMATYKA NR 6/2010, str. 30 - 33.

[4] D. Capanidis. Tribological testing of polioxymethylene (POM) – Tarnoform sliding composites. TRIBOLOGIA Nr 3/2004 (str. 25-33).

[5] E. Rusiński, J. Czmochowski, T. Smolnicki. Advanced finite element method in the load-bearing structures Publishing House of Wrocław University of Technology. Wrocław 2000.

[6] J. Stryczek. Cycloidal gears in design of gear pumps and motors. Habilitation Dissertation. Scientific Papers of the Institute of Machine Design and Operation of Wroclaw University of Technology. Series: Monographs nr 15. Wroclaw 1991.

[7] J. Stryczek, Gears of the hydraulic machines. Publishing House of Wrocław University of Technology.Wrocław2007.
http://www.dbc.wroc.pl/dlibra/docmetadata?id=20045&from=&dirids=1

Key Engineering Materials Vol. 490 (2012) pp 166-178
© (2012) Trans Tech Publications, Switzerland
doi:10.4028/www.scientific.net/KEM.490.166

Modelling the contact between the rolling elements and the raceways of bulky slewing bearings

Marek Krynke[1,a], Ludwik Kania[2,b], Eugeniusz Mazanek[3,c]

[1]Czestochowa University of Technology, Institute of Production Engineering al. Armii Krajowej 19 B, 42-200 Czestochowa, Poland

[2,3]Czestochowa University of Technology, Institute of Mechanics and Machine Design, ul. Dabrowskiego 73, 42-200 Czestochowa, Poland

[a]krynke@zim.pcz.pl, [b]ludwik@imipkm.pcz.pl [c]mazanek@imipkm.pcz.pl

Keywords: Slewing bearing, contact angle, contact zone, load distribution, equivalent characteristic of truss element, finite element method (FEM).

Abstract. In slewing bearings, a great number of contact pairs are present on the contact surfaces between the rolling elements and raceways of the bearing. Computations to determine the load of the individual rolling elements, taking into account the flexibility of the bearing ring, are most often carried out using the finite element method. Construction of a FEM full model of the bearing, taking into account the shape of the rolling elements and the determination of the contact problem for every rolling element, leads to a singularity of stiffness matrix, which in turn makes the problem impossible to solve. In FEM models the rolling elements are replaced by one-dimensional finite elements (linear elements) to simplify the computation procedure and to obtain an optimal time for computations. The methods of modelling the rolling elements in the slewing bearing, in which balls have been replaced by truss elements with a material non-linear characteristic located between the raceway centres of the curvatures in their axial section, are presented in the paper.

Introduction

At present rolling slewing bearings are the most often applied rotational connection between the body and support structure of working machines. The most important feature of rolling slewing bearings in comparison to standard rolling bearings is their ability to transfer axial load Q, radial load H and tilting moments M acting in planes crossing the longitudinal axis of the bearing (Fig. 1). In contrast to typical rolling bearings, the flexibility of support subassemblies, including the flexibility of the bearing rings and quantity of clearance in the bearing rows, has a decisive influence on the load-carrying capacity of the slewing bearings [1,2]. Although these types of bearings have been applied for eighty years, brand catalogues still contain simplified characteristics of their static load-carrying capacity. The existing state of technical knowledge is only apparently sufficient and still needs further investigation. The selection of a bearing, made on the basis of the above mentioned catalogue characteristics could lead to serious structural errors, such as a drastic reduction in the life of the bearing node and undcrusc of the bearing usability [3].

Knowledge of the values of the forces acting on the bearing rolling elements during work is the basis for estimating the bearing's ability to fulfil its function. The method of computation of internal load distribution is the main topic of papers dealing with slewing bearing problems [4,5]. Generally, the problem of determining the force distribution in a slewing bearing is repeatedly treated as statically indeterminate problem. The computational methods used to the determine the loads of individual rolling elements can be divided into classical methods and numerical methods with the use of FEM. In the classical methods the solution is determined by the assumption of the ideal rigidity of the rings, while the numerical methods allow the flexibility of the bearing rings to be taken into account.

The load-carrying capacity of slewing bearings is frequently calculated assuming non-deformability of the bearing rings [4]. But in certain circumstances, especially for bearings with large diameters applied in constructions which do not guarantee sufficient rigidity of the bodies with the bearing rings mounted on them, the above assumption can lead to overly large errors. This is why methods, which allow the flexibility of the bearing rings and mounted bolts to be taken into account as well as the flexibility of support structures of working machines, should be applied to determine the capacity of slewing bearings. Computational models of the bearings, which take into consideration the above mentioned factors, are built from the finite elements. Slewing bearings need special methods of modelling due to their specific features [6,7].

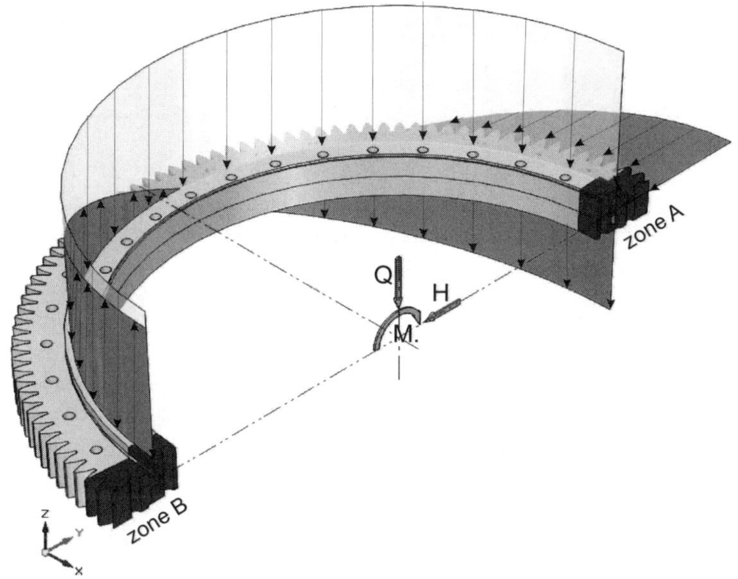

Fig. 1. An external load of a slewing bearing

Replacement rolling elements in the slewing bearings

Building of FEM models of the slewing bearings, due to their complex construction, offers many difficulties. An adequate modelling the rolling elements is one of them. FEM model of the contact zone of one rolling element with raceway at sufficient accuracy contain from several to tens of thousand degrees of freedom [1]. Due to large number of these elements (tens or more), construction of a full model of the bearing, taking into account the shape of rolling elements and the modelling the contact problem for each of them, is practically impossible. The difference in scale of systems *rolling element – raceway* and *support-construction* makes it difficult to obtain exact accuracy of computations and to obtain convergence of numerical solution due to too high differences in rigidities between individual finite elements. It should be noted that computations considering geometrical and physical nonlinearity are required to be carried out. So, to simplify computation procedures and to obtain optimal time of computations, the rolling elements in FEM models have been replaced by one-dimensional finite elements (linear elements). The most often applied finite elements replacing rolling elements in computational models of slewing bearings are:

- the truss elements with adequate unilateral non-linear material characteristic *stress – strain,*
- special elements (super-elements), mainly for ball bearings, on the basis of super-element proposed by T. Smolnicki in [1].

Replacement elements in the form of the truss elements are applied first of all to roller bearings where rollers are replaced by singular truss elements with the length equal to the diameter of roller. The system of two or more truss elements can be also used to substitute rollers what makes it possible to model asymmetrical transfer of the load by the roller (Fig. 2). This kind of replacement elements can be applied to computational models of ball bearings with constant contact angle equals 90°.

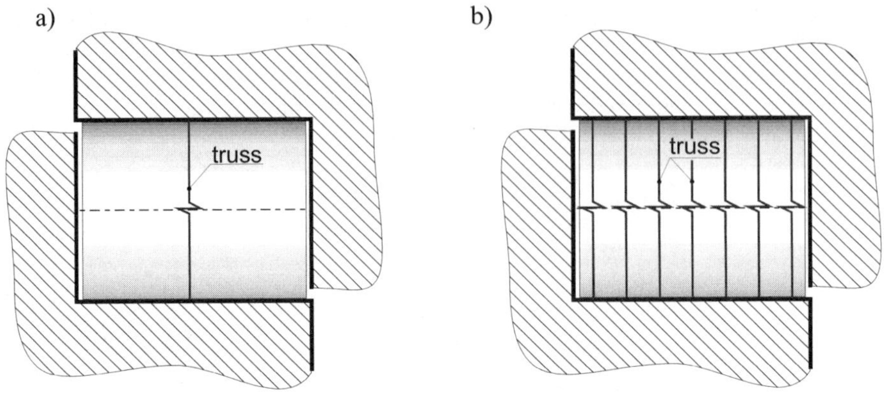

Fig. 2. Replacement truss elements for rollers: singular truss element (a), the system of the truss elements (b)

The truss element is of exactly definite length and cross-section as well as appropriately determined material characteristic $\sigma(\varepsilon)$. This element is performing a role of non-linear spring [8]. An assumed axial clearance s of the bearing or initial clamp Δ in FEM models can be directly determined by displacement of material characteristic of the truss element according to direction of strain ε as shown in Fig. 3 or by using additional contact elements with adjustable gap, e.g. elements of *gap* type.

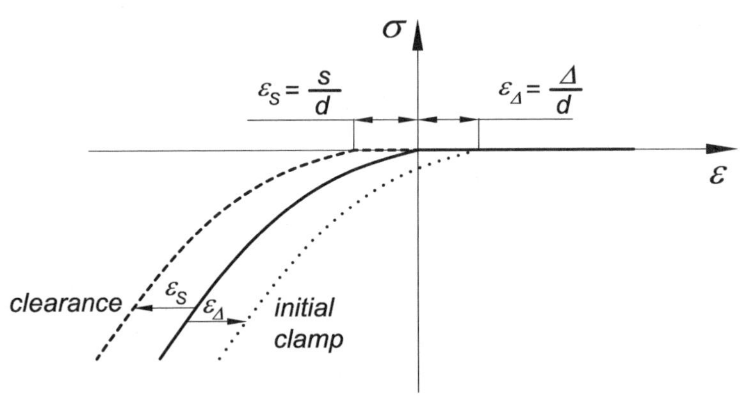

Fig. 3. Characteristic of stress – strain of the truss element replacing rollers in FEM models of the bearings

Introduction of super-element into numerical model of the bearing allows to take into account:

- the flexibility of support subassembly,
- variable contact angle of rolling elements, ,
- clearance (initial clamp) in the system *rolling element - raceway,*
- errors in realization of support subassemblies,
- predominance of rolling element (seeming negative rigidity).

Hence, super-element is mostly applied to model balls in slewing bearing with variable contact angle due to the loads carried out by the bearing.

Finite beam elements with joints, elastic element with non-linear characteristic *force – deflection* as well as the contact element have been used to build super-element according to T. Smolnicki (Fig. 4) [1].

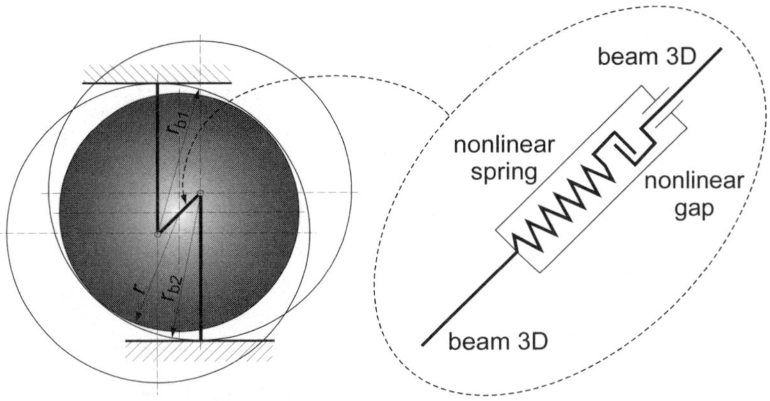

Fig. 4. Schematic diagram of super-element according to [1]

Numerical FEM model of slewing bearing, where rolling elements were replaced by special elements, has been presented in work [9]. Balls, likewise in the Smolnicki method, were substituted by non-linear springs placed between centres of curvatures of raceway in their axial intersection. The ends of non-linear springs were connected to appropriate raceways using the system of rigid beams. Nodes of beams were not connected directly to the nodes of ring mesh but with the use of indirect rigid shell elements of sizes corresponding to length $2a$ and width $2b$ of the contact ellipse of ball and raceway at maximal load of rolling element (Fig. 5) [10]. This method of modelling the rolling elements does not need additional correction of equivalent characteristic of the spring (elimination of II stage of determination of substitute characteristic) [12].

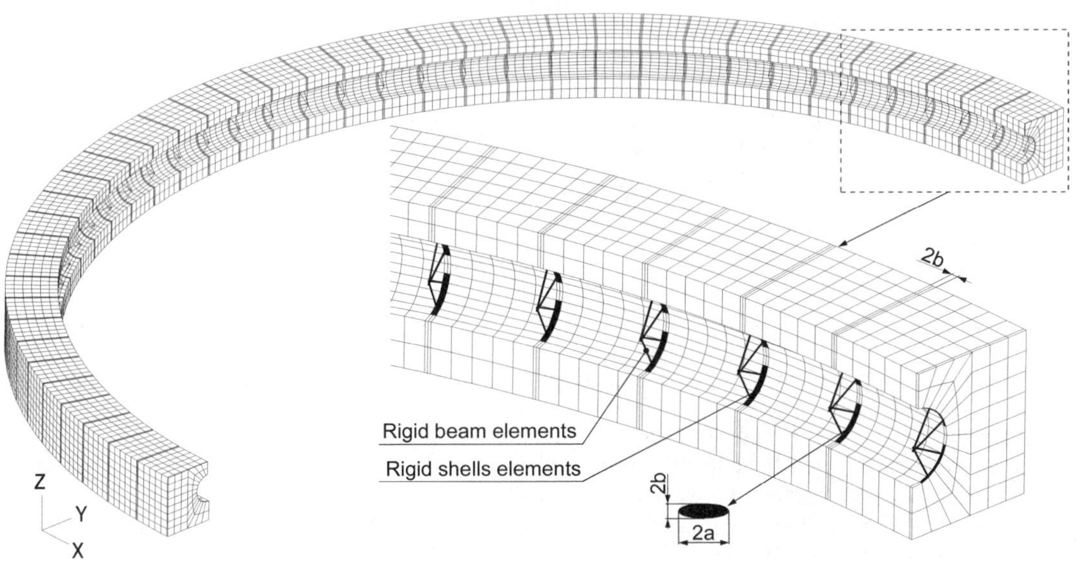

Fig. 5. Modelling the contact between the rolling elements and raceways – S2 type

Three types of super-elements, denoted as S1, S2 and S3 in slightly modified form in comparison to super-element proposed by T.Smolnicki, have been discussed in this paper. In all three cases of the proposed replacement elements authors resigned from the contact element and elastic element was replaced by the truss element with non-linear asymmetrical material characteristic. Modified super-element S1 consists of the system of two beam elements jointly connected in the place of centres of curvatures of raceway in their axial intersection with singular truss element. The remaining ends of beam elements are connected to the nods of mesh of the bearing rings (Fig. 6).

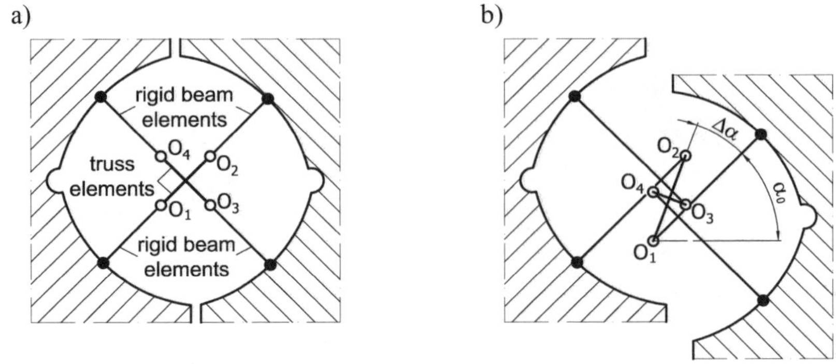

Fig. 6. Diagram of super-element S1 applied in a single-row ball bearing: in nominal state (a), under the load (b)

Essential difference in super-element S2 building results from its structure. Zones, digitized by shell elements, have been isolated in the raceways of the bearing rings. Rigidity of these elements is high comparing to rigidity of adjacent structures. The sizes of the zones correspond to the sizes of the contact area of rolling element at its maximal load. In the case of ball bearings it is the contact ellipse of the length $2a$ and width $2b$ [11]. So, in super-element S2 type, truss elements are connected to the bearing raceways using the system of two pairs of beam elements. Their remaining ends are connected to nods of the shell elements and finally to nods of mesh of the bearing rings (Fig 7.).

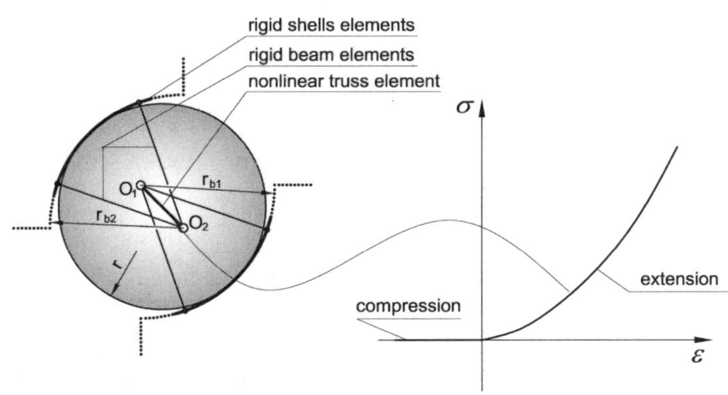

Fig. 7. Super-element S2 with non-linear asymmetric material characteristic for balls

Super-element S3 is the consecutive modification of super-element S2. In super-element S3, adequate system of rigid elements *RBE2* (Fig. 8) have been introduced in the place of rigid shell elements. The rigid connection between the nods can be defined using *RBE2* elements.

Independence of freedom degrees of nods connected by the those elements is determined in one point, where the freedom degrees of remaining nods *RBE2* are simultaneously determined. The central point of nod mesh connected by *RBE2* elements is the nod of the truss elements, while the remaining nods are the nods of the mesh of bearings rings. The bearing rings are digitized by body elements 8 – nodal. Meshing should be carried out in a way enabling to separate nods of elements in the contact zones of rolling elements and raceways. The sizes of these zones should correspond to the sizes of the contact area of rolling element at maximal load, similarly as in the case of super-element S2.

FEM software defined optional non-linear materials for the truss element. The definition of a material characteristic is based on introduction into the software particular points of this characteristic, it means strain ε and corresponding stress σ, in a tabular form. Therefore every strain of the truss element causes corresponding stress, what in turn allows the force carried by the truss element to be determined. Interruption of computation took place when deformation of the truss elements would exceed values given be material definition due to deficiency in convergence of the solution. Simulation of strain in the contact area of the contact pair is carried out using the characteristics described above. Super-element allows the displacement of real rolling element and the changes in the contact angle of the ball bearings to be determined. The real rolling element is subjected to external load transferred to rolling elements by the bearing rings.

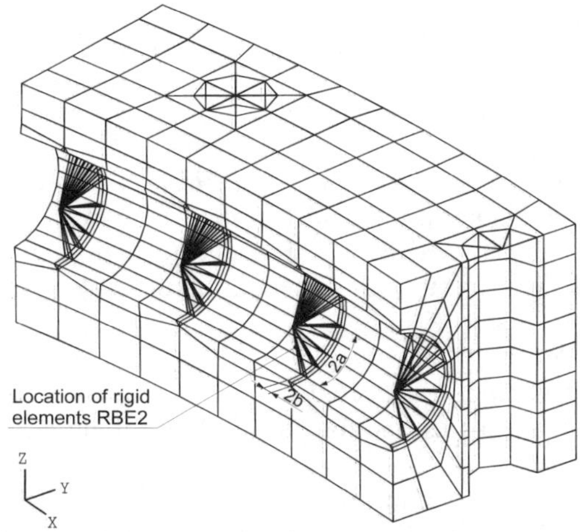

Fig. 8. Structure of super-element S3 in the external ring of single-row ball bearing

Examples of application of the truss elements to the modelling cross-roller and three-row roller bearings are presented in Figs. 9c and 9d, while application of super-element S2 to the modelling the balls in a single-row and a double-row ball slewing bearings are shown in Figs. 9a and 9b.

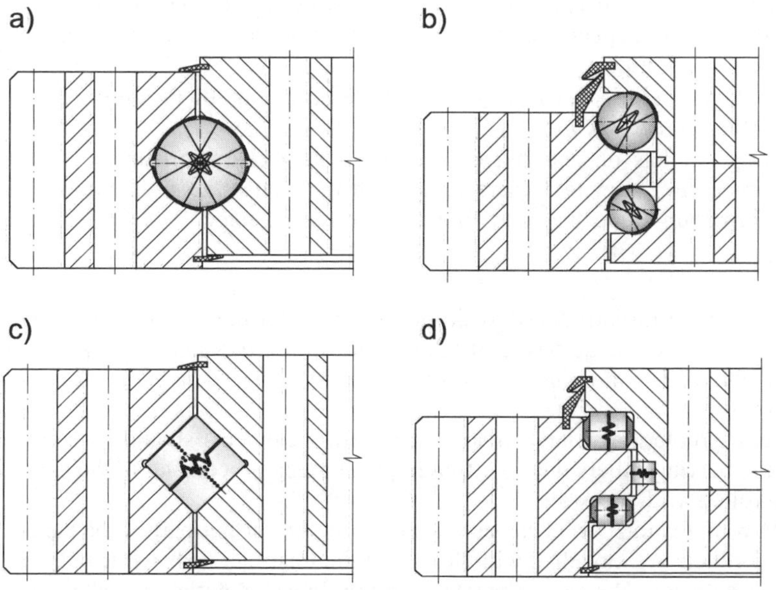

Fig. 9. Replaced rolling elements in: a) single-row four-point bearing, b) double-row ball bearing c) cross-roller bearing, d) three-row roller bearing

Equivalent characteristic of the rolling element

As it was mentioned before, rolling elements are replaced by linear elements in the modelling the slewing bearings. Thus, multiple modelling the contact problem can be avoided whereas appropriate determination of non-linear characteristics of such an element must be introduced. These elements should fulfil the following conditions [1, 8]:

- ensure the characteristic of the contact zone *deformation – load* like in the real condition of the bearing work,
- unidirectional carriage of the load,
- consideration of an influence of interaction of FEM mesh of the bearing rings, which is built of solid elements, on linear elements simulating mutual interaction of rolling elements in the bearings. In a different manner so called effect of concentrated force should be taken into account.

Asymmetrical, non-linear characteristic *force – deflection* of the system *rolling element – raceway* is an input data in building of the model of the whole bearing. This characteristic can be determined theoretically or experimentally [1]. But due to high degree of osculation of rolling element to the raceway and non-linearity of material, the method of finite element [8] should be used to determine this characteristic.

Advantages of FEM model are the possibilities to [6]:

- determine characteristic *force – deflection*,
- determine relationships between the force carried by rolling element and stress in it and in raceway,
- determine plastic deformation of raceways,
- consider expanding of raceways in the case of the bearings with unhardenable raceways,
- consider a wear of raceway.

The linear elements used in numerical models of the bearings must fulfil the conditions mentioned above. They could be ensured by computing adequate so-called equivalent characteristic of the rolling elements, the bearing rollers or balls. Because such a characteristic – apart from deflection of the system *rolling element – raceway* – should take into account the flexibility in FEM model of the bearing ring, it can be obtained in three stages of calculations [7]:

- first stage - it is analysis of the contact zone of rolling element with raceway and determination of dependence between force imposed on rolling element and reciprocal displacement of rings η_I,
- second stage - relies on analysis of mesh deformations of the bearing rings to determine deformation of raceway η_{II},
- third stage - relies on determination of equivalent characteristic η_z of the rolling part as a difference in deformation characteristics of the contact zone determined in the first stage and deformation of raceway determined in the second stage.

$$\eta_z = \eta_I - \eta_{II} \,. \qquad (1)$$

Solid model of the ring introduced again the flexibility of the bearing ring. It would be mistake not to correct this in the characteristic of the system *rolling element – raceway*. The flexibility of ring model for the load by a point force results from its shape, density of mesh of finite elements and their type.

Determination problem of characteristics of the truss elements should be considered separately for balls and the bearing rollers due to different phenomena proceeding in the contact zone of balls or the bearing rollers with raceways.

In the FEM computational models being analysed in the further part of the paper it was assumed that the bearing rollers are modelled with the use of the truss elements while balls are substituted by particular super-element. The truss element with non-linear material characteristic, determined on the base of equivalent characteristic, is an integral part of the super-element. These elements should be given equivalent value of the intersection to ensure their required rigidity. It should not be mistaken with real intersection, eg. of solid element, it is only computational quantity applied to simulation of deformation of rolling elements. The truss element modelling rolling elements should be elastic one, so its intersection can be computed using Hooke's law. After taking into account basic relationships applied to compute stresses σ and relative stresses ε, one can computed the required value of the intersection of the truss element. It was assumed this value is computed for the maximal load of rolling elements F_{max} and deformation caused by this load $\eta_{z\,max}$ [8]:

$$A_t = \frac{F_{\max}}{\eta_{z\max}} \frac{l_T}{E} \tag{2}$$

where l_T is the length of the truss element and equals:

- for the bearing rollers

$$l_T = d \tag{3}$$

- for balls after applying super-element

$$l_T = d\left(\frac{1}{k_p} - 1\right) \tag{4}$$

where k_p is the osculation of the ball and raceway.

Characteristic is most often defined in the tabular form as relationship *strain– stress* ($\sigma_i = f(\varepsilon_i)$) [13] in programmes for FEM software. It results from analysis of equivalent characteristic of the truss element that computed intersection A_i has different values for every points of this characteristic. Because the truss elements of the changeable intersection cannot be applied, individual points of non-linear material characteristic of equivalent element should be computed considering adequate correction factor k_A. So particular points of material characteristics are given by the following equation:

$$\sigma_i = \varepsilon_i E k_A \tag{5}$$

where:

$$\varepsilon_i = \mp \frac{\eta_{zi}}{l_T} \tag{6}$$

$$k_A = \frac{A_i}{A_t} \tag{7}$$

$$A_i = \frac{F_i}{\eta_{zi}} \frac{l_T}{E} \tag{8}$$

In the equation (6) sign „–" is applied when equivalent characteristic for the bearing rollers is determined , while „+" in the case of determination of equivalent characteristic for balls using super-element.

When applying S2 or S3 super-elements one can resign from deformation analysis of the mesh of the bearing rings to define deformation of raceway. Therefore the second and third stages are omitted in determination of equivalent characteristic of the truss element.

Computational example

The computation was carried out for single-row bearing with 84 balls of $d = 50$ mm in diameter, placed on the track diameter $d_t = 1400$ mm. The contact coefficient of the ball and raceway was equal to $k_p = 0.96$, and hardness of raceway - 54 HRC. It is the bearing of a standard construction with nominal the contact angle $\alpha_0 = 45°$. The bearing is fixed to the rings of enclosure by 36 M24 bolts of 10.9 class with corresponding initial stress $F_M = 139$ kN. The bearing model used in computations is presented in Fig. 10. Half of the bearing and rings of enclosure were assumed for analysis. Suitable boundary conditions resulting from model symmetry were introduced into the model. The bolts fastened the bearing, placed in the symmetry plane of the bearing, were modelled as beam elements with half core of a cross section of A_3 bolt. Initial stress equals ½ F_M of clamping screw initial stress was imposed to edge bolts with intersection of ½ A_3. Similarly, half of intersection A_t computed from equation (2) is assumed in the truss elements replacing nodal rolling elements. Adequate contact surfaces between the bearing rings and enclosure rings were defined. The friction factor equal to 0.15 was assumed in the contact zone of the elements.

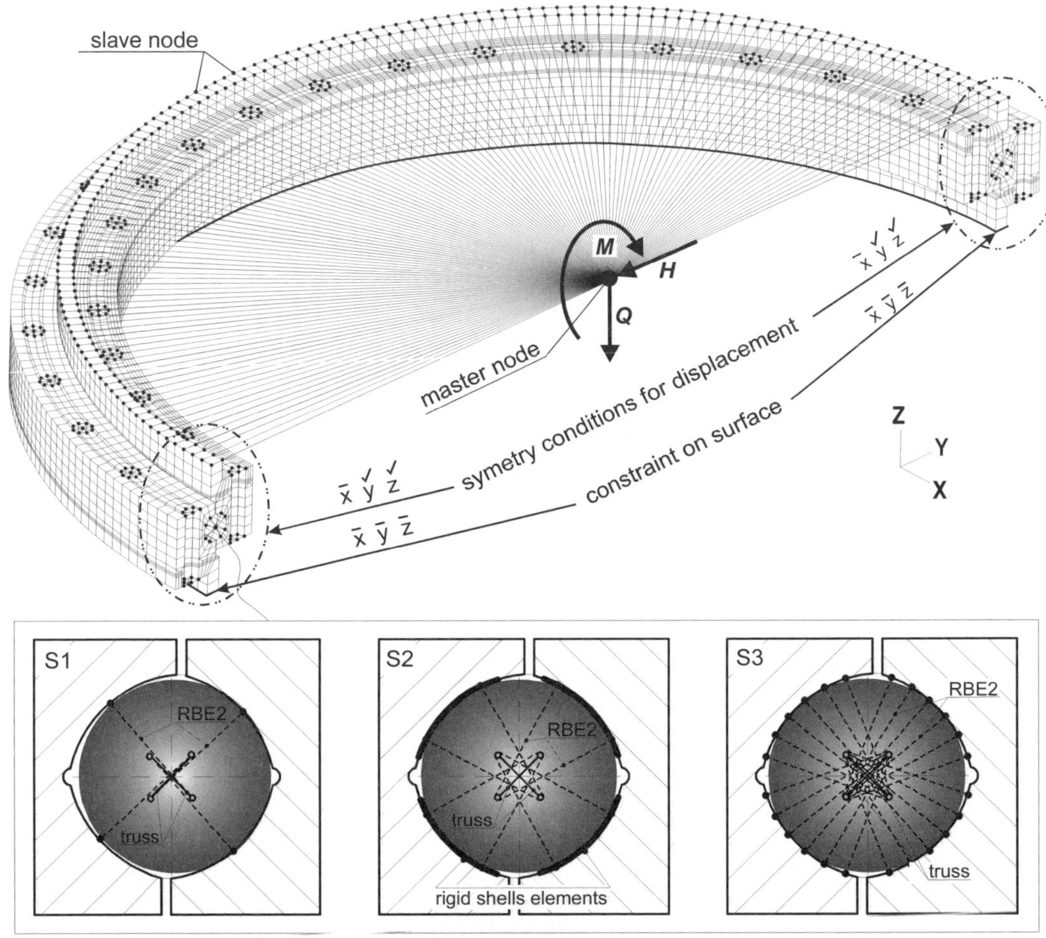

Fig.10. Boundary conditions

Under-surface of support ring, to which external bearing ring is fixed, was rigidly supported. Internal ring is fixed on the upper surface of enclosure ring. The constrains between nodes placed on this surface (*slave nods)* and central node (*master node)* placed in the centre of gravity of nodes *slave* was imposed on the upper surface of enclosure ring. Pairs of nodes, formed in such a way, create "rigid spider" with master node as a central point. External load in the form of concentrated force and concentrated moment is imposed on the master node. When model undergoes deformation, *slave* nodes are forced to displacement and rotation in such a way that distance between *slave* and *master* nodes stays constant and that rotations of *slave* nodes are identical as adequate rotations in the main node. The bearing load is generated by given tilting moment M, axial force Q and radial force H (Fig. 10).

Distribution of internal load and the load of clamping screws, obtained from computations for three different super-elements (S1, S2, S3) at the bearing loading by limited tilting moment M_{max} when $Q = 0$ and $H = 0$, are shown in Figs. 11 and 12.

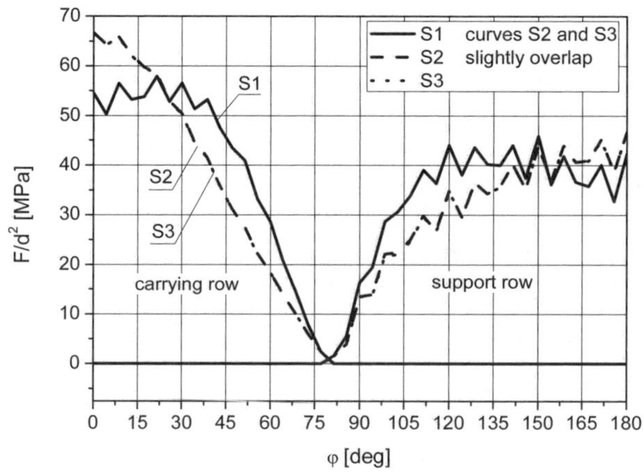

Fig. 11. Comparison of distribution of internal load in relation to applied super-element (S1, S2, S3) at the bearing loading by maximal tilting moment

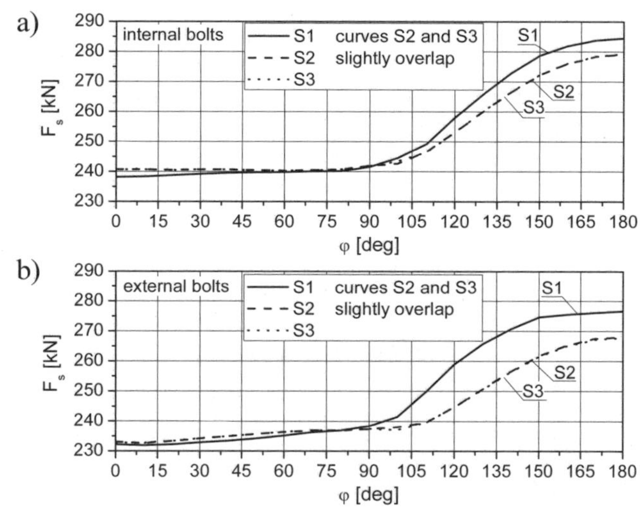

Fig. 12. Comparison of the load distribution of individual clamping screws in relation to applied super-element (S1, S2, S3): a) screws of internal ring, b) screws of external ring

Comparing the above diagrams one can notice that the method of modelling the rolling elements can have the influence on the obtained results. When applying super-element of S1 type, slightly smaller load of the rolling elements in relation to the loads obtained in the reminding cases (S2 and S3) was received. Thus, the usage of the first method of modelling the rolling elements applying super-element S1 can overstate computed the bearing capacity. Different distribution of the load obtained during application of substituted element S1 has the influence on the distribution of forces transferred by individual bolts fixing the bearing rings to support sub-assemblies (Fig. 12). Too low rigidity of connection between single beam element of super-element S1 and the nods of the bearing raceways is the reason of the above phenomenon. Beam element in the super-element structure is highly bended due to the change of direction of the force transferred by the rolling element in the ball bearing [14]. Thus, the rigidity of one-node connection of the beam element - in the case of super-element S1 with the bearing ring – becomes insufficient. The rigidity of such a connection also depends on density of the mesh of finite element in the connection zone with super-element. If the bearing rings are divided into higher number of smaller finite elements, then the application of super-elements of S2 or S3 type is recommended.

The characteristic of static bearing capacity considering the flexibility of the bearing rings, obtained by computation of numerical model using super-element S2 or S3 (curve 1), is presented in Fig. 13. The curve of the bearing limiting capacity obtained assuming non-deformability of its ring is placed in this figure for the comparison (curve 2) [4]. The following curves in the diagram (3÷8) are related to the right axis of the diagram and refer to axial displacement ("z" direction in Fig. 10) and to radial displacements and radial displacement ("y" direction in Fig. 10) of central point, at which external load is imposed. Curves 3 and 4 depict axial and radial displacements applying super-element S1, respectively, while curves 5 and 7 and curves 6 and 8, practically overlapping, depict the same displacements but using super-elements S2 and S3.

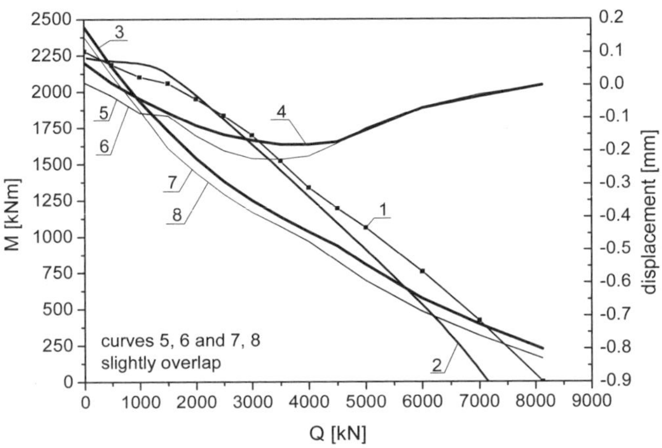

Fig. 13. Characteristic of the bearing capacity (1 – FEM model, 2 – analytical model) and the displacements of central point of internal ring: axial displacements η_z (3 – η_z for S1, 7 – η_z for S2, 8 – η_z for S3) and radial displacement η_r (4 – η_r for S1, 5 – η_r for S2, 6 – η_r for S3)

The following positions of internal bearing ring for individual points of individual points of its limited capacity are presented in Fig. 14.

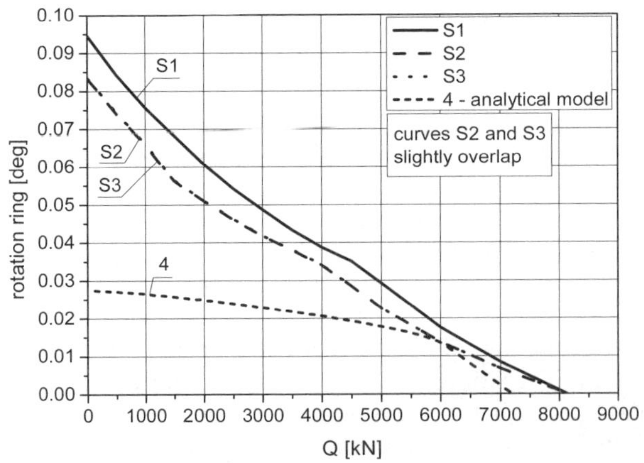

Fig. 14. Rotation of the bearing rings in the plane of tilting moment action for the maximal load of the bearing by the tilting moment

Additionally, in this diagram the changes in a rotation angle of internal ring of the bearing in the plane of tilting moment action (rotation along "x" axis in Fig 10) with application of three substitute elements (S1, S2, S3) are presented. The curve 4 describing the location of the bearing rings at the assumption of infinite rigidity of the bearing rings is also inserted in this diagram. It results from this diagram that displacements of the bearing internal ring are higher when substitute element S1 was used in relation to displacements in the remaining cases (S2 and S3). Thus, the results indicate the same reasons described above concerning the lower rigidity of connection between replacing elements of S1 type with the bearing raceway.

Summary

Computational models of slewing rolling bearings must undergo simplifications due to their complexity. The same goes for the computation of the bearing capacity using the finite element method.

The presented analysis justifies the usage of numerical methods in practical computations. The analysis allows the force distribution in the individual rows of the bearings to be exactly determined, which is impossible using analytical methods. The analytical methods are reduced for the bearing computations by the assumption of the non-deformability of the bearing rings. Computational modelling allows all constructional and technological changes introduced into the geometry of the real bearing to be quickly verified, and also allows the geometrical characteristic of mounting structure of the bearing in machine to be taken into account. The method of modelling the rolling elements, which are replaced by special elements (super-elements), is the key stage in the presented computational methodology. Both the structure of the super-element and its substitute characteristic can have a profound influence on the obtained results of computations. To obtain an adequate characteristic of the bearing capacity one must take into account both phenomena proceeding in the contact zone of the rolling elements and raceways, and the influence of the method of rolling element modelling in the FEM numerical model of the bearing.

On the basis of the obtained results one can state that it is essential to connect substitute elements (super-elements) with the nods of the bearing raceway in the numerical FEM models of slewing bearings. Due to the carried loads, the change in direction of the forces carried by the rolling elements as a result of the change in the contact angle that has taken place in the ball slewing bearings. This leads to the bending of beam elements in the structure of the super-element. This is the reason why one should pay special attention to the method of connecting the centres of raceway curvature to the nods of the bearing rings.

On the basis of the analysis carried out on the three types of substitute elements, the following conclusions can be formulated:

- the application of super-elements S2 or S3 in dependence on the usability of the FEM program is recommended to model balls in the slewing bearings,
- the super-elements S2 and S3 do not need additional correction of the equivalent characteristic *force – deflection*, which must, however, be carried out in the case of the super-element S1 due to the single point transfer of the force from this element to the bearing rings,
- the super-element S1 can be applied to model balls in the rolling bearings with a contact angle equal to 90°, and the characteristic *force – deflection* must be appropriately corrected in dependence on the density of the finite element mesh for the bearing raceway.

Acknowledgement

The study has been carried out within the framework of Research Project No. N N502 266237 awarded by the Polish Ministry of Science and Higher Education.

References

[1] T. Smolnicki, Physical aspects of coherence of large-size rolling bearings and deformable support structures, Wrocław: Wrocław University of Technology Publishing House, 2002 (*in Polish*).

[2] R. Kunc, I. Prebil, Numerical determination of carrying capacity of large rolling bearings, Journal of Materials Processing Technology, 155–156 (2004) 1696–1703.

[3] E. Mazanek, Constructional and strength problems of large-size rolling bearings, Monografie nr 105, Wydawnictwa Politechniki Częstochowskiej, Częstochowa 2005 (*in Polish*).

[4] J.I. Amasorrain, X. Sagartzazu, J. Damián, Load distribution in a four contact-point slewing bearing, Mechanism and Machine Theory, 38 (2003), 479-496.

[5] S. Zupan, I. Prebil, Carrying angle and carrying capacity of a large single row ball bearing as a function of geometry parameters of the rolling contact and the supporting structure stiffness, Mechanism and Machine Theory, 36 (2001), 10: 1087-1103.

[6] T. Smolnicki, E. Rusinski, Superelement-Based Modeling of Load Distribution in Large-Size Slewing Bearings, Journal of Mechanical Design, 129, (2007), 4: 459-463.

[7] L. Kania, Modelling of rollers in slewing bearing calculations with the use of finite elements, Mechanism and Machine Theory, 41 (2006), 11: 1359-1376.

[8] L. Kania, Analysis of internal load of slewing bearings in respect of their load carrying capacity, Seria monografie nr 111, Wydawnictwa Politechniki Częstochowskiej, Częstochowa 2005 (*in Polish*).

[9] Z. Chaib, A. Daidie, A. Ghosn, 3D simplified finite elements analysis of load and contact angle in a slewing ball bearing, Journal of Mechanical Design, 130 (2008) 8: 082601 (8 pages).

[10] Z. Chaib, A. Daidié, D. Leray, Screw behavior in large diameter slewing bearing assemblies: numerical and experimental analyses, Int J Interact Des Manuf (2007) 1:21-31.

[11] J. Brändlein, P. Eschmann, L. Hasbergen, K. Weigand, Ball and Roller Bearings: Theory, Design and Application, Willey, (1999).

[12] L. Kania, Substitute characteristics of the rollers in modelling ring roller bearings, Przegląd Mechaniczny, 57 (2003), 7-8: 35-40 (*in Polish*).

[13] ADINA, Theory and Modeling Guide, Volume 1: ADINA. ADINA R&D, Inc., Watertown 2007.

[14] L. Kania, M. Krynke, E. Mazanek: The variation of the contact angle in a flexible four – point slewing ball bearing, 24th Scientific Conference „Trends in The Development of Heavy Duty Machines", Zakopane 2008: 211–218 (*in Polish*).

Key Engineering Materials Vol. 490 (2012) pp 179-186
© (2012) Trans Tech Publications, Switzerland
doi:10.4028/www.scientific.net/KEM.490.179

Modernization of the Trailer for Tyre/Road Rolling Resistance Measurements

Ryszard Wozniak[1a], Stanislaw Taryma[1b] and Grzegorz Ronowski[1c]

[1] Gdansk University of Technology, Faculty of Mechanical Engineering, ul. Narutowicza 11/12, 80-233 Gdansk, Poland

[a]rwozniak@pg.gda.pl, [b]staryma@pg.gda.pl, [c]gronowsk@pg.gda.pl

Keywords: Road Measurement, Tyre, Rolling Resistance, Aerodynamic Drag

Abstract. In the article the ways of defining tyre rolling resistance are presented. The advantages of the laboratory tests of tyre/road rolling resistance and the advantages and the disadvantages of on the road tyre/road rolling resistance tests are described. The description of the special trailer used for tyre/road rolling resistance measurements designed and constructed in Faculty of Mechanical Engineering at Gdansk University of Technology is presented also. The trailer during it's modernisation was equipped with special test systems which compensate the impact of disturbance factors such as: aerodynamic drag and inertia force acting on the tested tyre, slope of the road, tilt of the trailer and vibrations of the measuring arm. This article contains the description of only one compensation system applied in the measuring trailer which eliminates the aerodynamic drag. The conclusions which came from the measurements performed using this compensation test system are included.

Introduction

Rolling resistance is one of the main drags of a moving vehicle. In a moving car, depending on its movement conditions it presents from 25.2 to 46.6% of the total drags of moving vehicle [1, 2]. Such a large share of the rolling resistance in total drags of movement translates to it's significant impact on fuel consumption by a moving vehicle. Studies conducted in Norway [3] show that for a typical mid-size passenger car moving in European cycle [2, 4] reducing rolling resistance by 10% causes a decrease in fuel consumption by 3%.

The effect of lower fuel consumption in addition to energy savings is lower cost of transport. In addition to lower fuel consumption it reduces exhaust emissions of environmentally harmful substances such as CO_2, CH_4, NH_3 and N_2O.

Vehicle rolling resistance is dominated by tyre/road rolling resistance. It is defined as the tangential reaction force evoked during tyre rolling, which movement occurs as a reaction on force applied to the middle of the wheel. Torque is not applied to the tyre. Assuming the above conditions causing the movement of the tyre is assumed not to be in encyclical deflection. In addition all lateral loads acting on the tyre are ignored. Typically in such conditions is carried out to tyre rolling resistance forces both on the road or in the laboratory.

Tyre/road rolling resistance drag is defined as the value of the rolling resistance coefficient C_{RR}. It is the ratio of the above force to the normal surface reaction force acting on the tyre. Sometimes the term of rolling resistance power P_t is used which is the product of rolling resistance force F_t and linear speed V ($P_t = W_t \cdot V$). Still other researchers [5] determine the rolling resistance of the amount of energy lost (in Nm) during the tyre rolling at the distance of one meter. Usually the rolling resistance coefficient C_{RR} is used.

In order to reduce rolling resistance special investigations are carried out on designing and manufacturing of modern tyres and road surfaces. Essential condition for this type of work is the ability to access the appropriate methods of measurement. Laboratory methods for rolling resistance investigations are very well suited to examine the influence of construction and maintenance on rolling resistance. Those methods are characterized by higher accuracy and better repeatability of measurements, as well as the ability to control and maintain constant test conditions. Road rolling resistance test methods are better for the impact of road surface on rolling resistance. Unfortunately they are characterized by lower accuracy and repeatability due to the influence of such factors as weather conditions and method of driving.

In experts opinion genuine tyre rolling resistance measurements can be performed only in real conditions - it means on the real road. Study of this type can only be done using special test equipment such as a dynamometer trailer equipped with high sensitivity measurement system.

The Measurement Trailer Constructed in Gdansk University of Technology

Faculty of Mechanical Engineering in Gdansk University of Technology has begun work related to investigate tyre rolling resistance. As part of this work the special measurement trailer was designed and manufactured. This trailer (see Fig. 1) with electronic instrumentation is used for rolling resistance measurements of tyres for passenger cars and vans in traffic conditions on real road surfaces.

Figure 1. The measurement trailer for tyre/road rolling resistance research.

The principle of measuring rolling resistance using the measuring trailer is shown in Fig. 2. The trailer is connected with the towing vehicle by the triangular hitch. The steering system of the measuring trailer makes possible to turn back on a public road and automatic return of the steering wheels to the straight position (self-steering capability of the trailer). The hydraulic inertia type braking system of the trailer, operating on the front wheels, provides the possibility of braking during transport and during the tests. The trailer is equipped with three wheels: two self-steering wheels are placed at the front, one test wheel with tested tyre is placed at the rear of the trailer. That is why the required stability of the longitudinal, lateral and directional motion of the trailer was achieved. Front independent suspension based on the double transverse arms was made on the basis of passenger car front suspension. The motorcycle elastic-absorbing columns were applied consisting the springs and the shock absorbers. The rear wheel suspension of the tested tyre is based on the vertical measuring arm 4 (see Fig. 2). Front and rear suspension are linked together through the trailer frame 1. The axis of rotation 3 of the frame 1 is positioned at the height of the geometric axis of rotation of the front wheels. The other end of the frame is jointedly connected to the measuring arm 4 on which the wheel with the tested tyre 5 is installed. The axis of rotation 3 is the axis of the swing arm 2 with weights as well. The swing arm 2 is rested on the frame 1 by the elastic-damping column 7. On the swing arm 2 the pod with the weights 6 has been mounted to obtain the proper vertical load of the tested tyre 5.

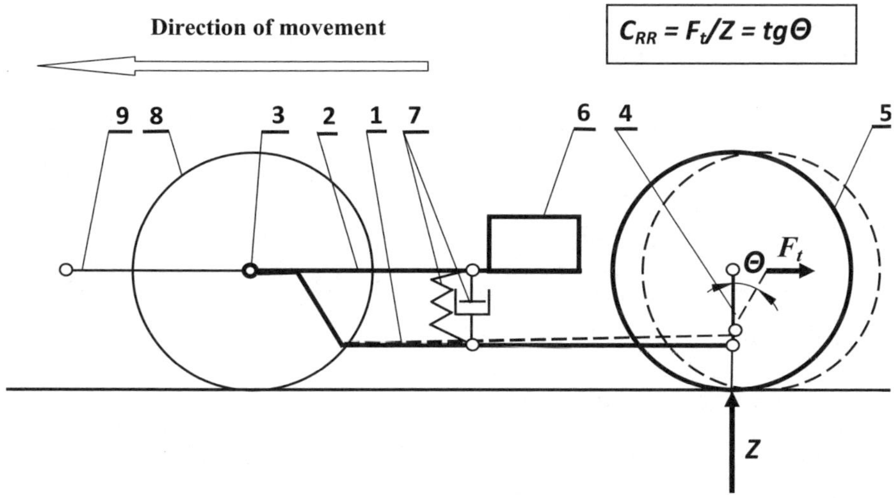

Figure 2. The rolling resistance measuring principle using the measuring trailer:
1 - frame of the trailer, 2 - swing arm, 3 - axis of rotation of the frame and swing arm with weights, 4 - measuring arm, 5 - tested tyre, 6 - pod with weights, 7 - elastic-damping column, 8 - front wheel, 9 - trailer hitch.

This design solution makes the mass of swing arm 2 with the weights 6 the suspended mass of the trailer. This results in a reduction of the vertical dynamic load of the tested tyre. Change of the static load of the tested tyre 5 is obtained by changing the number of weights in the pod 6. During the movement of the trailer with the towing vehicle on smooth, preferably horizontal surface, rolling resistance force acting on the tested tyre 5 deflects measuring arm 4 at angle Θ. Determination of the tyre rolling resistance on the road surface is performed by measuring this deflection (Θ angle).

For an accurate and reliable measurements the trailer must be equipped with compensation systems for the impact of disturbance factors such as: aerodynamic drag and inertia force acting on the tested tyre, slope of the road, tilt of the trailer and vibrations of the measuring arm.

Elimination of Disturbance Influence of the Aerodynamic Drag

During the road tests on the tested tyre which is uncovered by chamber, an aerodynamic drag F_{air}, described by formula (Eq. 1) is acting.

$$F_{air} = 0.5 \cdot C_{xtyre} \cdot A_{tyre} \cdot \rho \cdot V^2. \tag{1}$$

where:

C_{xtyre} - frontal air drag coefficient of the tested tyre,
A_{tyre} - frontal area of the tyre,
ρ - density of air,
V - tyre speed relative to air.

This force causes additional deviation of the measuring arm which disturbances measurements. Therefore, in the algorithm for determining the tested tyre rolling resistance coefficient the correction was used liquidating the impact of this disturbance. The algorithm was presented in [4]. In equation (1) frontal air drag coefficient of the tested tyre C_{xtyre} is unknown. In order to determine this coefficient the following measurements were done. The reference tyres A, B, C and E (their description was placed in [6] and in Table 1) were tested on the drum facility on the steel surface and on the very rough APS-4 surface.

Table 1. Description of the reference tyres used in the measurements.

Symbol of the tyre	Producer	Size	Tread
A	AVON	185/65R15 88H	ZV 1
B	AVON	185/65R15 88T	CR 322
C	AVON	185/65R15 88T	CR 65 M+S
E	MICHELIN	185/65R15 88H	E3A Energy

Two internal pressure values in the tyres were performed: 180 kPa and 350 kPa. Test speeds: 20, 30, 50, 70, 90, 110 and 130 km/h were applied. The vertical load Z of the tested tyres was fixed at 2972 N. The measurement results are shown in Fig. 3.

The rolling resistance coefficient of reference tyres on both surfaces was constant in practice in all range of test speeds. Also the rolling resistance coefficient of tyre E was measured in road conditions on the SMA 0/12.8 surface at internal pressure of 350 kPa. Vertical load of the tested tyre was fixed at 2972 N. The measurement results are shown in Fig. 4.

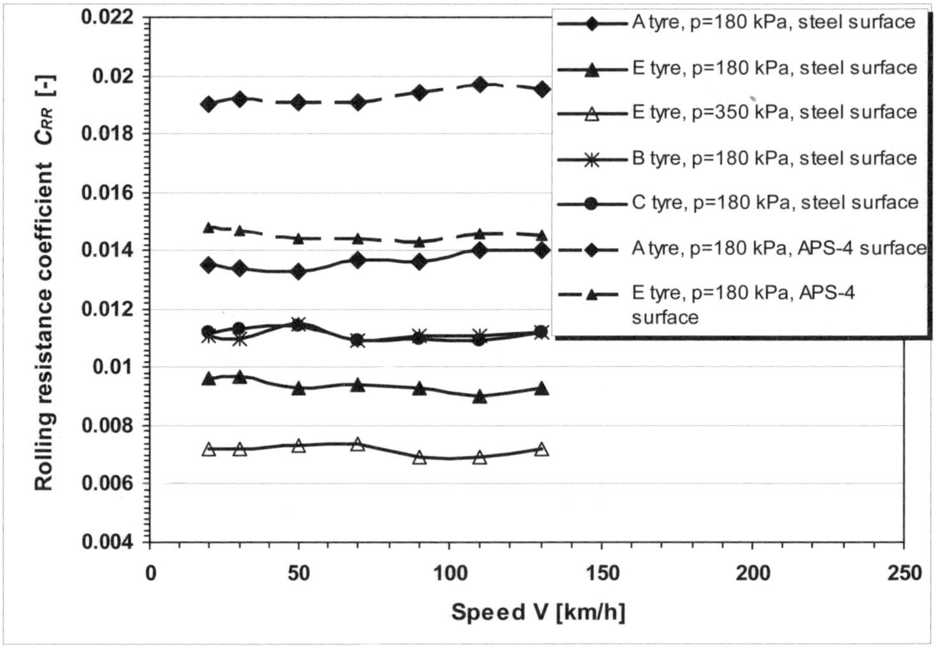

Figure 3. Dependence of the rolling resistance coefficient of reference tyres: A, B, C and E as a function of speed on the steel and the APS-4 surfaces.

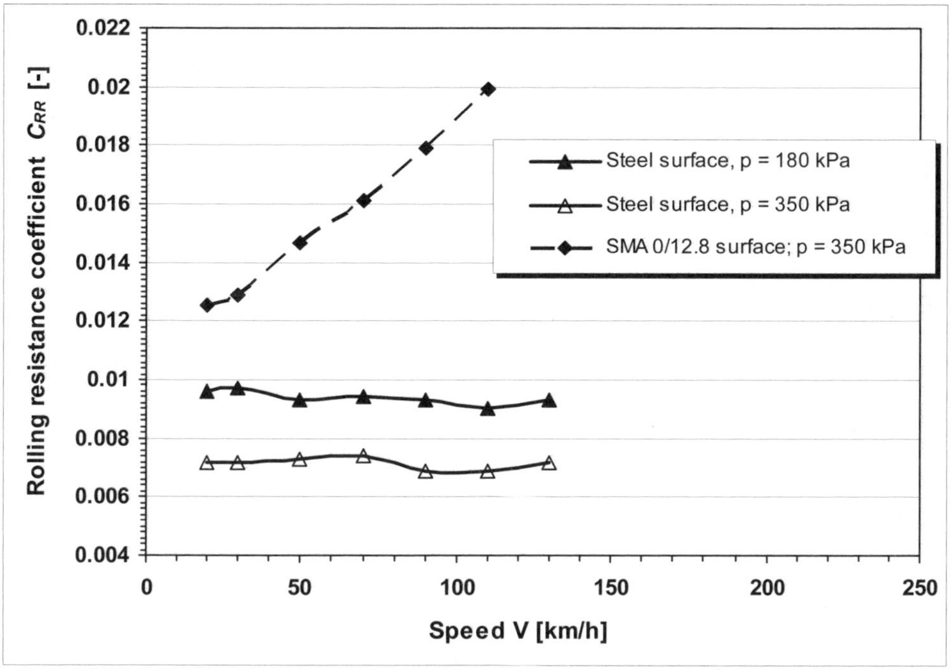

Figure 4. Dependence of the rolling resistance coefficient of reference tyre E as a function of speed on the steel and the SMA 0/12.8 surfaces.

Assuming that on the SMA 0/12.8 road surface the rolling resistance coefficient of E tyre was constant in whole measured speed range, measured increase of the rolling resistance coefficient proportional to the square of the speed was caused by the aerodynamic drag F_{air} acting on the tested tyre. Comparing this measured F_{air} force with measured growth it was possible to determine frontal air drag coefficient of the tested tyre; $C_{xtyre} = 0.38$.

As far as the C_{xtyre} value is concerned to be sure that this value was correct it was decided to perform additional tyre/road rolling resistance measurements of E tyre on the SMA 0/12.8 road surface. The measurements were done for two speeds: 70 and 90 km/h on two different road sections and for two directions of movement for each road section and the same internal pressure and vertical load parameters of the tested tyre as before. The measurement results without the aerodynamic drag correction are shown in Fig. 5. Fig. 6 presents the same measurements results with the aerodynamic drag correction applied using $C_{xtyre} = 0.38$.

It is noteworthy that in formula (Eq. 1) there are parameters which accurate determination can be difficult, and in turn the changes of their values can be a source of errors. This applies especially to air density and to speed of the tested tyre relative to air (affected by the wind direction and speed).

Therefore to be sure of tyre/rolling resistance measurements performed using this measuring trailer it was decided to eliminate the influence of aerodynamic drag of the tested tyre by closing it in a special aerodynamic chamber (see Fig. 7a and Fig.7b). When the aerodynamic chamber in the measuring trailer was ready the additional road tests were performed to investigate the tyre/road rolling resistance coefficients on the SMA 0/16 road surface for two directions of movement for three tyres described in Table 2. The internal pressure of the tyres was fixed at 210 kPa and the vertical load Z of them was fixed at 4002 N. The measurements were done only for two speeds: 50 and 80 km/h (speeds above 80 km/h were rejected because of the road traffic regulations). The results are shown in Fig. 8.

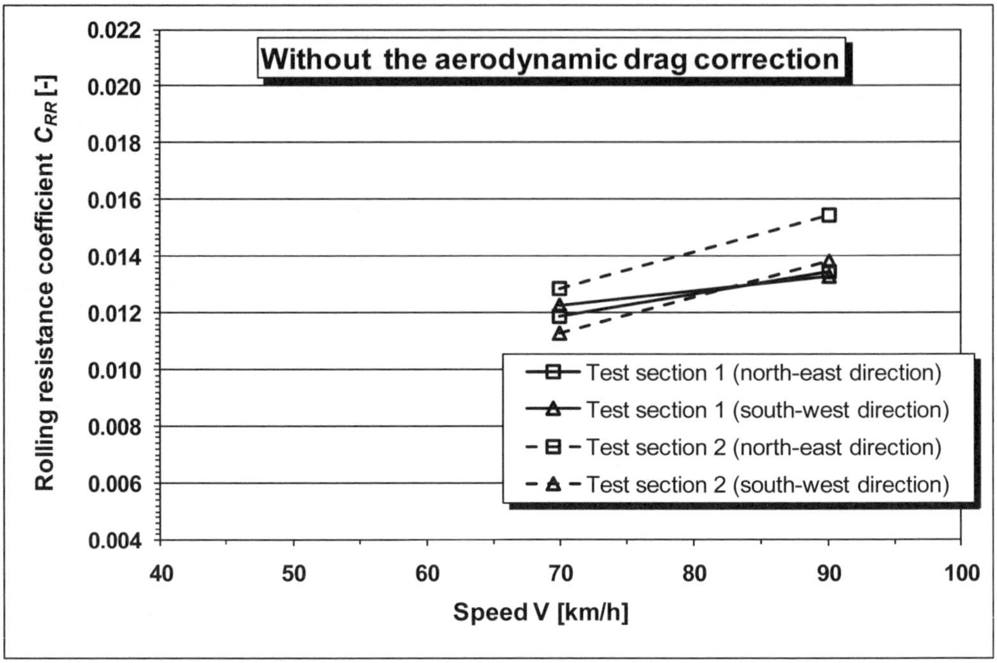

Figure 5. Dependence of the rolling resistance coefficient of reference tyre E as a function of speed on the SMA 0/12.8 surfaces (for two test sections and for two directions of movement) without the aerodynamic drag correction.

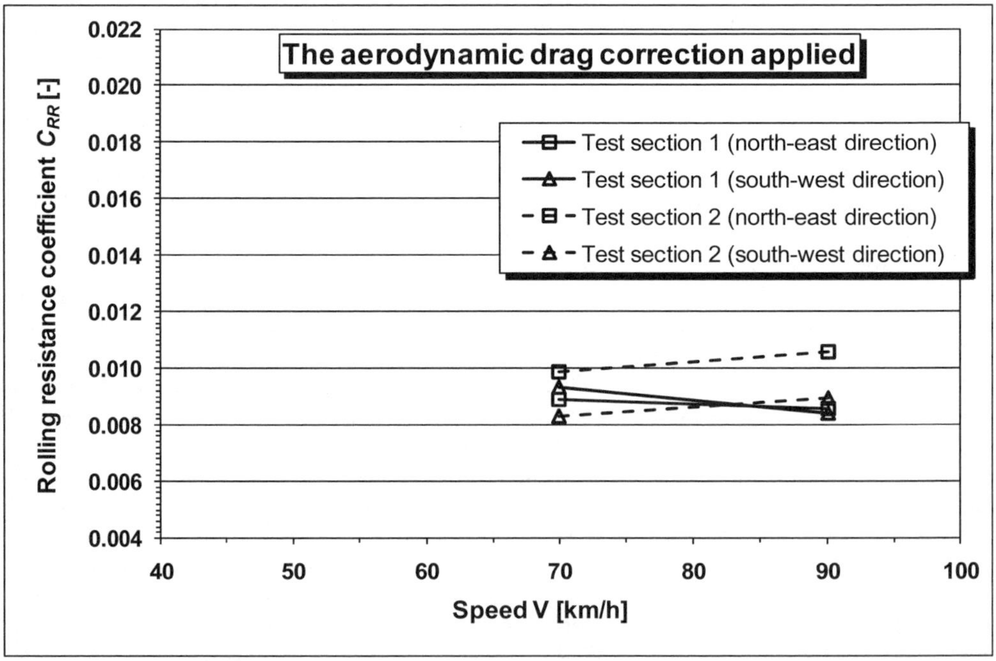

Figure 6. Dependence of the rolling resistance coefficient of reference tyre E as a function of speed on the SMA 0/12.8 surfaces (for two test sections and for two directions of movement with the aerodynamic drag correction applied.

a) b)

Figure 7. The tested tyre in the measurement trailer:
a) without the aerodynamic chamber, b) enclosed in the aerodynamic chamber.

Table 2. Description of the tyres used in the additional road tests.

Symbol of the tyre	Producer	Size	Tread
SRTT	UNIROYAL	P225/60R16 97S M+S	TIGER PAW
MCPR	MICHELIN	225/60R16 98V	PRIMACY HP
A	AVON	185/65R15 88H	ZV 1

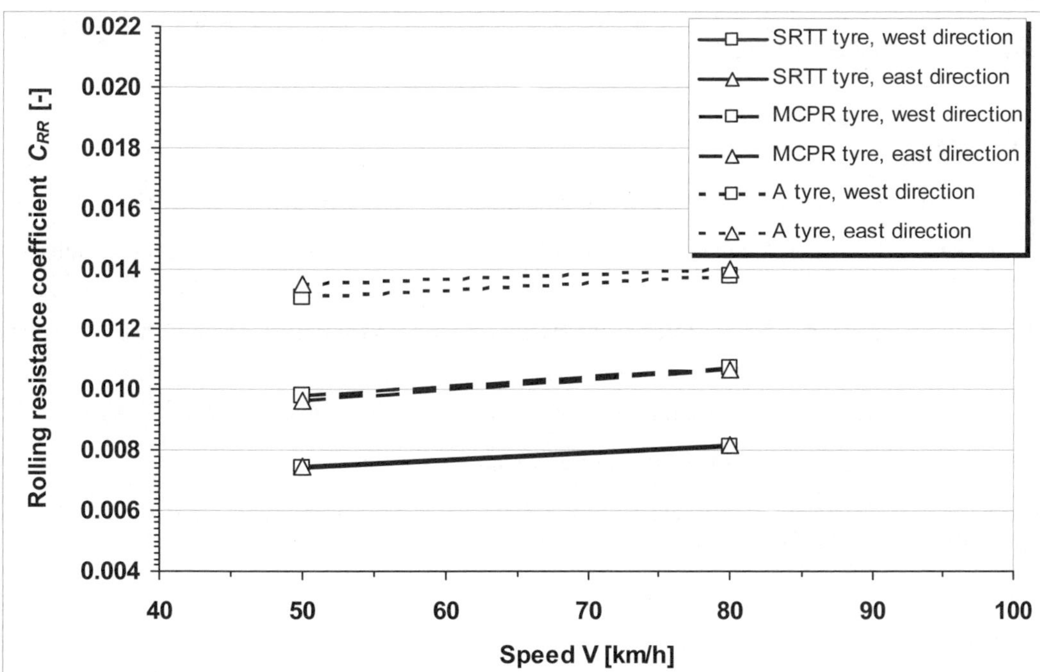

Figure. 8. Dependence of the tyre/road rolling resistance coefficient of tyres: SRTT, MCPR and A (enclosed in the aerodynamic chamber in the measurement trailer) as a function of speed on the SMA 0/16 surface for two directions of movement.

These presented measurement results are typical for great number of tyre/road rolling resistance tests performed using this measurement trailer. Additional test speeds in the range of 50 and 80 km/h (it means 60 and 70 km/h) were not performed because of the financial reasons. The authors intend to perform these additional test speeds during the measurements in the future.

Notice that on this surface (SMA 0/16) the tyre/road rolling resistance coefficient for MCPR tyre is 31% higher in comparison to SRTT tyre for both speeds (average value for both directions of movement). For A tyre the tyre/road rolling resistance coefficient increases 78% for 50 km/h and 71% for 80 km/h in comparison to SRTT tyre (average value for both directions of movement). As a matter of fact A tyre is the tyre of very high rolling resistance coefficient in comparison to rolling resistance coefficient of great number of car tyres.

Conclusions

True assessment of tyre/road rolling resistance can only be done performing measurements in real conditions - it means on the real road.

The research of this type can only be done using a special device such as the measurement trailer equipped with high sensitivity measurement system.

To perform accurate and reliable road tests the measurement trailer has to be equipped with special systems which compensate the impact of disturbance factors for example such as aerodynamic drag acting on the tested tyre.

Correction of this disturbance factor can be done either programmatically (in the algorithm for determining of the tyre/road rolling resistance coefficient) or by eliminating the source of disturbances by the special device described by the authors. In their opinion much more effective is the method of eliminating the sources of disturbances by closing the tested tyre inside the aerodynamic chamber (see Fig. 7b).

References

[1] J. Lanzendoerfer, Vehicles Testing, Communication Publishing House, Warsaw, 1977.

[2] J. Reimpell, J. Betzler, Vehicles Chassises - Bases of Constructions, Communication Publishing House, Warsaw, 2001.

[3] J. R. Bang, Influence of Tire Rolling Resistance on Fuel Consumption, Bang Consult, Sandnes, 1999.

[4] Z. Chlopek, Natural Environment Protection, Communication Publishing House, Warsaw, 2002.

[5] K. S. Clark, Mechanics of Pneumatic Tires, Washington, 1978.

[6] S. Taryma, R. Wozniak, Tyre/Road Rolling Resistance Measurements in Road Conditions, Asphalt Pavements. 2 ($2008_{(14)}$) 6-9.

Key Engineering Materials Vol. 490 (2012) pp 187-194
© (2012) Trans Tech Publications, Switzerland
doi:10.4028/www.scientific.net/KEM.490.187

Multicaterpillar track chassis of big machines - identification of loads

Tadeusz Smolnicki[1,a], Paweł Maślak[2,b]

[1] Wroclaw University of Technology, Faculty of Mechanical Engineering, Poland

[2] Wroclaw University of Technology, Faculty of Mechanical Engineering, Poland

[a] tadeusz.smolnicki@pwr.wroc.pl

[b] pawel.maslak@pwr.wroc.pl

Keywords: multicaterpillar track chassis, open cast machines, traction force, steering force

Abstract: Multicaterpillar track chassis are used in the biggest machines. This kind of chassis is commonly used in open cast mines for mobile transfer conveyor or for bucket wheel excavators. In each machine the chassis consist of 6 to 12 caterpillar tracks, which some are steered. Existing models used for analytical calculations are based on simplified relations taken from quasistatic equations of stability and geometrical connections. The complex research was conducted on mobile transfer conveyor, which is used in open cast mine. The aim of measurements was to identify the value of force that acts on elements of chassis during the steering and driving the machine. The measurements were taken during the drive in different directions, with different positions of center of gravity. The strain gauges were used to measure the deformation on each caterpillar track, trusses and drawbar.

Introduction

The aim of the research was to identify the loads acting on multicaterpillar track chassis. The object of study was the open cast mine machine - mobile transfer conveyor, shown in figure 1. The biggest problem in such kind of machines is to diagnose properly the state of machine and to remove it from operation before any dangerous accident for people and machine will take place [1,2].

Fig. 1. Open cast mine machine: - mobile transfer conveyor

Multicaterpillar track chassis

In open cast mines there were many accidents caused by unexpected damage of some elements of machine. One of it was the damage of the driveshaft, which broke and the steered set of caterpillar track immersed into the ground, what is shown in figure 2.

Fig 2. Destruction of driveshaft of steered set of caterpillar tracks

The mobile transfer conveyor A_2RsB 12500 was designed in early 70' last century, on the basis of TGL Standards. There were many indefinite problems, which are still unknown. Nowadays we cannot answer the question how long the machine will work without any failure, therefore the assessment of actual technical condition need to be done.

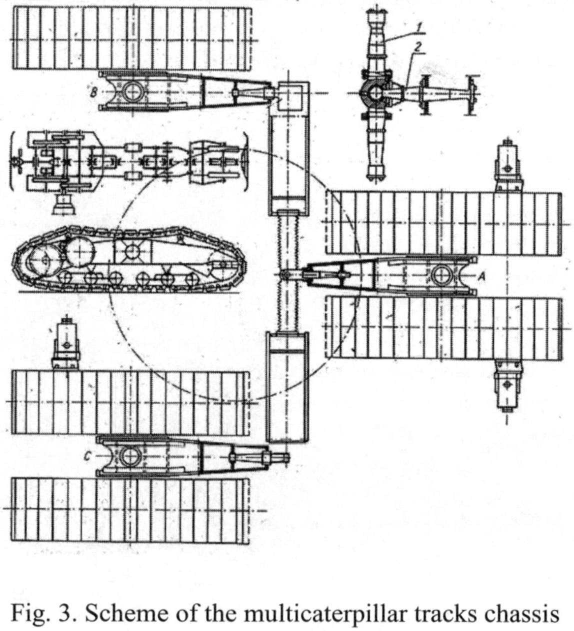

Fig. 3. Scheme of the multicaterpillar tracks chassis

Multicaterpillar track chassis consist of 6 caterpillar tracks which are connected by axis and the steering drawbar. The scheme of the chassis is shown in figure 3. Two caterpillar tracks are connected into one set, where there are 3 sets. Two of sets are fixed and do not have the possibility to turn, one set is steered by driveshaft and has an possibility to change the relational angle between the portal of the machine and the main axis of set. Tracks are powered by electric motor, one caterpillar track by one motor.

Boundary conditions

Nowadays standards state that the friction is held at the value of 0.6. It is the maximum value that is taken from the updates standards [6, 7]. Several cases of loads and boundary conditions are used for strain calculations. According to the standards, loads cases do not include all of the possibility of real loads. In the presented paper the most popular cases of used boundary conditions are shown. In the figure 4 the boundary conditions used for calculations of caterpillar track are shown [7, 8]. The caterpillar track is loaded on each arm or on drive-wheel and tension-wheel. The caterpillar track is fixed on its axis' bushing. The force F1 and F2 are generated by the reaction with the ground, the force Q is the result of the interaction with the ground during the turning, force Fb comes from the reaction of electric motor.

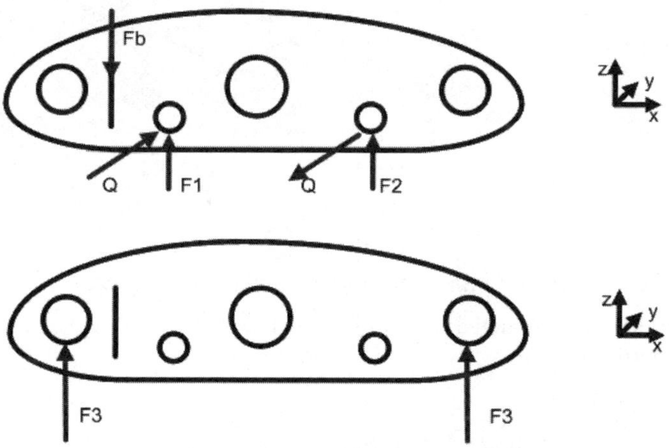

Fig. 4. Scheme of the load on caterpillar track

The most appropriate for the calculations of steering drawbar case of boundary conditions is to fix one end of the steering drawbar and apply force at the other end in the horizontal direction, that is perpendicular to the maim axis of the drawbar and the vertical force applied to the place, where the ball joint is.

Measurements

Measurements scheme is shown in figure 5. The points, where the strain gauges were located are shown in the scheme. There are 6 points on caterpillar tracks' where the traction forces were measured. The strain gauges were located on the support of caterpillar track, where the motion is fitted. The other 2 strain gauges were located on each truss and the last one was located to the drawbar. The measurements were taken during the drive in different directions: forward and backwards, with different positions of center of gravity, with different positions of the load carrying structure and the chassis. The measurements were taken during turning machine to the right and to the left, riding forward and backwards with and without the overburden. During the measurements over 30 sets were recorded.

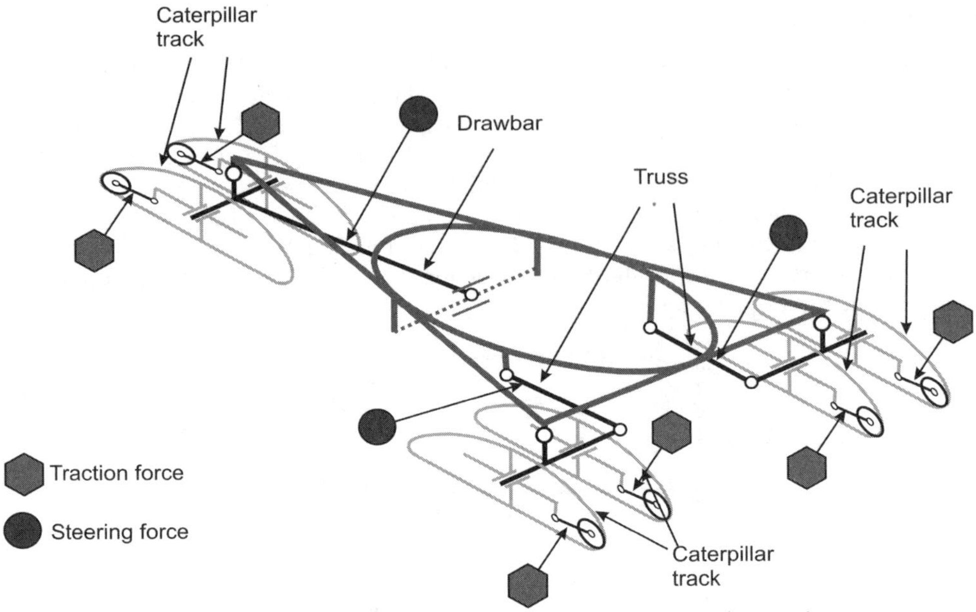

Fig. 5 Scheme of caterpillar track chassis and points of locating strain gauges

The data acquisition was conducted by multichannel recorder. All of the signals were recorded parallel from each strain gauge.

Calibration of measurements

The measurements were calibrated by using the finite element method [4]. The support of caterpillar track was loaded by the known force and then the stress results were obtained.

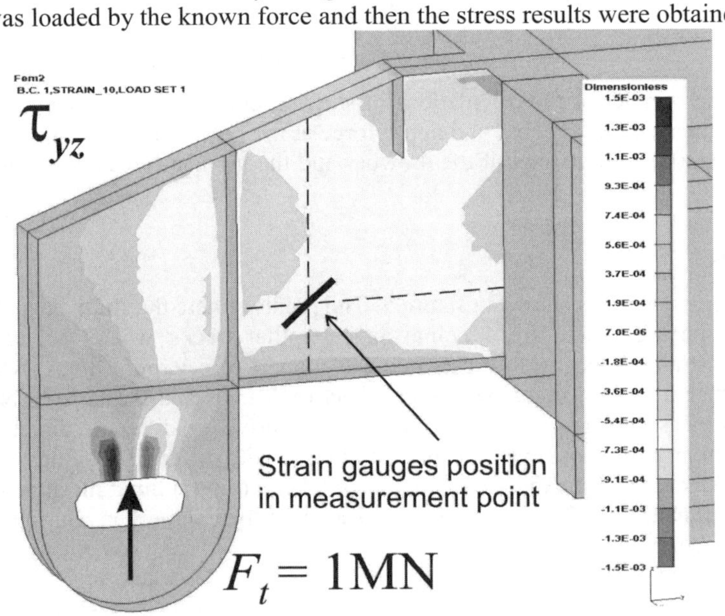

Fig. 6. Results of numerical stress calculation of caterpillar track

In figure 6 the stresses along Z direction are shown, obtained with use of finite element method calculations. The stresses along other directions were used to get full calibration of the strain gauges measurements. The measurements were taken by the half - Wheatstone bridge strain gauges. This kind of strain gauges allowed to measure the component along Z and Y coordinates [3].

Analytical calculations of traction and steering force

Analytical calculations were based on the equation 1 [5]. Equations used for calculations of torque and traction force are linear and do not include any stochastic relations between chassis elements and interaction with ground. The equation 1 shows that only linear equations are used to calculate the traction force.

$$F_t = R \frac{p_{av}}{2Lp_0} k_z \qquad (1)$$

Where:

F_t - traction force [N],
R - reaction for each caterpillar track [N],
p_{av}- average pressure between ground and caterpillar track [Pa],
L - length of caterpillar that takes part in interaction with ground [m],
p_0 - ground deformability modulus [Pa/m],
k_z - factor that takes 1 for rectangle load and 3 for triangle load of caterpillar track [-].

Equation 1 shows that the calculation of caterpillar track chassis is very simplified. Load is not static and in real conditions the change in very high range. The measurements on the real object are necessary to be conducted before new caterpillar track chassis will be designed. Calculations were done for 7 different types of ground on which the mobile transfer conveyor drives. The results of calculations are shown in table 1.

Table 1. Traction force Ft calculated for different types of ground

Type of ground	**Traction Force *Ft* (rectangle load) [kN]**	**Traction Force *Ft* (triangle load) [kN]**
Mud, peat	26.7	80.1
Soft clay	12.8	38.5
Compacted sand, sandy clay	9.1	27.5
Hard, plastic clay	5.8	17.5
Semi hard clay, marl, moist loess	3.3	10.1
Heavy clay, semi hard marl, dry loess	2.5	7.6
dry loam-clay ground	1.9	5.8

If the ground is wetter, the traction force is bigger. On the dry ground the machine does not fall into the ground so the forces are smaller than on wet or not compacted ground. Drive on mud or on peat is very destructive for the entire machine with multicaterpillar track chassis. The equations used for analytical calculations of traction force use very simplified load that comes from interaction with ground. The load is not linear and should not be calculated rectangle or triangle load. The load is much more stochastic. The results of calculations shown in table 1 are 3 times bigger for triangle load then for rectangle load.

Value of the steering force calculated for the steering set bases on equation 2 shown below. The equation shows the relations between: p_{av} – average pressure between ground and caterpillar track, μ_{av} – average factor of friction between ground and caterpillar track, L – length of caterpillar track, b – width of caterpillar track and r – distance between caterpillar track axis set and axis of steering screw. Value of factor of friction equals μ_{av} = 0,6 based on the standard PN-G-47000-2:2005 for calculating open cast machines.

$$F_S = \frac{\frac{\mu_{av} p_{av}}{24} 2Lb\sqrt{L^2+b^2} + L^3 \ln\left(\frac{b+\sqrt{L^2+b^2}}{L}\right) + b^3 \ln\left(\frac{b+\sqrt{L^2+b^2}}{L}\right)}{r} \quad (2)$$

Calculated steering force is at the level of 170 kN. Calculated force depends on static variables and is not related to the change of angle of steering set. Measurements shows that the force changes its value during the steering and riding the entire machine.

Results of the measurements

Results of the measurements show that the real traction force is not linear, but very stochastic. Traction force was gained during riding forward and backwards. Traction force for forward drive is shown in figure 7. The average value is at the level of minus 50 kN with amplitude at the level of 70 kN.

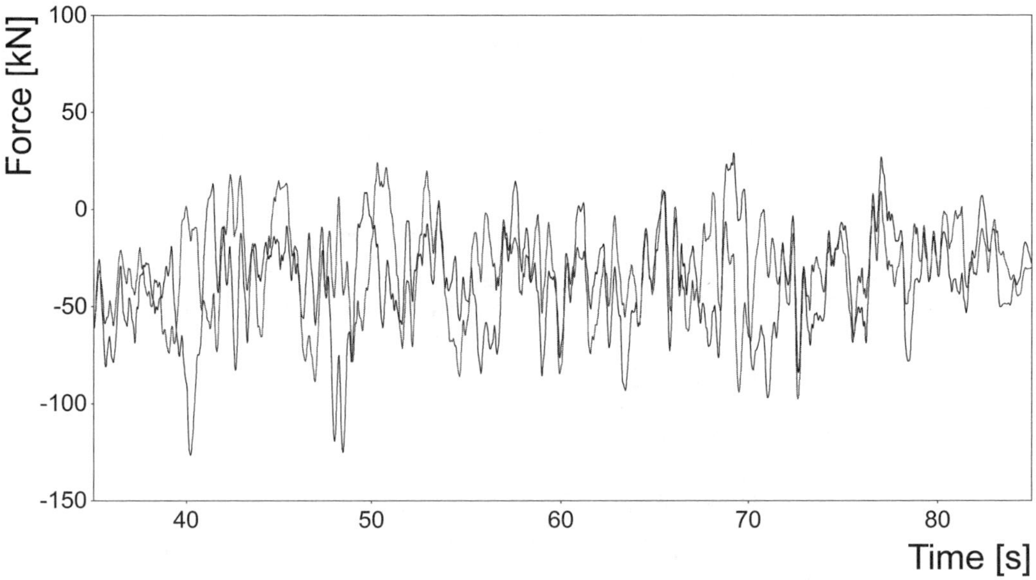

Fig. 7. Traction force of steering caterpillar track set - forward drive

For backwards drive the average value is with opposite sign, but the absolute value is at the same level of 50 kN and the amplitude is equal. The steering caterpillar track set snakes during the drive, so the value of measurement on each caterpillar track oscillates around the average value, what is shown in figure 8. The snaking of the caterpillar sets is a normal situation during the drive. It is caused by the geometry of each caterpillar track, size of each piece of caterpillar track belt and the movement of the hole body of the machine. The entire machine need to be flexible and also stiffness have to be enough high to resist all operational forces.

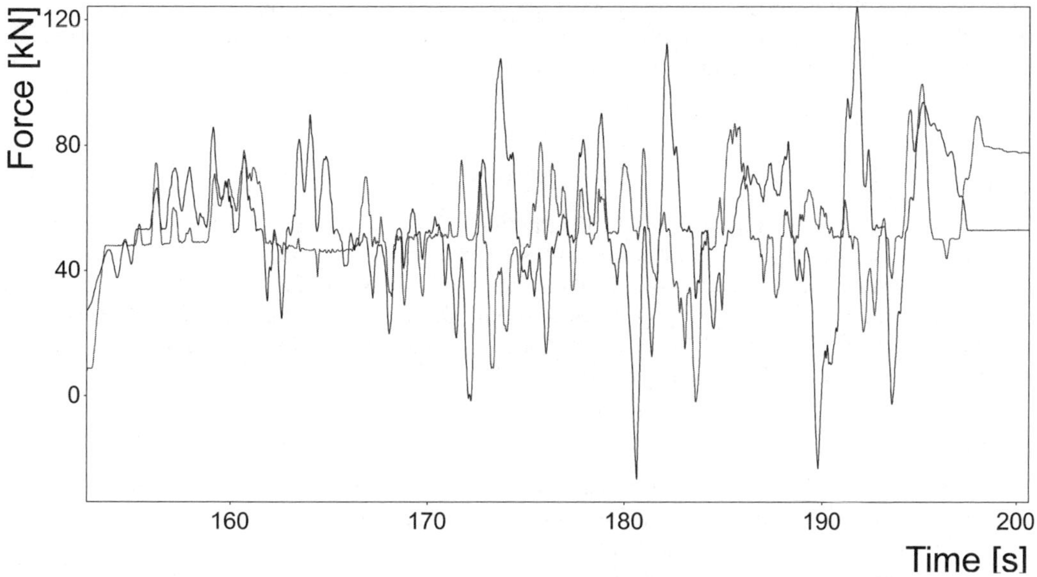

Fig. 8. Traction force of steering caterpillar track set - backwards drive

The measurements of steering force are shown in figure 9. The steering forces are shown as the function of steering angle. The most important thing in this measurement is to show how the force changes its value during turning the steering caterpillar track set. Steering force changes from 0 kN to 650 kN. Force value rise during the increasing angle from 0 to 13 degrees of steering caterpillar track set. Tendency lines for graphs that illustrate changes of steering forces are also shown in the figure 9. Measured steering force is over 3 times higher than the force calculated with use of analytical method.

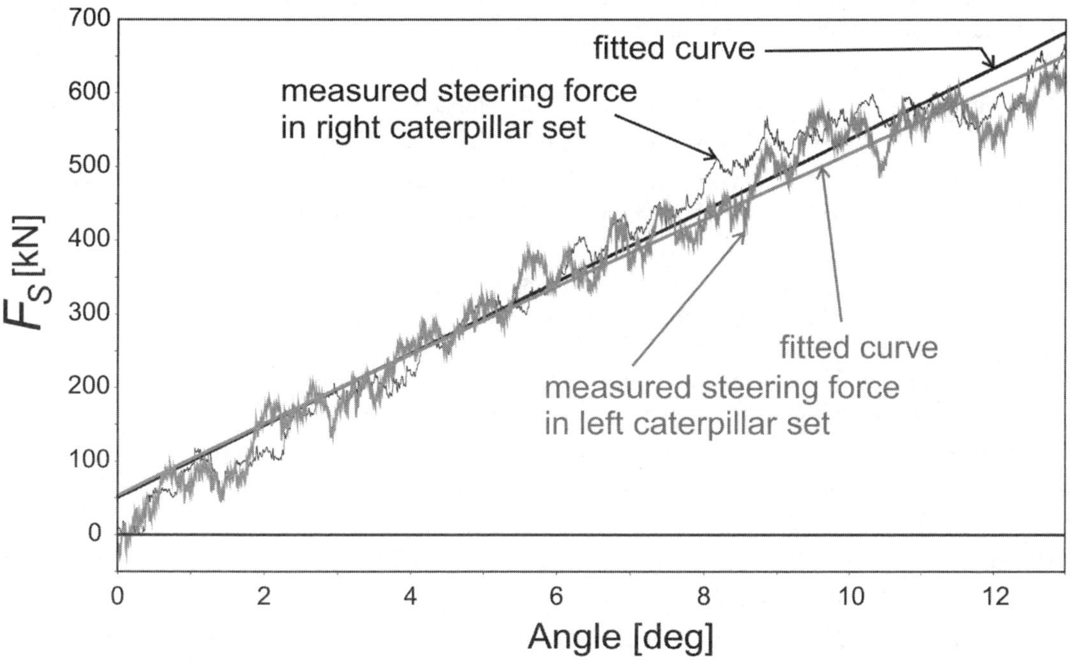

Fig. 9. Steering force in trusses – turning right from 0° to 13°, forward drive

The equations that describe the fitted curves are shown below. Equation 3 and 4 are related to the trusses, equation 3 for the right truss and equation 4 for left truss. The equations were found with use of exponential function method of approximation of the graphs.

$$F_S(\alpha)=50{,}22(1+\alpha)^{0{,}98} \quad (3)$$

$$F_S(\alpha)=53{,}25(1+\alpha)^{0{,}94} \quad (4)$$

Equations 3 and 4 give opportunity to calculate the value of steering force on steering caterpillar track set related to the angle of steered set. In those equations value of the steering force depends on two constant values and a one variable. In those equations the value of angle of steered caterpillar set is the variable. Equation 3 and 4 are dedicated only for mobile transfer conveyor A_2RsB 12500. For other kinds of such machines it is necessary to conduct the research and to acquire the measurements of the traction and steering forces.

Summary

There are many indefinite problems with the loads on chassis of open cast mining machines. Existing equations do not take into consideration any stochastic loads. The measurements show that the traction force changes very rapidly and has a very high value of amplitude. Currently there is lack of guidelines to receive the appropriate boundary conditions and operational loads. It is necessary to conduct more research on open cast mining machines to define more properly the loads acting on multicaterpillar track chassis. The measurements should be taken in different conditions: during the winter and summer, when the ground is frozen, wet and muddy or dry and sandy. Research should be done in the different positions of the entire machine and with or without transporting the overburden.

References

[1] S. Bosnjak. M. Arsić, N. Zrnić, M. Rakin, M. Pantelić, Bucket wheel excavator: integrity assessment of the bucket wheel boom tie-rod welded joint, Engineering Failure Analysis 18 (2011) 212-222

[2] E. Rusiński, P. Harnatkiewicz, M. Kowalczyk, P. Moczko, Examination of the cause of a bucket wheel fracture in a bucket wheel excavator, Engineering Failure Analysis 17 (2010), 1300-1312,

[3] T. Smolnicki, P. Maślak, Experimental identification of loads in multi-caterpillar mechanism of the drive of dumping conveyor, 6th International Conference Mechatronics Systems and Materials, MSM 2010,

[4] T. Smolnicki, D. Derlukiewicz, M. Stańco: Evaluation of load distribution in the superstructure rotation joint of single-bucket caterpillar excavators. Automation in Construction. vol. 17, nr 3, 2008, pp 218-223

[5] K. Pieczonka, Working machines engineering, Part 1 Fundamentals of working, driving, turning and lifting. Oficyna Wydawnicza Politechniki Wrocławskiej, Wroclaw, 2007, (In Polish)

[6] W. Szepietowski, Exploitation research on stress oft wist mechanism of multi-crawler traversing gears of stacking machines with belts, Czasopismo Techniczne z. 1-M/2005, Technical University of Cracov, 2005, 377-385, (in Polish)

[7] Standard PN-G-47000-2:2005 Surface mining – Bucket wheel excavators and stacker - Part 2: Calculation foundations, (in Polish)

[8] Standard DIN 22261-2 – Excavators, stackers and supplementary machines in open coast mining (in German)

Key Engineering Materials Vol. 490 (2012) pp 195-205
© (2012) Trans Tech Publications, Switzerland
doi:10.4028/www.scientific.net/KEM.490.195

New satellite pumps

Paweł Sliwinski

Gdansk University of Technology, Department of Mechanical Engineering,
Team of Hydraulics and Pneumatics, 11/12 Narutowicza Str., 80-233 Gdansk

Keywords: satellite pump, oil, water, noise

Abstract. Stosowanie Maszyn company from Katowice was initiator and originator of newest original high pressure satellite pumps series of types PSM. Construction solution, research and development of these pumps have been conducted by Team of Hydraulics and Pneumatics of Gdansk University of Technology in cooperation with this company within the confines of research project nr R0300103 (source of funds: Ministry of Science and Higher Education). Prof. A. Balawender was the manager of this project. Working medium for these pumps may be oil, emulsion HFA-E and water. Results of laboratory research of the pump would be presented in the article. Results of research contain: volumetric losses, hydraulic-mechanical losses, efficiencies of those pump and their characteristics of noise. For all types of pumps from PSM series characteristics are similar.

Introduction

Satellite pumps PSM series have emerged in 2010 as a result of research project conducted by the Team of Hydraulics and Pneumatics of the Gdansk University of Technology [1]. Construction of satellite pumps was created basing on the construction of satellite motors SM series [2,3]. The new design was an object of patent applications nr P.391422, P.392623 and P.392624 [4,5,6]. Cooperation with Polish companies Stosowanie Maszyn and Hydromasz form Katowice allowed, in 2010, to start production of constant displacement pumps in sizes: 5,3; 8,6; 12,4; 18,6 cm^3/rev. The general view of satellite pump is presented in figure 1.

Fig 1. General view of satellite pump [1].

The new pumps showed good features like: nominal working pressure 25 MPa, peak pressure up to 32 MPa, nominal speed 1500 rpm, power density exceeding 2,5 kW/kg, ability to work in self-sucking mode. Working medium for satellite pumps are: oils, emulsions and tap water. The nominal working pressure of these pumps for water is the same as for oil. The low dynamic viscosity and poor lubricant property of water results in high volumetric losses and higher mechanical losses than in pump working with oil. In effect, total efficiency of the pump working with water is lower than total efficiency of the same pump working with oil [7,8,9,10,11,12,13,14,15].

Structure of satellite pump

The construction of satellite pump is showed in figure 2. The working mechanism of pump is gear mechanism (figure 3) and consists of: rotor (4 humps), curvature (6 humps) and gear wheels (satellites). The number of satellites is the sum of humps of rotor and humps of curvature (10 satellites in this case).

The principle of operation of satellite mechanism (working mechanism) is as following. When the rotor rotates, then the space between two satellites is changing its capacity. In this way, when capacity increases then it is suction process and when capacity decreases then it is press process. On one turn of shaft there are 24 working cycles. Spaces in satellite' mechanism are closed by two commutation plates (item 6 in figure 2). The commutation plates 6 (figure 2 and figure 3) play the role of the compensation plates simultaneously. In this way the satellite pump includes axial clearances compensation. In plates 6 are the inlet and outlet holes (commutation holes). The shape of those holes are non circular [5,6].

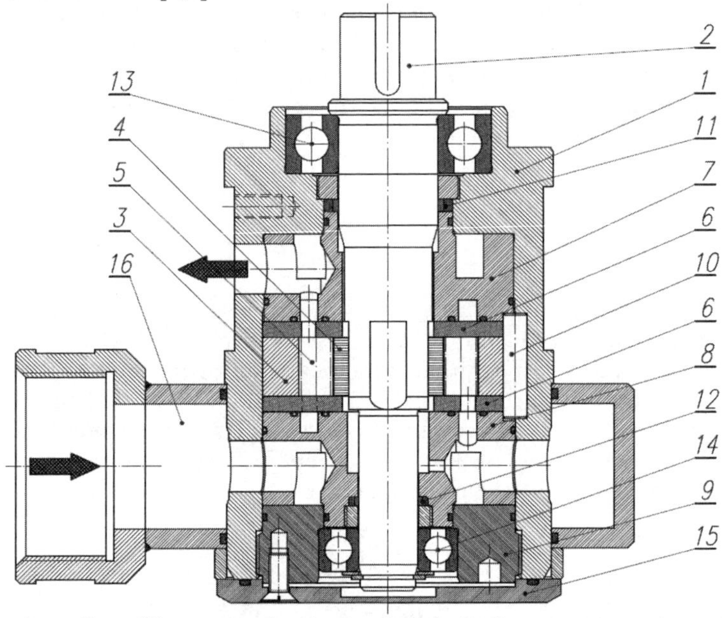

Fig. 2. Cross section of satellite pump: 1 – body, 2 – shaft, 3 – curvature, 4 – rotor, 5 – satellite, 6 – compensation plates (commutation plates), 7 – exhaust manifold, 8 – suction manifold, 9 – clamp, 10 – dowel pin, 11 and 12 – shaft seals, 13 and 14 – bearings, 15 – cover, 16 – suction flange [1,4].

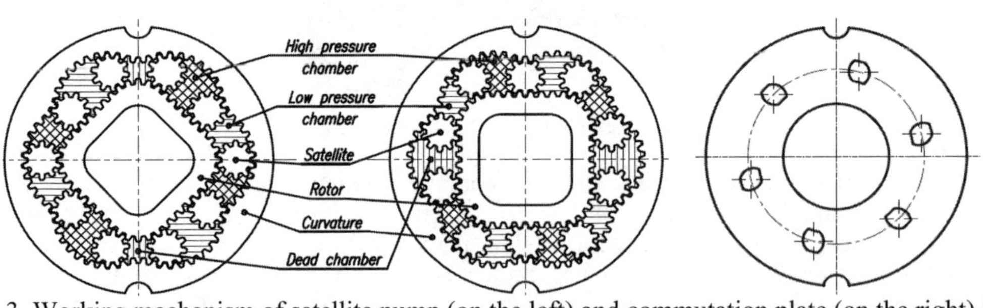

Fig. 3. Working mechanism of satellite pump (on the left) and commutation plate (on the right) [6].

The idea of principle of operation of axial clearances compensation is illustrated in figure 4. On the one side of the plate acts pressure from working chambers and on the other side acts pressure from exhaust manifold channel (load of pump). The force of compensation P_c is bigger than the force given by working chambers ($P_{wc}+P_{gr}+P_{gs}$) and then the compensation plates is bending down. In this way total axial clearances are lower, internal leakage is lower and volumetric efficiency higher in effect.

In all satellite pumps, satellites and rotor have value of 5±1 μm lower than curvature (figure 4a) parameter A).

Between the rotor and compensation plates there are no leek stoppers. Thus, existing leakage goes from working chambers to the chamber of shaft. Then the leakage drains to suction manifold channel.

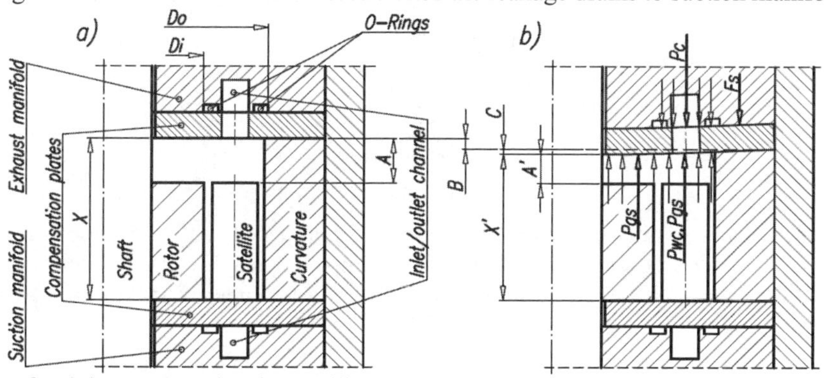

Fig. 4. Idea of axial compensation in satellite pump: a) state before tighten up of clamp (X – high of curvature, A – axial clearance of satellites and rotor, D_i – inside diameter of compensation area, D_o – outside diameter of compensation area); b) state after tighten up of clamp and under load (B – value of decrease of high of curvature after tighten up of clamp, A' – real axial clearance in loaded pump, C – value of bend of compensation plate, F_s – force of press down curvature generated by clamp, P_c – pressure in exhaust manifold (load of pump), P_{wc} – pressure in working chambers, P_{gs} – pressure in satellite gaps, P_{gr} – pressure in rotor gap) [9].

Depending on size of the pump, the teeth module of working mechanism are: m=0,4mm (q=5,3cm^3/rev), m=0,5mm (q=8,6cm^3/rev), m=0,6mm (q=12,4cm^3/rev) and m=0,75mm (q=18,6cm^3/rev).

The compensation plates are made of sintered carbide. The satellite mechanism is made of special tool steel and sulfonitrided. Therefore satellite pumps can work not only with oil but also with other liquid like water. Application of liquid with poor lubricant properties doesn't cause scuffing and seizing of working mechanism.

All satellite pumps can also work submerged in liquid. Then the suction flange 16 (figure 1) is not necessary.

Tested pumps and measurement parameters

In laboratory of Team of Hydraulics and Pneumatics of Gdansk University of Technology following satellite pumps have been tested:

- PSM-0,4/15 (q=5,3 cm^3/rev),
- PSM-0,5/25 (q=8,6 cm^3/rev),
- PSM-0,6/25 (q=12,4 cm^3/rev),
- PSM-0,75/15 (q=18,6 cm^3/rev).

Tests of pumps were carried out using the following liquids:

a) oil Total Azolla 46 (tests in oil temperature 43°C, kinematic viscosity ν=40cSt);
b) HFA-E emulsion prepared on base of Isosynth VX110BF concentrate (1% concentrate in water) (tests in emulsion temperature 30°C, kinematic viscosity ν=0,86cSt);
c) pure tap water (tests in water temperature 30°C, kinematic viscosity ν=0,80cSt);

During testing of the pump were measured following parameters (figure 5):

- delivery of pump Q in [l/min];
- pressure p_1 and p_2 in [MPa];
- rotational speed of shaft n in [rpm];
- torque on pump shaft M in [Nm].

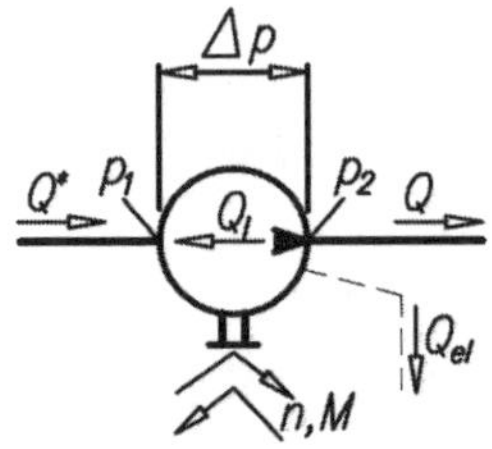

Fig. 5. Measured parameters in pump.

In the next parts of this paper characteristics of pump PSM-0,75/15 supplied with oil and water will be presented. Characteristics of this pump for poor 1% emulsion are similar to characteristics for water [1]. Characteristics for other satellite pumps are similar to characteristics of pump PSM-0,75/25.

Volumetric losses

Following factors influencing volumetric losses Q_l in satellite pumps [1,7,8,11,12,14,15,17]:

a) construction parameters:
   - height of axial gaps in working mechanism before assembly (A parameter in figure 4),
   - torque of tighten up of clamp 9 (figure 2) and thus reducing the high of curvature about value B (figure 4),
   - deformation of compensation plate under load about value C (figure 4);
   - clearances G and T_c (figure 6) between teeth of satellites and rotor and between teeth of satellites and curvature (satellites dislocate in clearances between teeth – when the speed of working mechanism increases then the volumetric losses also increases),
   - value of underlap X in commutation (figure 6);
   - stiffness of elements of working chambers (especially stiffness of teeth).

b) liquid parameters:
   - kinematic viscosity;
   - density;
   - compressibility of liquid;

c) load of the pump.

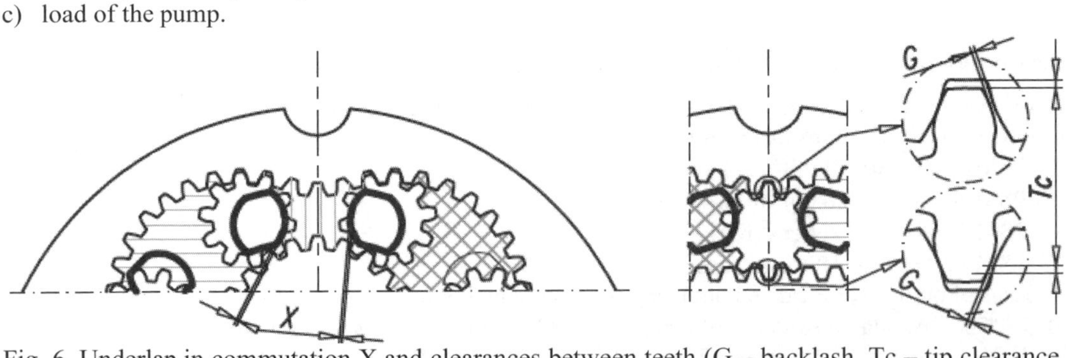

Fig. 6. Underlap in commutation X and clearances between teeth (G – backlash, Tc – tip clearance, X - underlap).

The results of the research on satellite pumps and motors showed that the dominant component of volumetric losses are [1]:

a) leakage in flat gaps in working mechanism (depends on differential pressure in working chambers and depends on pressure in shaft chamber),
b) leakage in clearances between teeth of satellites and rotor and between teeth of satellites and curvature (figure 6).

Other factors (above mentioned) have less influence on volumetric losses.

The flow of low viscosity liquid in flat axial clearances is not fully turbulent practically in all range of Δp in pump. Therefore the flow rate of low viscosity liquid (like water and emulsion) to oil is not 40 (for laminar flow of oil and water) but less than that. The ratio of flow of low viscosity liquid to flow of oil in axial clearances can be described by the following [1,11,12,13,15,17]:

$$\frac{Q_{lv}}{Q_O} = 38\frac{\rho_o}{\rho_{lv}^{4/7}}\frac{\nu_o}{\nu_{lv}^{1/7}}\frac{l^{3/7}}{h^{9/7}}\Delta p_i^{-3/7}$$

Where:

- Q_{lv} – flow rate of low viscosity liquid,
- Q_o – flow rate of oil,
- ρ_{lv} – density of low viscosity liquid,
- ρ_o – density of oil,
- ν_{lv} – kinematic viscosity of low viscosity liquid,
- ν_{lv} – kinematic viscosity of oil,
- l – length of gap,
- h – high of gap (axial clearance),
- Δp_i – different of pressure in working chamber.

Flow rate of water and emulsion in flat gaps in working mechanism of satellite pump are a dozen or so times higher than the flow rate of oil at the most.

In second case the flow rate in clearances between teeth of satellites and rotor and between teeth of satellites and curvature can be described by [1]:

$$\Delta q_l = \left(\frac{\Delta Q_l}{\Delta n}\right)_{\Delta p=const}$$

The Δq_l is approximately constant in all range n with Δp=const and increases when Δp increases (with n=const.).

For pump PSM-0,75/15:

a) for oil Δq_l=0,74,
b) for water Δq_l=1,15.

That is Δq_l depends on kind of liquid and their viscosity.

Volumetric losses in pumps are calculated from:

$$Q_l = q \cdot n - Q$$

In all satellite pump leakages are draining to the suction channel, that is $Q=Q^*$ (figure 4).

Characteristics of volumetric losses in satellite pump PSM-0,75/15 working with water and emulsion are showed in figure 7.

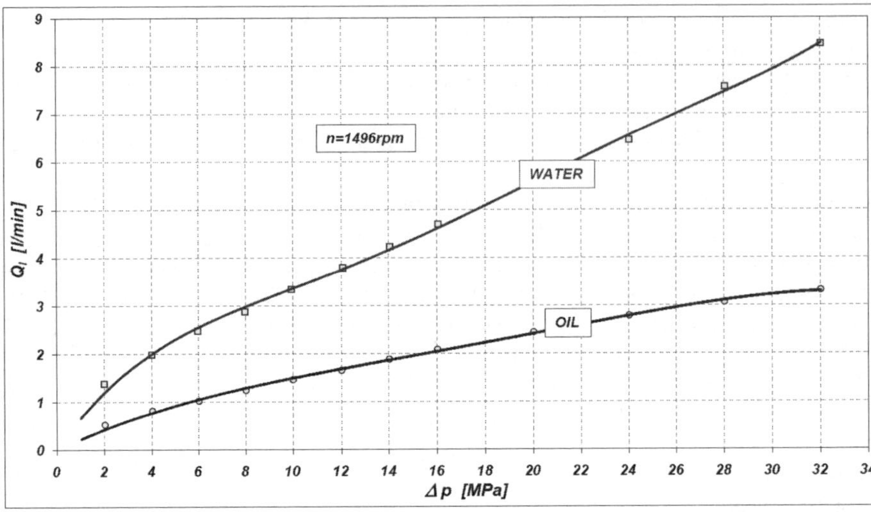

Fig. 7. Characteristics $Q_l=f(\Delta p)_{n=const}$ of volumetric losses Q_l in function of differential pressure Δp in pump PSM-0,75/15. Working medium: oil and water.

Mechanical losses

Torque of mechanical losses M_{ml} in pump depends on:

a) load of pump Δp,
b) lubricant properties of liquid,
c) rotational speed of n (if the pump can be driven with variable speed).

Torque of mechanical losses can be calculated by:

$$M_{ml} = M - \frac{q \cdot \Delta p_i}{2 \cdot \pi}$$

This torque is effect of mechanical friction forces, forces of inertia of working mechanism elements and forces of inertia of liquid convicted in working mechanism chambers, compressibility of liquid in dead chambers, compressibility of liquid in spaces between teeth and viscous friction in liquid.

Places where mechanical friction can occur in satellite pumps are all kinematic pairs, that is:

a) between rotor, satellite and compensation plate,
b) between teeth of satellites and rotor and curvature,
c) in bearings,
d) in seals.

The water demonstrates poor lubricant properties. Better lubricant properties demonstrates HFA-E emulsion. The mineral oil has the best lubricant properties [10,16,17]. Therefore the torque of mechanical losses is the higher in pumps supplied with water.

Characteristics of torque of mechanical losses in tested PSM pump supplied with oil and with water are shown in figure 8.

The occurrence of mechanical losses requires the bigger torque of the shaft of pump.

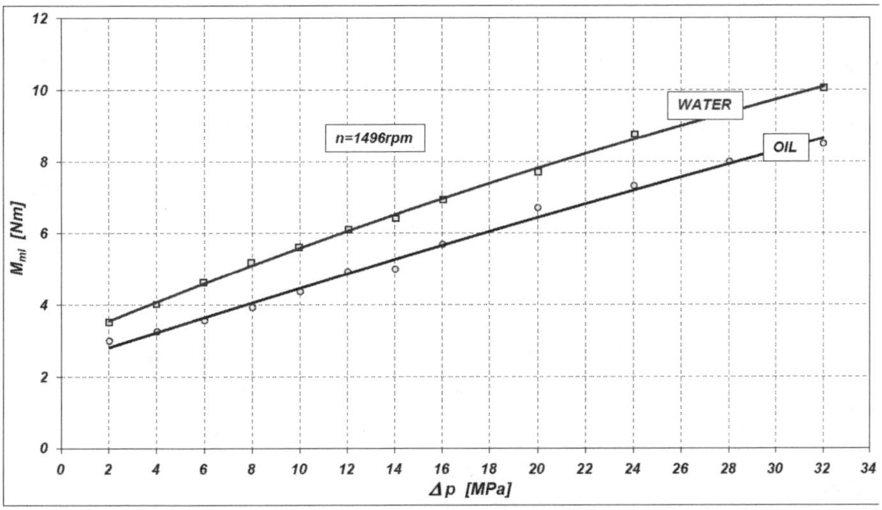

Fig. 8. Characteristics $M_{ml}=f(\Delta p)_{n=const}$ of torque of mechanical losses M_{ml} in function of differential pressure Δp in pump PSM-0,75/15. Working medium: oil and water.

Pressure losses

Pressure losses Δp_{ich} in internal channels (suction and outflow channels) of pump directly depends on [10,11,12,14,17]:

- liquid parameters, that is viscosity ν and density ρ;
- the flow rates Q and Q^*;

Indirectly load of pump Δp also influences on Δp_{ich}. When load of pump increases then delivery of pump decreases and Δp_{ich} decreases. That is:

$$\Delta p_{ich} = f(Q, \nu, \rho, \Delta p)$$

In non loaded pump working with n=1496rpm the drop pressure totals:

a) in suction channels:
   - about 0,2 bar for oil,
   - about 0,17 bar for water and emulsion.
b) in outflow channels:
   - 1,1 bar for oil,
   - 0,9 bar for water and emulsion.

Thus pump PSM-0,75/15 is self section pump for n=1500rpm. Other satellite pumps are self-section too.

Results of the pumps' test prove that pressure losses in the internal channels are larger for oil than for the water and HFA-E emulsion. The main cause of this phenomenon is the consequence of a laminar flow component in the general flow rate Q in internal channels. That is:

$$\Delta p_{ich} = C_l \cdot \nu \cdot Q + C_t \cdot \rho \cdot Q^2$$

where:

- $C_l \cdot \nu \cdot Q$ – laminar flow component;
- $C_t \cdot \rho \cdot Q^2$ – turbulent flow component;
- C_l and C_t – constants mainly dependent on geometrical dimensions of internal channels of pump.

The pressure losses in pump contribute to bigger value of torque on the pump shaft.

Total efficiency

The pressure, mechanical and volumetric losses, which exist in pump have influence on total efficiency. Pump working with water has the worse total efficiency because the water characterizes in the lowest viscosity and poor lubricant property. Characteristics of total efficiency of pump PSM-0,75/15 are presented in the figure 9.

For n=1496rpm and for nominal working pressure $p_2 \approx \Delta p$=25Mpa the efficiency totals:

- 81% for oil,
- 68% water and HFA-E emulsion.

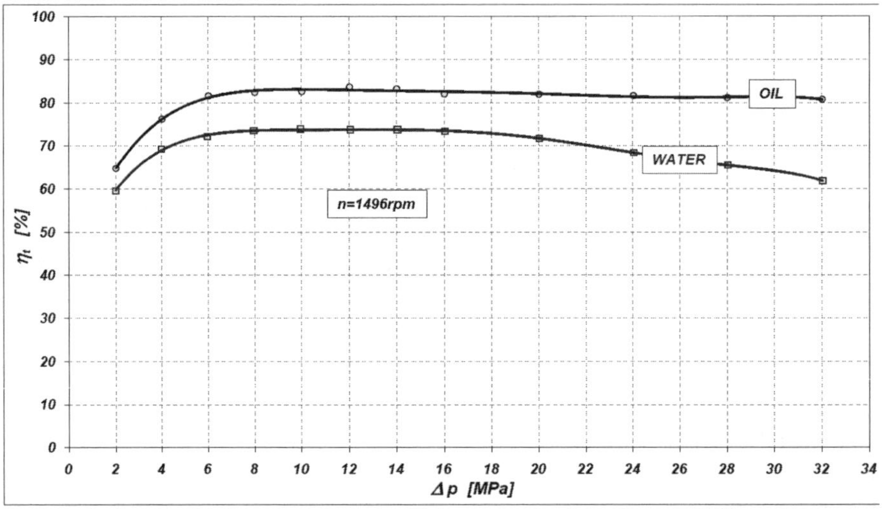

Fig. 9. Characteristics η_t=f(Δp)$_{n=const}$ of total efficiency η_t in function of differential pressure Δp of pump PSM-0,75/15 supplied with oil and water.

The total efficiency of pump working with water and emulsion is 13% less than the efficiency of the same pump working with oil.

Level of noise

The main reasons of noisiness of positive-displacement pump are:

- fast, of short duration pressure check during the movement of the working chambers from suction phase to pumping phase;
- pressure fluctuation in outflow channels in pump,
- cavitation phenomena,
- another hydraulics phenomena, characterized for given construction.

The noise is transferred by construction elements of pump and elements of hydraulic system, by liquid and air.

The measurements of noise level were realized in condition of soundproofing electric motor and coupling of pump (multilayer shield with elastics layer absorbed of noise). But the pump, hydraulic couplings and tubes and overflow valve emitted noise to surroundings. Furthermore in measurement place there was a reflection of noise waves from near walls, ceiling and floors. As a result, it accepted unfavorable system for noise testing. But this system was near to exploitation condition.

During the measurement of noise, the pump was driven by electric motor with speed of 1450rpm. The pump sucked in the oil and water from tank by the elastic tube with ~1,5m length and 30mm inside diameter and pumped the oil and water by the elastic tube (with ~1,5m length) to the overflow valve (setting of load of pump). The microphone was standing at tested pump axis level, within the distance of 1m from pump.

For measurements of noise high-class apparatus Bruel&Kjaer (B&K) was used. This equipment contained:

- analyser B&K 2144,
- calibrator B&K 3541;
- intensity probe B&K 3584.

The tests of noise of pump were conducted under different working pressure. The bigger level of noise was registered under biggest load of pump. Characteristics of noise spectrum generated by the pump and overflow valve under working pressure 25 MPa are presented on figure 10.

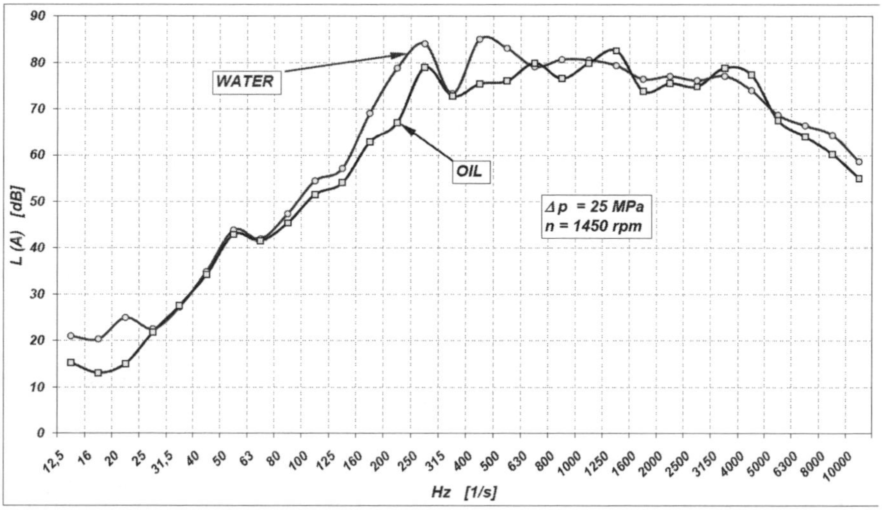

Fig. 10. Spectrum of noise L(A) emitted by loaded pump PSM-0,75/15 and overflow valve in self section conditions of oil and water (H_z – frequency of noise). Load of pump Δp=25MPa and speed n=1450rpm.

The results of noise test obtained for unit pump and overflow valve shown that the level of noise don't cross the level required for devices, which is 85 dB(A). Furthermore, it is smaller than characteristic level of noise of hydraulic system. Some of the pumps have generated noise higher than 90 dB(A). That is the level of noise emitted by loaded unit pump and overflow valve that can be accepted as satisfied.

Conclusion

Satellite pumps PSM are modern construction. The application of sintered carbide and special sulfonitrided tool steel on elements of working mechanism are accurate and allow to use satellite pumps with different liquids like mineral oil, different emulsions and pure tap water. Furthermore, one of satellite pumps was applied in petrol pumping system. Applications of different liquid for PSM pump haven't required interfere in construction of this pump. At an angle of working parameters satellite pumps can be compared with classic gear pumps. But the speed of satellite pump cannot be higher than 1500 rpm. Above this speed satellite pumps required initial feed.

The low dynamic viscosity and poor lubricant property of water results in high volumetric losses and higher mechanical losses than in pump working with oil. In effect total efficiency of pump working with water is lower than total efficiency of the same pump working with oil.

It is necessary to say that characteristics of volumetric, mechanical, pressure losses and characteristics of efficiency for other satellite pumps have not many different courses that characteristics of pump PSM-0,75/15. But efficiencies of pumps with smaller module of teeth insignificantly decrease [1].

The life of satellite pumps are unknown. In further perspective live tests of PSM pumps supplied with oil and pure tap water are planned. Detailed tests of those pumps working with vegetable oil are also planned. If results of research will be positive, than satellite pumps may be used in hydraulics systems for food industry devices.

Literature

[1] A. Balawender, P. Sliwinski and others, Report of research project nr R0300103: Developmental research of hydraulic satellite motors and satellite pump with small geometrical displacement supplied with water, emulsion and oil. Faculty of Mechanical Engineering, Gdansk University of Technology. (2010)

[2] T. Szwajca, Polish Patent nr 200588, Hydraulic circulating motor. (2009)

[3] T. Szwajca, Patent application nr P.389444, Hydraulic satellite motor. (02.11.2009)

[4] Pawel Sliwinski and Piotr Sliwinski, Patent applications nr P.391422, Satellite pump. (07.06.2010)

[5] Pawel Sliwinski, Piotr Sliwinski and T. Szwajca, Patent applications nr P.392623. Inflow and outflow channels in side plates of satellite working mechanism of displacement machine. (11.10.2010)

[6] Pawel Sliwinski, Piotr Sliwinski and T. Szwajca, Patent applications nr P.392624. Inflow and outflow channels in side plates of satellite working mechanism of displacement machine. (11.10.2010)

[7] A. Balawender, Physical and mathematical model of losses in hydraulic motors, Developments in mechanical engineering, Gdansk University of Technology Publishers. Gdansk, Poland. (2005)

[8] Z. Paszota, Power of energetic losses in hydrostatic drive system elements – definition, relationships, ranges of changes, energetic efficiencies. Part 1 – hydraulic motor. Napędy i Sterowanie, nr 11/2007. (2007)

[9] P. Sliwinski, Methodology of developmental research of new type satellite motors and pump. International Scientific and Technical Conference: Drives and Hydraulic and Pneumatic Controls 2009: Domestic brand in condition of turbulent market. Wrocław, 07-09 October 2009.

[10] P. Sliwinski, Comparison of lubricant property of water, oil in water emulsion HFA-E and oil Total Azolla 46 as working liquid in hydraulic systems. XXX Polish-wide Tribological Conference ADVANCED TRIBOLOGY, Naleczow. Tribology: Theory and Practice, nr 3=225. (2009)

[11] P. Sliwinski, Power balance in hydraulic satellite motors supplied with oil and HFA-E emulsion. Proceedings of the 5th FPNI – PhD Symposium on Fluid Power, Krakow University of Technology. (2008).

[12] P. Sliwinski, Influence of oil and HFA-E emulsion on losses and power balance in hydraulic satellite motors. Chapter in monograph CYLINDER 2008 "Research, construction, manufacturing and eksploitation of hydraulic systems". Centrum Mechanizacji Górnictwa KOMAG, Gliwice 2008.

[13] P. Sliwinski, Model of volumetric losses in hydraulic satellite motors supplied with oil and HFA-E emulsion. Conference Mechanics 2007. Gdańsk, Faculty of Mechanical Engineering, Gdansk University of Technology. (2007)

[14] P. Sliwinski, Influence of oil and HFA-E emulsion on flow characteristics in internal channels of hydraulic satellite motors. Hydraulika i Pneumatyka nr 4/2007.

[15] P. Sliwinski, Influence of oil and HFA-E emulsion on flow characteristics in gaps of hydraulic satellite motors. Hydraulika i Pneumatyka nr 5/2007.

[16] P. Sliwinski, Lubricant property of working liquid of hydraulic systems. Oil Total Azolla 46, HFA-E emulsion and water. Hydraulika i Pneumatyka nr 6/2007.

[17] P. Sliwinski, Doctoral dissertation: Comparison of phenomena in hydraulic satellite motors supplied with water in oil emulsion and oil. Faculty of Mechanical Engineering, Gdansk University of Technology. Supervisor: Balawender A. (2006)

Key Engineering Materials Vol. 490 (2012) pp 206-215
doi:10.4028/www.scientific.net/KEM.490.206

Novel Design of an Ocean Wave Power Device

Czeslaw Dymarski[1, a], Pawel Dymarski[2,b] Wojciech Litwin[3,c]

[13]Gdansk University of Technology; ul. Narutowicza 11/12; 80-233 Gdansk, Poland
[2]Ship Design and Research Centre; Waly Piastowskie 1; 80-958 Gdansk, Poland
[a] cpdymars@pg.gda.pl , [b] padym@wp.pl, [c] wlitwin@pg.gda.pl

Keywords: renewable energy, wave power device

Abstract. The paper presents a novel device for capturing the energy of sea-waves and the construction of its model for conducting test in a towing-tank. The paper also presents the research program and general analyses on selecting the hull form and systems of anchoring and propulsion. In addition, it includes remarks and observations gathered following initial tests and measurements of the device operating in regular wave conditions.

Introduction

The last few decades have witnessed a dramatic growth in demand for energy which has been accompanied by decreasing fossil fuel resources and increasingly stringent environmental protection requirements. Such situation has resulted in an ever growing interest in renewable sources of energy. One of them is the widely available energy provided by the rolling waves of seas and oceans. While the energy potential this source is immense, it is also spread over a huge area and for this reason has remained practically unexplored. For the European Union it is estimated at roughly 120 – 190 TWh/year for open waters and additional 34 – 46 TWh/year for coastal zones [1]. The average wave of the Atlantic Ocean carries with itself approximately 70 KW per meter of width. The waves of the Baltic carry a somewhat lower value of energy - it is estimated that their technically feasible capturable power amounts to 30 kW/m [1]. It may be assumed that even for calmer marine regions of lower energy potential it is realistically possible to obtain 10 – 20 kW/m.

However, it should be pointed out that due to its scattered nature, as well as the widely unpredictable character of marine conditions, the of this energy is extremely challenging in its technical aspect, and therefore also quite expensive. Despite that the constantly rising energy prices, as well as ever improving technology, materials and control systems have resulted in the creation of more conceptual designs, prototype devices and facilities for capturing the energy of sea waves. Numerous solutions are already in operation [2÷8 and 10÷15]. Initially, these consisted of power plants located ashore or in shallow coastal waters but recently also of floating structures created to operate in the open sea.

The paper presents construction solutions of a power buoy prepared at the Faculty of Ocean Engineering and Ship Technology of the Gdansk University of Technology based on the general concept contained in the patent application by Mr. Sławomir Klukowski [5]. The main idea is based on utilizing the buoy's motion on the rolling sea-wave to power a water turbine placed under the buoy, as illustrated in Fig.1. In addition, the analysis of its operational-technical conditions is presented with construction variations and anchoring systems. The paper also includes required measuring systems meant for testing the model buoy in a towing-tank and the scope of planned research.

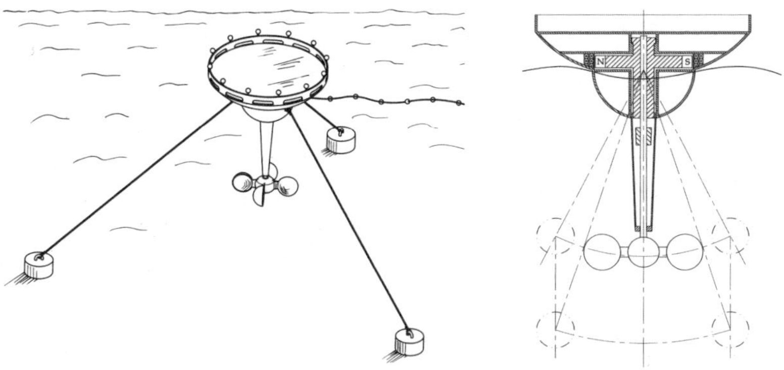

Fig. 1. Schematic illustration of the power buoy in perspective and axial section views [5].

Design requirements assumptions

Prior to starting to design any novel structure, especially one mean to work under marine conditions, it is necessary to conduct analysis of the forecasted atmospheric and sea conditions which may occur during operation with emphasis on extreme weather scenarios. Such analysis allows for assessing the character and value of acting loads, as well as of the various hazards and technical problems, allowing for greater accuracy in establishing technical requirements assumptions of the structure and its equipment.

In case of the presented power buoy it was decided that the prototype would be meant to operate on the Baltic Sea in the Polish coastal zone at the depth of about 20 meters. It was also determined that the design and construction of the prototype should be proceeded by model based research in a towing-tank. The decision was made to build at least two models of the buoy at the scale of 1:5.The scale was selected based on the possibility of conducting research tests of the models on the large towing-tank located in the Ship Design and Research Centre in Gdansk, where it is possible to create model conditions close to those found at sea. This fact has significant impact on generalizing the measurement results and employing them for the verification of calculations. The towing-tank has the dimensions of L x B x H = 260 x 12 x 6 m and is equipped with a wave generator, as well as a towing carriage with measuring equipment. The tank allows for conducting tests for a wide range conditions with waves up to 0.7 m in height and 14 m in length for regular wave and even greater height for irregular waves. The large towing-tank dimensions significantly decrease the influence of wall deflection and creation of a shallow-water wave, allowing for conducting longer measuring sessions.

Taking into account the above factors, the following general model buoy design requirements were set:

- transverse buoy hull dimensions – 2.0 to 2.5 m;
- total draught of turbine column – up to 3.5 m;
- anchoring system for constant direction of incoming waves;

Construction solutions

Prepared buoy construction solutions differed mostly in the shape of the hull, power transmission system, the type of power generator and the buoy anchoring system. The two variants presented below were selected for model construction.

Variant I

In the initially prepared version of the model buoy it was assumed that an axially symmetrical composite hull would be built, as was specified in the mentioned patent application. However, the hulls differ in shape geometry, as can be seen by comparing the buoy shape in Fig.1 with the shape presented in Fig.2.

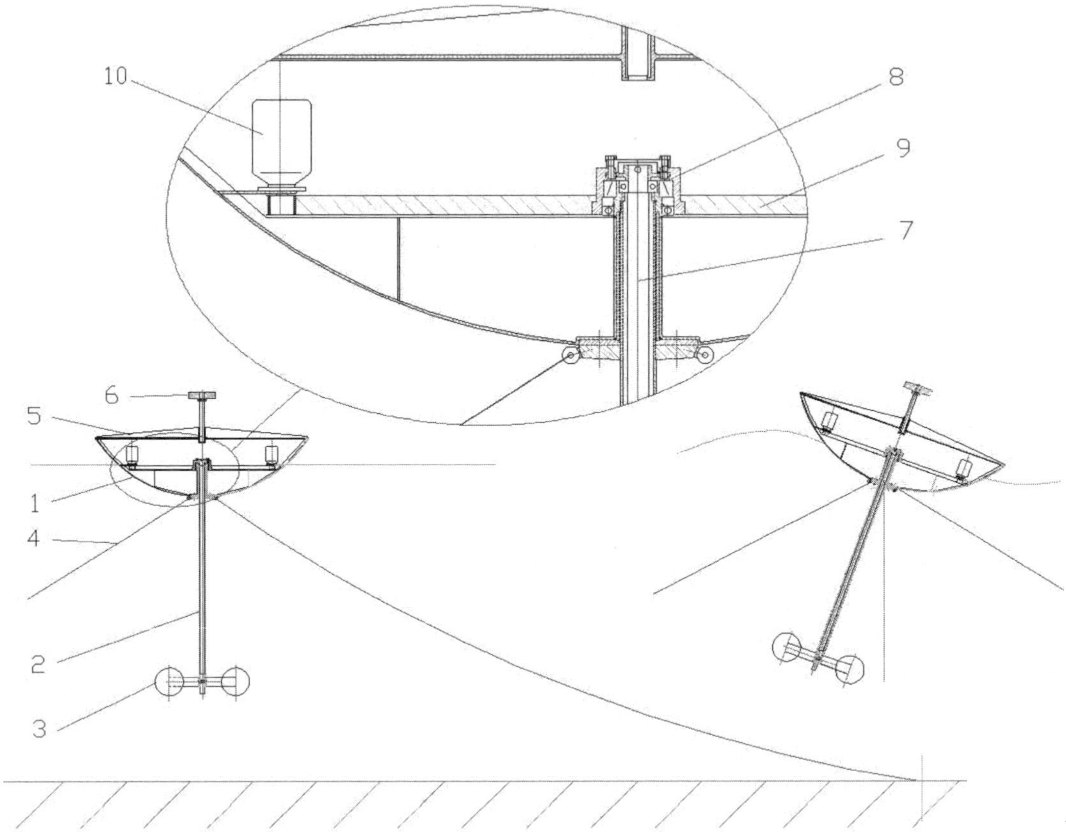

Fig. 2. Conceptual design of the buoy model for towing-tank tests: 1 – buoy body , 2 – column in form of tube, 3 – turbine, 4 – anchoring chain or line, 5 – buoy body cover, 6 – slidable weight for controlling buoy's centre of gravity, 7 - shaft, 8 – unidirectional clutch with bearing system, 9 – main gear , 10 – generator.

The buoy model's hull (1) is spherical in form thanks to which the forces of water resistance are significantly lower than for a hull shape with variable curve radius, and even more so for hull shape of abruptly variable form. The anchoring system is of a star type, not tighten, with chain strands (4) attached to the hull close to its axis of symmetry. Such a solution allows for the buoy to work regardless of wave direction with only minor dampening of the buoy's rolling motion. Thanks to this the displacement of the lower end of the column (2) with turbine (3) may be larger, which means a greater power effect. The upper part of the column (2) is rigidly fixed to the hull (1) on the axis of its symmetry. The shaft (7) is placed inside the column on roller bearings. Attached to the lower end of the shaft is the turbine (3). The blades of the turbine are semi-spherical in shape, thanks to which any lateral motion of the turbine will cause it to spin in only one direction. Placed below the lower shaft bearing are two sealing rings which protect the inside of the column from water. The upper end of the column is rigidly connected to a head placed in the hull (1) on which the main gear flywheel (9) is positioned on roller bearings The torque from the shaft (7) is transferred to the main gear (9) through the unidirectional clutch (8) which is necessary for protecting the turbine, when changes in wave direction occur, and the speed of the rotor in relation to water diminishes. The large main gear (9) drives the smaller gears placed on the shafts of two small permanent magnet generators (10) .The significant mass and moment of inertia of the main gear ensure relatively constant shaft speeds of the generators despite varying torque on the turbine shaft. Measurements of generator shaft speeds and parameters of the current allow for establishing the value of acquired power for specific simulated wave conditions acting on the tested model.

Rigidly attached to the cover of the buoy hull on its symmetry axis is a rod, on which a slidable weight (6) is placed. Through changing its position, the buoy's moment of inertia may be controlled which allows for investigating its impact on the parameters of model's rolling motion, and therefore on the effects of capturing wave energy.

Variant II

The second variant of the buoy was created at a later date following a more extensive theoretical analysis of the behavior of objects atop a wave, as well as after taking into account realistic technical and financial capabilities of the research. In comparison to the previous solution a number of significant changes were introduced. The axially symmetrical hull shape was abandoned in favor of form somewhat similar to "Salter's duck" with a markedly elongated front and cylindrical rear part, as can be seen in Fig. 3. It was decided that such a shape allows for better use of waves' kinetic and potential energy. This stems from the fact that the significant power of the hydrostatic lift of the water below acting on the elongated bow part of the hull should increase the angle of buoy's deflection. In addition, after the crest of the wave passes the bow part above the water surface will increase the torque of the returning motion due to the longer arm of its centre of gravity in relation to the axis of the roll.

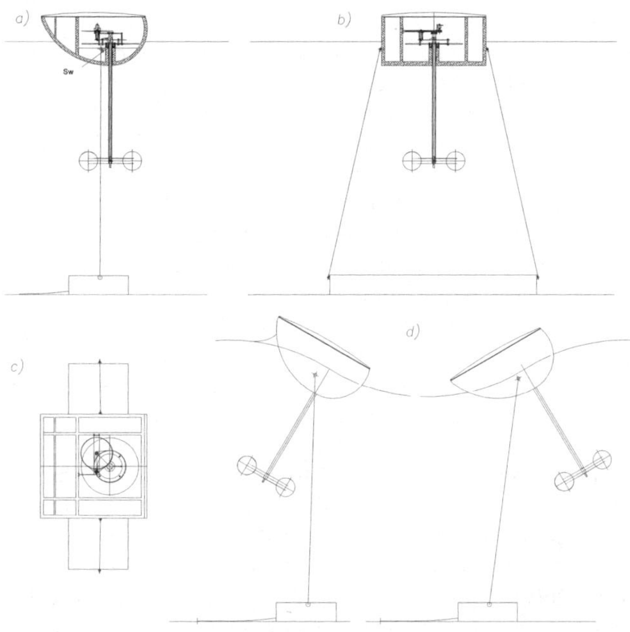

Fig. 3. Variant II of the power buoy: a) – cross-section of side view of the PS surface plane, b) – cross-section of front view, c) – view from above with cover removed, d) – outermost positions of the buoy atop wave

The new buoy hull also required changes to the anchoring system. Two anchoring strands were attached to the flat sidewalls of the hull. Proper selection of the attachment point of the anchoring strands was of high importance. In view of the authors' of this paper the attachment point should be located somewhat in front of and at the same time above the buoy's uplift centre. Such location is characterized by a very favorable course of the righting moment in the function of the buoy's angle of deflection. This moment is very small for low deflection angles due to the short arm of the force acting in the anchoring strand and does not interfere with hull's deflection. It increases to large values with high deflection angles, as a result of the uplift center and roll axis shifting away from the point of anchoring strand attachment, which to some degree protects the buoy from turning over and facilitates its returning motion. The anchoring system differs from conventional ones also by the method of its attachment to the bottom. In this solution the lower ends of anchoring strands are attached to a sinker resting on the bottom which in turn is connected by a separate strand to the bottom anchoring element. The sinker has the form of a box filled with heavy elements. The width of the sinker is substantially larger than the buoy width which protects the turbine from catching onto the anchoring strands during its rolling atop waves. The weight of the sinker in the water is appropriately smaller than total hull displacement, so that in the event of a very high wave the buoy would remain on the surface lifting only the sinker. The length of the anchoring strands was set as the approximate straight line distance between the points of their attachment to the buoy and the sinker. This means that in a certain foreseen range of wave height, the buoy not being able to float freely on the water surface will as the waves swell increase its draught. In doing so it will create increasingly large resistance, taking on growing amounts of wave energy which will manifest itself in increased amplitude of its roll which translates into greater efficiency.

The power transmission system was also modified. It is more sophisticated, allowing for controlling the torque and obtaining power directly from the turbine shaft. For clarity its construction is presented in Fig.4 and Fig. 5.

The upper end of the turbine shaft (2) is placed on a tapered roller bearing (4) which transfers transverse and longitudinal forces from its and the turbine weight to the head of the column (3) and the buoy hull (1). The shaft is topped with a journal connected to the torque meter (8). The torque meter body is protected against turning by a bracket (9) which is rigidly fixed to the column head. The torque meter is connected to the unidirectional clutch (10). The casing of the clutch contains two horizontal rods (11) whose ends are placed loosely in pivot recesses (12) rigidly connected to the flywheel (13). This way of connecting the turbine shaft with the flywheel ensures advantageous, flexible transfer of the torque in only one direction. The flywheel is rigidly attached to the hub (7) which is placed on tapered roller (5) and ball (6) bearings in relation to the head (3). The flywheel is rigidly connected by bolts (14) to the larger 1st stage chain pulley (15) of 1:10 gear ratio with chain. The smaller chain pulley (16) is placed on the lower journal of the shaft (17) resting on two ball bearings (18) in the sleeve (19) rigidly connected to the transmission casing (20) attached to the walls of the buoy hull. Attached to the upper shaft journal (17) is the larger pulley of 2nd stage gear of the same ratio and geometrical parameters. The smaller pulley (22) is placed on the shaft (23) resting on two ball bearings (24) in the casing (20). The shaft (23) is connected by clutch (25) with the generator shaft (26) attached with a bracket (27) to the transmission casing (20).

In the presented solution there appeared a challenging technical problem of running the measuring cable out of the torque meter located between revolving elements to the buoy hull. The problem was solved by making a pulley (15) in the form of a ring so that the free space inside it was sufficient to fit the cable running from the torque meter casing to a specially machined channel on the column head (3) through which the cable was taken to the free space below the roller bearing unit. The cable was run further along the internal walls of the hub to the hub cover from which together with the generator and other measuring devices cables were run the outside

It is planned that in the case of the buoy prototype anchored at sea, the cables running from the hull will be loosely connected to one of the anchoring strands and taken ashore through a shallow groove in the sea bottom covered with loose rubble.

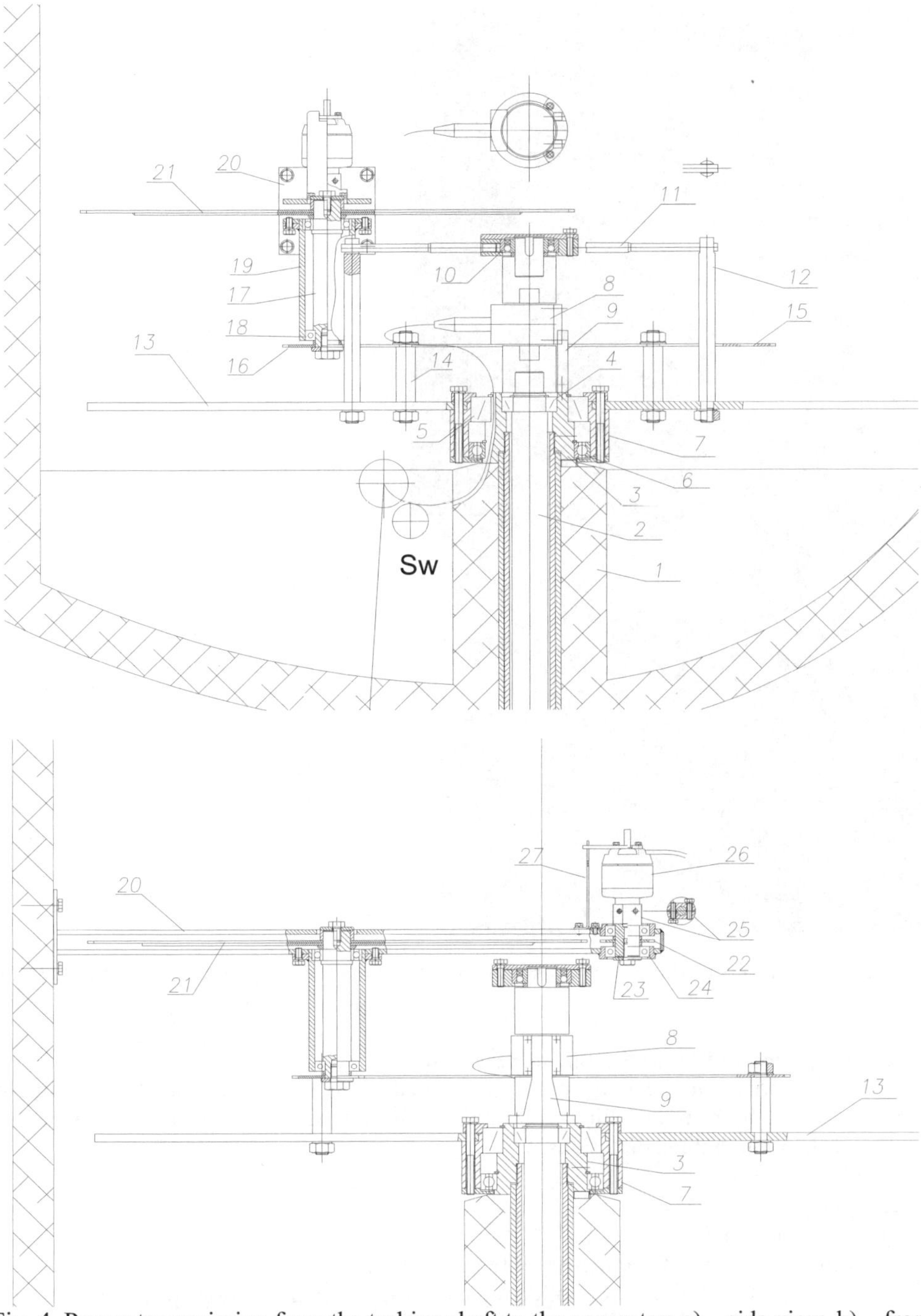

Fig. 4. Power transmission from the turbine shaft to the generator: a) – side view, b) – front view. Sw – centre of uplift on calm water

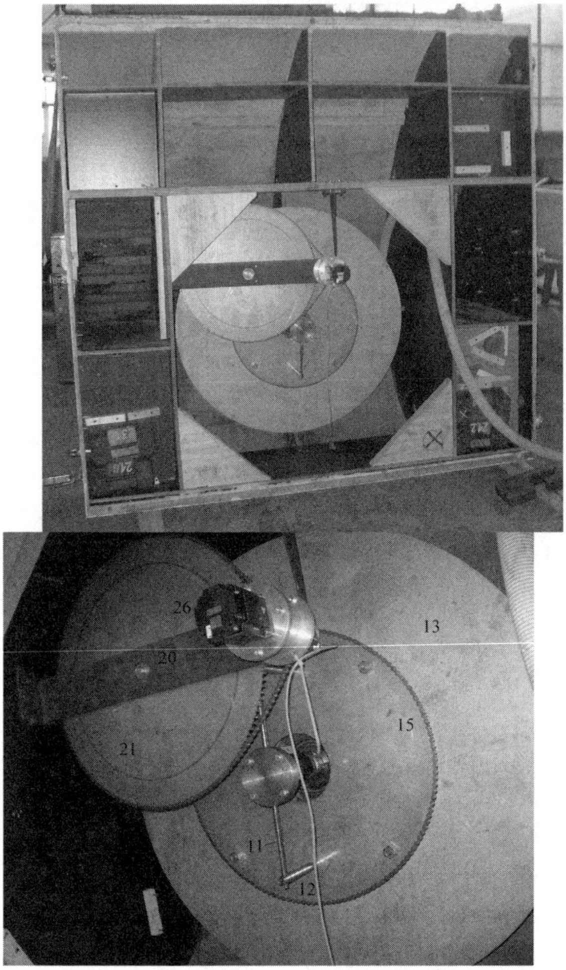

Fig. 5. Photographs of the power transmission mechanism during assembly inside the bouy

Measuring possibilities and scope of research

The second variant of the buoy model presented above was tested in the towing-tank. The apparatus installed on the model allowed for measuring and recording the following parameters:

- linear acceleration in three directions,
- buoy rolling angles,
- heaving height,
- torque on turbine shaft,
- rotation speed and current parameters of the generator.

The measuring system allows for collecting data with constant rotation speed and variable torque or with constant torque and variable rotation speed of the generator.

The prepared program encompassed the following research tasks:

- Establishing favorable shape of the buoy hull depending on the chosen variant of defined or undefined wave direction.
- Preparing efficient anchoring system for both of the above mentioned variants of wave motion.
- Establishing the impact of the location of buoy's centre of gravity and moment of inertia on its rolling characteristic and power generation efficiency.

- Investigating the impact of the moment of inertia of the rotating elements of the system transmitting power from the turbine to the generator, as well as the influence of the gyroscope effect on the effectiveness and parameter stability of the generated electrical power..
- Setting outermost wave parameters:
  - the minimum necessary for starting the process of power generation,
  - the maximum where danger of mechanical or electrical damage appears - if possible to obtain in the towing-tank.
- Setting the general efficiency characteristic of generating power as function of wave parameters.
- Preparing a mathematical model and calculating program for simulating operation of the power buoy.

  Verification of calculation results with data obtained during the model-based research phase.

First model test results

As it has already been mentioned the built model was tested in the large towing-tank of the Ship Design and Research Centre in Gdańsk. While this paper was being written two sets of measurements have already been carried out with the buoy model on a regular wave for various wave values and frequencies . These series are not quite complete as they are time consuming and were started with the smallest wave parameters. Some model tests results converted to fool scale are shown in Fig.6.

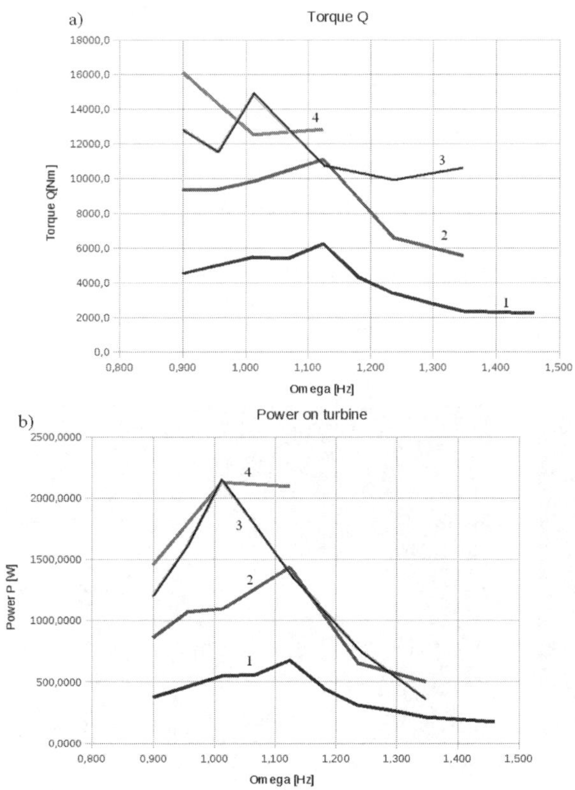

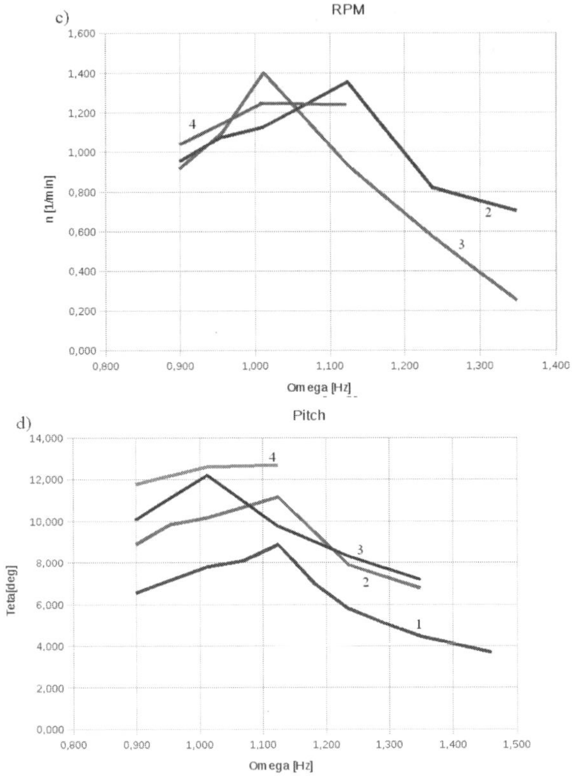

Fig. 6. Some model tests results converted to fool scale. Nomenclature: Height of wave: 1 – 1,0 m, 2 – 1,5 m, 3 – 2,0 m, 4 – 2,5 m.; Omega – angular frequency of wave

Certain observations have already been made and based on them it can be ascertained that qualitatively its behavior atop wave and operation of the power transmission system have satisfied expectations. Two photographs of the model buoy are presented in Fig. 7 below - before launching in the towing-tank and during tests in regular wave conditions.

Fig. 7. Photographs of the power buoy model - on the left during preparation for launching, on the right during tests in the towing tank.

Final remarks

It should be pointed out that as early as the design phase a number of calculation based analyses and simplified model tests had been carried out, especially on the hull, the turbine and the anchoring system. They were carried out with the aim of reducing to the necessary minimum the number of variants tested in the large towing tank. In case of the hull, wood was chosen as construction material so that after testing the first variant of non-symmetrical shape it would be relatively simple to create the shape of the second variant by attaching additional sponsons. Similarly, in turbine's case, prior to making the final decision on the number and placement radius of the blades, relatively simple and inexpensive tests were performed on a larger scale model turbine for constant water flow. The obtained results were taken into account in the construction of the rotor prepared for the buoy model.

The values of the power recived during the first tests are not high, but in authots opinion increasing efficiency of the the devices is possible. It can be obtained by proper choise of the technical and exploitational parameters.

It is also our hope that following the completion of model based research, regardless of the operational effectiveness of the presented power buoy, the obtained results will constitute interesting research material and will be published.

References

[1] H. Bernhoff, E. Sjostedt, M. Leijon: Wave energy resources in sheltered sea areas: A case study of the Baltic Sea. The Fifth European Wave Energy Conference, 17-20 September 2003, In Cork, Ireland.

[2] T. Setoguchi, S. Santhakumar, M.Takao, T. Kim, K. Kaneko: A modified Wells turbine for wave energy conversion. Renewable Energy. Vol.28, Issue 1, January 2003.

[3] J. Cieśliński, J. Mikielewicz: Unconventional sources of energy. Niekonwencjonalne źródła energii. Gdansk University of Technology Press. Gdansk 1996.

[4] A. Leao Rodrigues, A. Steiger Garcao: Pelamis Converter for Wave Power Extraction. Congresso Luso-Mocambicano de Engenharia. Maputo, 2-4 Setembro 2008.

[5] S. Helkin, C. Velez: Novel Design of an Ocean Wave Power Device Utilizing a Bi-directional Turbine. University of Central Florida Department Mechanical, Materials, and Aerospace Engineering. 4000 Central Florida Blvd., Orlando, FL 32816 USA

[6] McCabe, A.P.; Aggidis, G.A.[a]; Widden, M.B: Optimizing the shape of a surge-and-pitch wave energy collector using a genetic algorithm. Engineering Department, Renewable Energy Research Group, Lancaster University, Lancaster LA1 4YR, UK

[7] Falnes J.: Ocean waves and Oscillating System. Cambridge University Press,2002

[8] Price A.: New Perspectives on Wave Energy Converter Control. A thesis submitted for the degree of Doctor of Philosophy. The University of Edinburgh. March 2009.

[9] S. Klukowski: Dispositif pour collecter l`energie de la houle (Converter for Sea Wave Power Extraction). Patent application in France - Paris No 08/00951. 22/02/2008 (INPI - Institut National de la Propriete Industtriell)

[10] http://mg.biz.pl/?title=Renewable_energy

[11] http://mg.biz.pl/?title=Wave_farm

[12] (http://www.pelamiswave.com/).

[13] http://www.esru.strath.ac.uk/EandE/Web_sites/01-02/RE_info/wave%20power.htm

[14] http://www.wavegen.co.uk/what_we_offer_limpet_islay.htm

[15] http://www.malibuwater.com/OceanWaveEnergy.html

Key Engineering Materials Vol. 490 (2012) pp 216-225
© (2012) Trans Tech Publications, Switzerland
doi:10.4028/www.scientific.net/KEM.490.216

Numerical analysis of crack propagation in silicon nitride

Waldemar Karaszewski

Technical University of Gdansk, ul. Narutowicza 11/12, 80-233 Gdansk, Poland

walkaras@pg.gda.pl

Keywords: crack propagation, silicon nitride, rolling contact, boundary element method

Abstract. The properties of ceramics, specifically low density, high hardness, high temperature capability and low coefficient of thermal expansion are of most interest to rolling element manufacturers. The influence of ring crack size on rolling contact fatigue failure has been studied using numerical fracture analysis. Such cracks are very often found on ceramic bearing balls and decrease fatigue life rapidly. The numerical calculation are based on a three dimensional model for the ring crack propagation. The stress intensity factors along crack front are analyzed using a three-dimensional boundary element model. The numerical analysis is verified by experimental studies.

Introduction

Catastrophic fracture failure of engineering structures is caused by cracks that extend beyond a safe size. Cracks, either as a result of manufacturing fabrication defects or damaged in service, may grow by mechanisms as fatigue, stress corrosion or creep. The crack growth leads to a decrease in the structural strength. Fracture, the catastrophic damage that takes place very rapidly, is preceded by slow crack growth during normal service conditions, mainly by fatigue due to cyclic loading.

Silicon nitride is one of the most important materials for rolling element bearings applications [1]. However, the difficulties of sintering and machining the material may result in surface defects. Surface defects as ring cracks are difficult to detect during high volume production process. Another surface cracks caused by blunt impact load have also ring shape and are the most common type found on ceramic ball surface. These defects decrease the rolling contact fatigue performance considerably [2-5] so it is important to understand the failure mechanisms caused by such cracks.

Boundary element method

Linear elastic fracture mechanics can be used in damage analyses to describe the behavior of cracks. The fundamentals postulate of linear elastic fracture mechanics is that the crack behavior is determined by the stress intensity factors which are the function of the applied load and the geometry of the cracked structure. So stress intensity factors play a fundamental role in linear fracture mechanics applications.

Crack growth processes are simulated with an incremental crack extension analysis. In general numerical methods must be used for the evaluation of the stress intensity factors in structures. The finite element methods has many application in fracture mechanics [6-8] but a disadvantage of the method is connected with the need for continuous remeshing to follow crack extension.

The boundary element method (BEM) is a numerical computational method of solving linear partial differential equations which have been formulated as integral equations. It can be applied in many areas of engineering and science including fluid mechanics, acoustics, electromagnetic. The dual boundary element method is also used in fracture mechanics in incremental analysis of crack extension problems [9]. The method is utilized in the present study.

The dual equations, on which the dual boundary element method is based, are the displacement and the traction boundary integral equations [9]. The crack is represented by two elements with the same physical location, each element representing a face of the crack. The displacement and traction boundary equations are on the crack surface.

In the paper the Beasy software is used to the numerical analysis of ring crack propagation in silicon nitride balls [9].

Stress intensity factors and criterion of crack growth

In a fatigue analysis the distance that each mesh point will grow for a given crack growth is determined from the fatigue crack growth law. In a BEASY software three different values for the stress intensity factors, one for each of the crack modes (opening, shearing and tearing) are computed. For the mixed-mode problem [10], the stress intensity factor Kmax is replaced by the effective stress intensity factor Keff which is defined as follows:

$$K_{eff} = \sqrt[1/2]{(K_I + |K_{III}|)^2 + 2K_{II}^2}$$

The criterion of crack propagation is based on the Paris model, which defines the rate of change of crack length with respect to the loading cycles and is expressed as follows:

$$\frac{da}{dn} = c\left(\Delta K_{eff}\right)^m$$

where a is the crack length, n is the number of load cycles. In the Paris equation c and m are the materials constants and for silicon nitride c=12-18 and m=1.023E-21 [11]. A constant c=12 was used in the numerical analysis presented in the paper.

The equation 1 gives a relationship between the individual stress intensity factors and effective stress intensity factor. However in the analysis it is necessary to compute ΔKeff which is defined:

$$\Delta K_{eff} = K_{eff}^{max} - K_{eff}^{min}$$

In the rolling contact the original crack propagation is connected with the stress intensity factors arising from the Hertzian stress field. The analysis was carried out with the maximum Hertzian contact pressure of 6.21 GPa. The crack front will propagate if the stress intensity factors on the crack front are higher then the threshold value. The threshold value of crack propagation is from 2 to 4 $MPa*m^{1/2}$ for hot isostatically pressed silicon nitride material and the fracture toughness value is 6 $MPa*m^{1/2}$ [10].

In order to predict crack growth the direction of the crack increment extension along the crack front needs to be defined. For each increment of the analysis, in which the crack extension is modeled with new boundary elements, the dual boundary element method is applied for the stress analysis. The incremental analysis is based on a prediction-correction technique to define the direction of the crack-extension increment. There have been several criteria to describe the direction of mixed-mode crack growth: minimum strain energy density, maximum principle stress, maximum energy release. The first criterion is adopted in the numerical analysis.

Calculation results

The original crack model is shown in Fig.1. The original crack was modeled based on the experimental data of pre-crack made by impact load (see description in Experimental methods chapter later in the paper). Fig. 2 shows views of pre-crack made by impact load, fig. 3 shows dimensions of pre-crack and fig. 4 shows meshing of pre-crack inside ceramic ball.

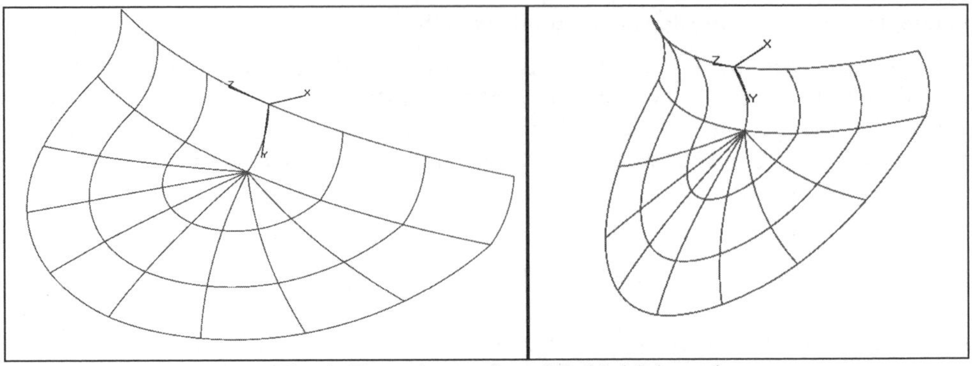

Fig. 1. Two views of modeled initial crack

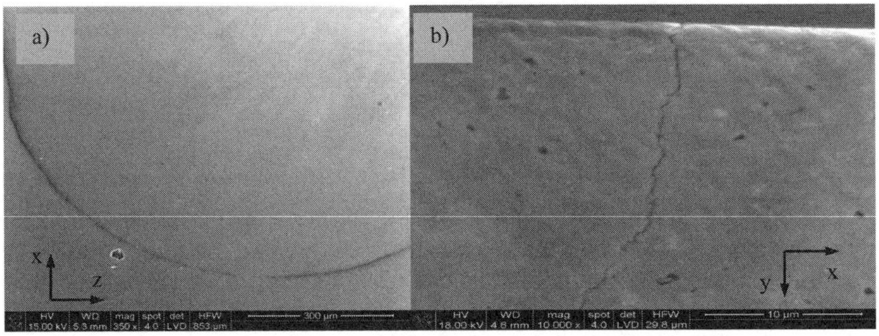

Fig. 2. Surface (a) and section (b) views of pre-crack

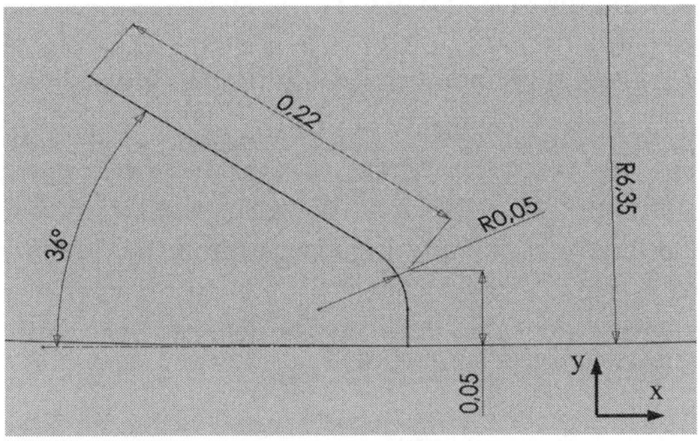

Fig. 3. Dimensions of pre-crack

Fig. 5 depicts the variations in stress intensity factors of initial crack. Crack face position from 0 to 1 is defined as presented in Fig. 4b. The stress intensity factors are the function of the position on the crack front. The mode I stress intensity K1 is the largest for crack face position 0.18 and 0.83. These positions are shown in Fig. 4a and are on the radius of crack face.

The comparison of effective stress intensity factors after different stages of crack extension is shown in Fig. 6. The highest value of effective stress intensity factor is for initial crack, and the Keff decreases with crack front propagation. This is agreed with experimental observation where cracks propagate only to a certain extension size.

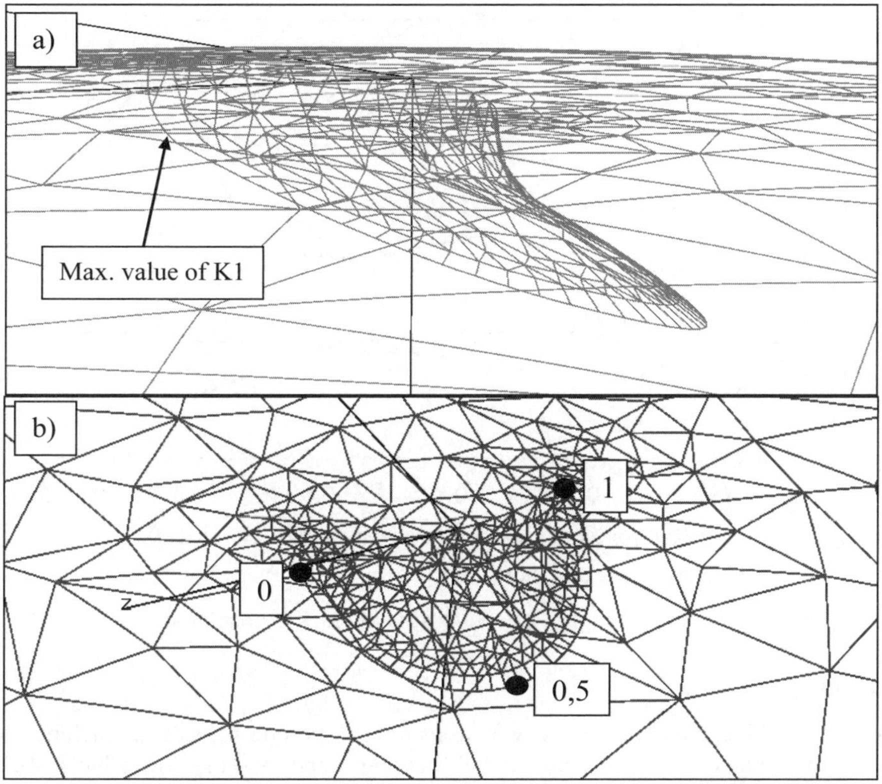

Fig. 4. Meshing of pre-crack inside a silicon nitride ball

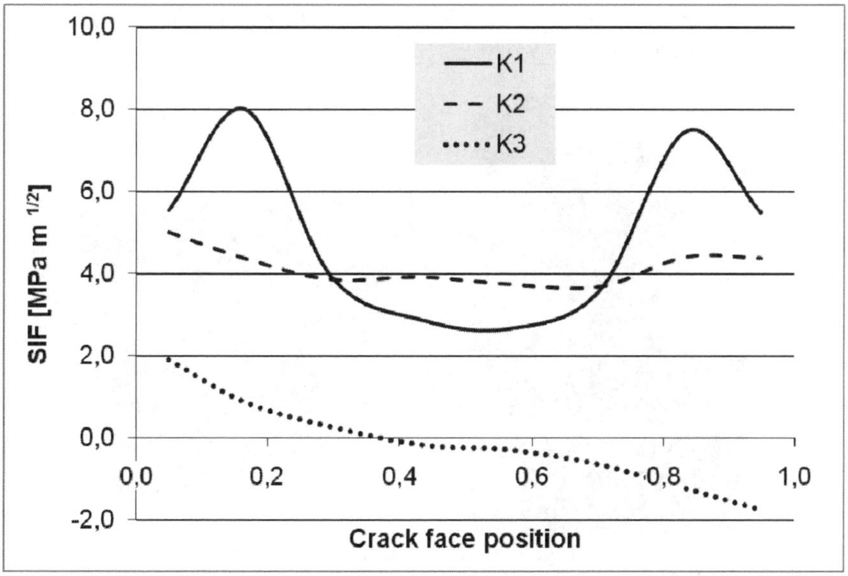

Fig. 5. Plot of SIFs in initial crack propagation

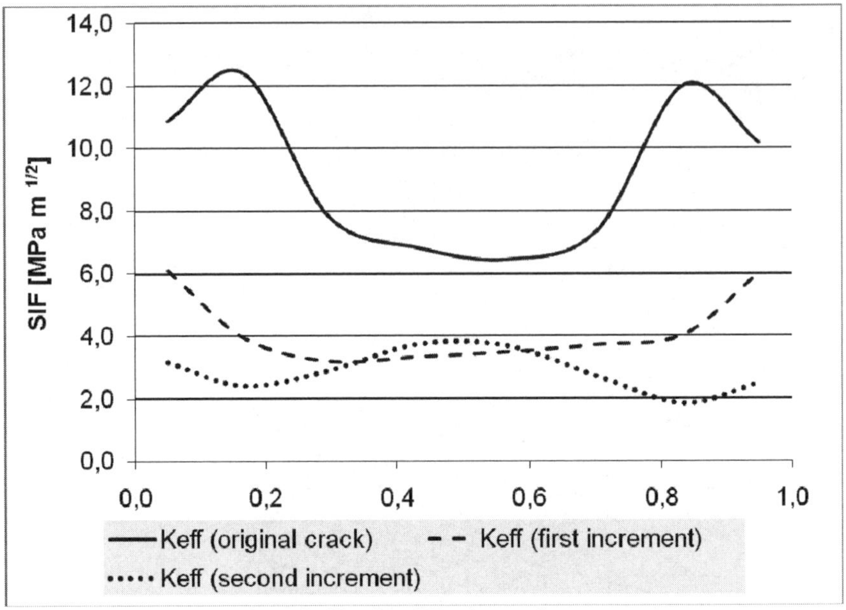

Fig. 6. Plot of effective SIFs in crack propagation

Experimental methods

A four-ball machine, shown in Fig.3, was used to test ceramic balls. The machine consists of elements which simulate an angular contact ball-bearing: upper ball (1), lower balls (2), steel cup (3). The assembly is loaded via a piston (5) below the steel cup. The upper silicon nitride ball is assembled to a drive shaft (4) and contacts three silicon nitride lower balls. Upper and lower balls are immersed in lubricating oil. All the tests were conducted at a maximum Hertzian contact pressure of 6.21 GPa. The upper ball speed was set at 1500 rpm. The upper ball is the silicon nitride test specimen with a pre-existing ring crack. Four-ball machines have been used to examine rolling contact by various researchers [2, 3, 15, 16].

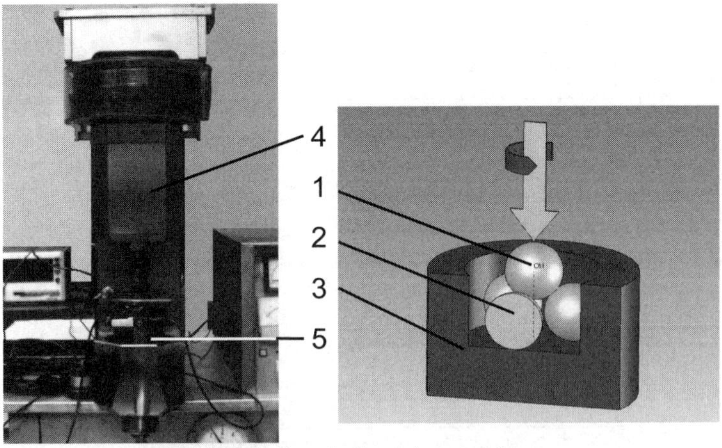

Fig. 7. A four-ball testing machine
(1-upper ball, 2-lower balls, 3-steel cup, 4-spindle/drive shaft, 5-load piston)

The silicon nitride balls were manufactured as 12.7mm diameter balls. Table 1 describes properties of silicon nitride material used in the research. Silicon nitride balls were manufactured by Comballs Materials Co. The roughness (R_z) of silicon nitride balls is 0.08 μm, the ball diameter tolerance is ± 2.5 μm and the ball roundness tolerance is 0.6 μm. This data is based on the manufacturer's information.

Table 1. Silicon nitride properties

Chemistry	99.9 % Si_3N_4
Compressive strength [MPa]	2353
Thermal conductivity [W/m*K]	29.3
Coefficient of expansion [1/K]	1.78 x 10E-6
Young modulus [GPa]	324
Hardness [HV]	1800
Density [g/cm^3]	3.2

The ring cracks were created on the upper silicon nitride balls surface by blunt impact loads. This method was chosen to create similar cracks which can be created during falls of the balls or exploitation as an effect of impact load. A silicon nitride ball was placed in a ball holder and tungsten carbide ball (12.7 mm diameter) was placed on top of the silicon nitride ball (Fig 4). The position of the silicon nitride ball holder was defined to make the ring crack on a silicon nitride ball in place of a 4-ball contact [9].

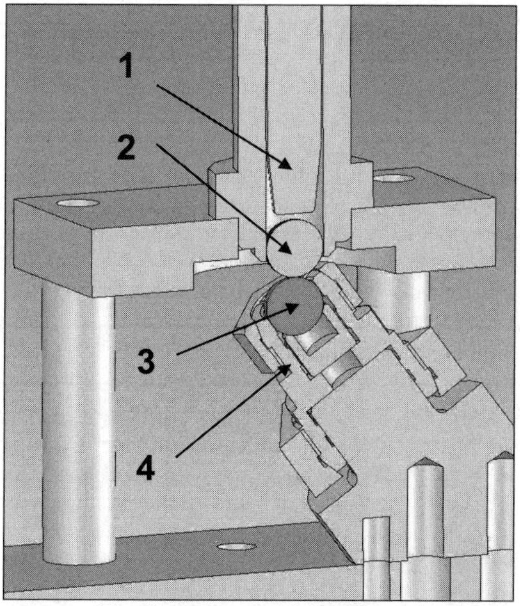

Fig. 9. A ring crack device maker
1 – dropped weight, 2 – tungsten carbide ball, 3 – silicon nitride ceramic ball, 4 –upper ball holder placed onto drive tribometer shaft

The tolerances of tungsten carbide balls are the same as those of the silicon nitride balls. Table 2 describes the properties of tungsten carbide material. Typical ring cracks were produced by a 400 g impact load, dropped from 500 mm.

Table 2. Tungsten carbide properties

Chemistry	93,5 - 94,5 % WC 5,5 – 6,5 % Co
Compressive strength	5200 [MPa]
Tensile strength	1520 [MPa]
Maximum useful temperature	430 [^{O}C]
Young modulus	640 [GPa]
Hardness	2200 [HV]
Density	13,2 [g/cm^3]

Table 3. Base lubricant properties

	Base lubricant
Trade name	T-9
Producer	Nynas
Properties:	
Kinematic viscosity (40°C) [mm^2/s]	9,13
Sulfur content [mass %]	0,01
Aromatic hydrocarbons content [mass %]	9,0

Experimental results and discussion

Figure 10 shows experimental results for pre-cracked rings made by defined impact energy. The tests were repeated five times. The time for failure was recorded and the tests were stopped at a high vibration level. A specific sound at high vibration level appeared at the moment of surface failure. Variations in length tests is connected with material properties and differences in pre-cracks length made by impact load.

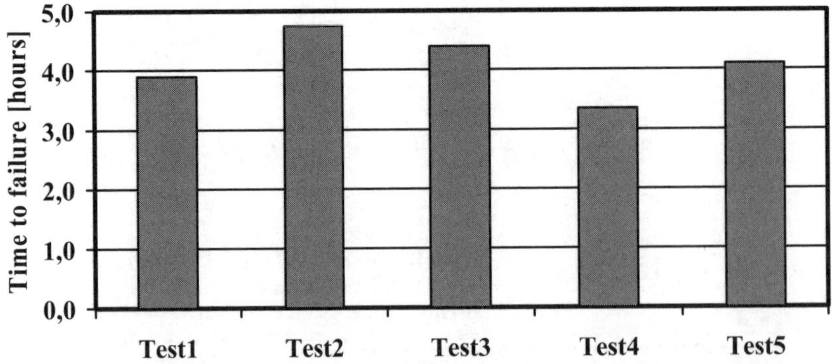

Fig. 10. Plot of time to failure versus test number

To describe the mechanism of a ring crack failure, experimental observations of the fatigue process are important. A typical crack propagation process is shown in Fig. 11. New surface cracks are found beside the original ring cracks. These new crack can be called as secondary surface cracks [10]. These new cracks are found after about 3 hours of testing. The ball is sectioned to examine of crack propagation. The subsurface image of silicon nitride ball is shown in Fig.12. Fig. 12a shows ring cracks on the ball surface which propagate away from the ball surface. These cracks are named secondary cracks.

The secondary surface cracks play an important role in the process of surface fatigue failure. The secondary cracks propagate away from the surface and stop at some depth. This phenomena was approved by the numerical analysis where effective stress intensity factor decreases with crack front propagation. In the same time subsurface cracks appear between ring cracks which is shown in Fig. 12b. Generated subsurface cracks between ring cracks are connected with large material removal between ring cracks. Fig. 13a presents initial crack which was open as an effect of material removal. Next secondary ring cracks are found on the ball surface (fig. 13b and 13c) near open initial crack. The time between formation of secondary cracks and large material removal is short an takes about 5-20 minutes during tests. The formation of subsurface cracks were also reported by other authors [2, 13, 14].

A detailed analysis of crack propagation was also presented in other papers [3, 15, 16].

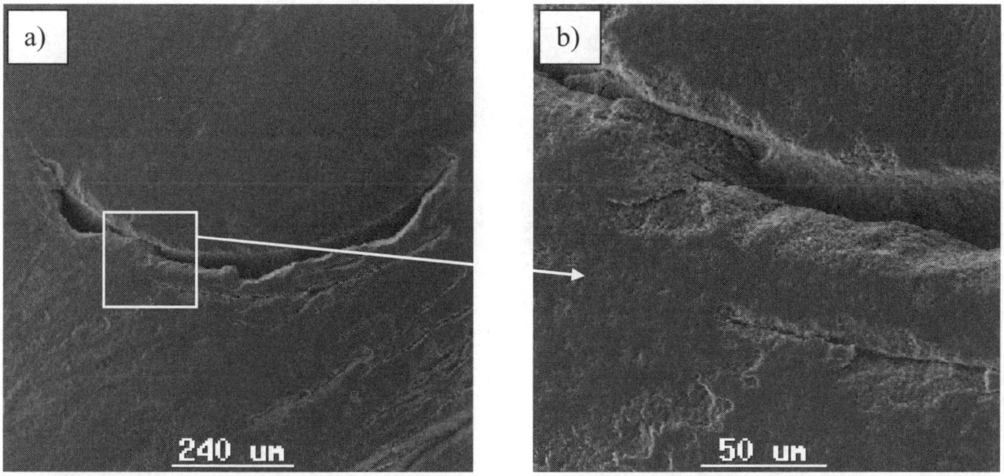

Fig. 11. Surface observations of silicon nitride ball with pre-crack after 3 hours of testing

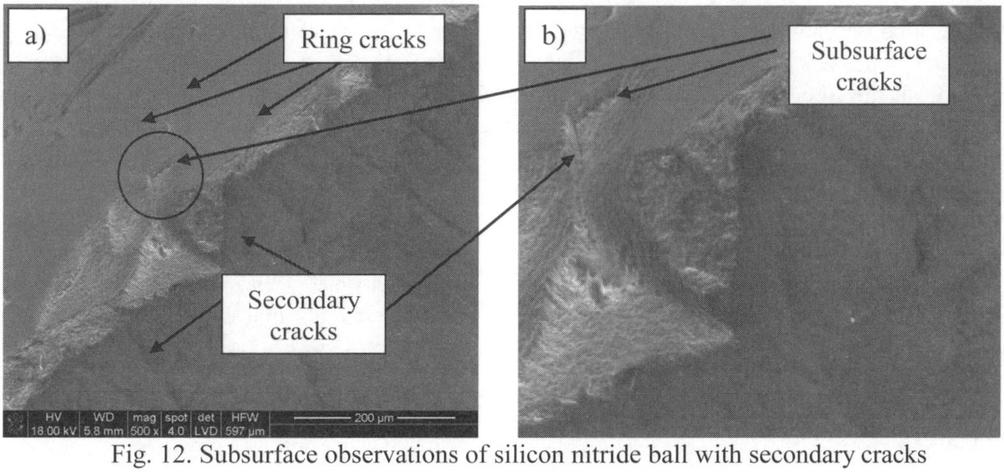

Fig. 12. Subsurface observations of silicon nitride ball with secondary cracks

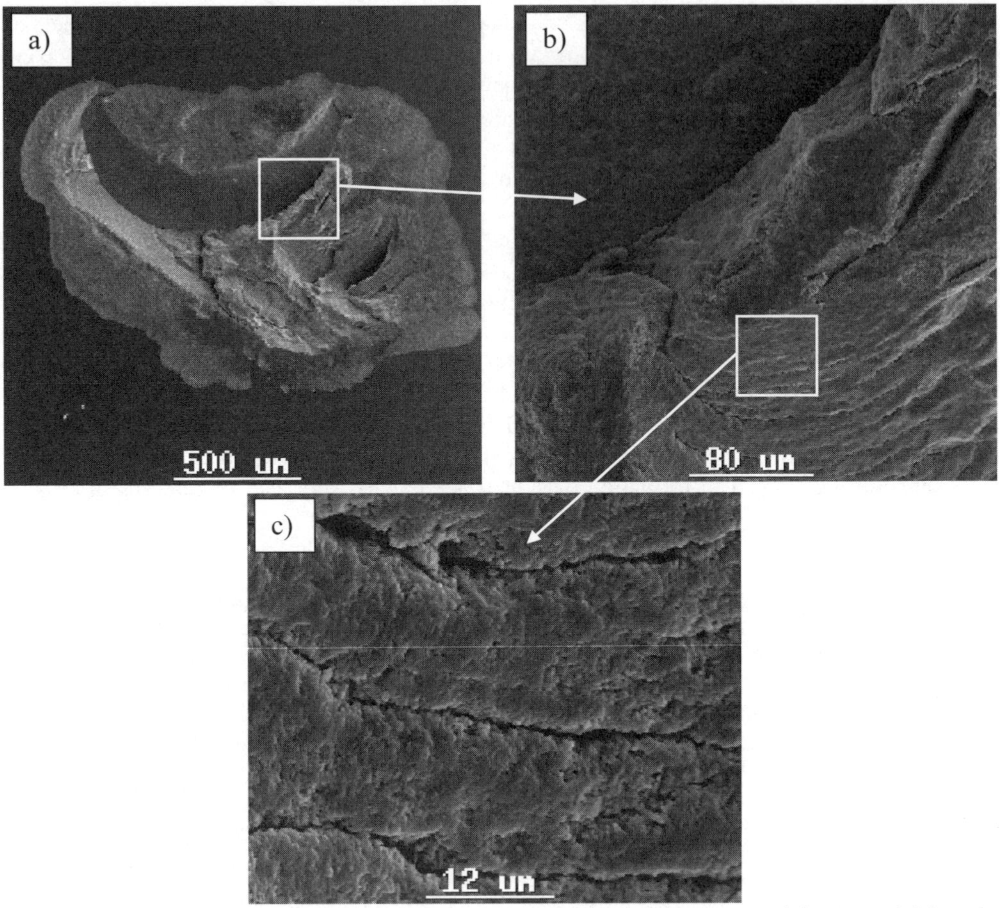

Fig. 13. Surface observations of silicon nitride ball with a large material removal (a) and secondary cracks (b, c)

Conclusions

The boundary element method is used for a crack propagation analysis of a ring crack on a silicon nitride ceramic ball. The SIF numerical analysis is based on a three-dimensional model of ring crack growth. The crack geometry and dimensions used in the model are based on the experimental observations of the surface ring cracks. The following conclusions may be drawn:

- The process of fatigue failure of silicon nitride with surface ring cracks is influenced not only by the pre-test crack propagation. Generated subsurface cracks play an important role in rolling contact fatigue performance and failure modes.
- A large number of material removal are connected with subsurface cracks which are generated between initial and secondary crack and later between the next two secondary cracks.
- The secondary cracks propagate conically away from the surface and stop at some depth. This phenomena was approved by the numerical analysis.
- The numerical analysis of ring crack propagation reveals that effective stress intensity factors along the crack front decrease with next increments of crack propagation.

This study was funded by the Polish Ministry of Science and Higher Education under contract no. 4TO7C03528.

References

[1] R.T. Cundill, Rolling element bearings into 21st century, IMechE, Seminar, London, November, 1990.

[2] Y. Wang, M. Hadfield, The influence of ring crack location on the rolling contact fatigue failure of lubricated silicon nitride: experimental studies, Wear 243 (2000), 157-166.

[3] W. Karaszewski, The influence of oil additives on spread cracks in silicon nitride, Tribology International, vol.41, issue 9-10 (2008) 889-895.

[4] Y. Wang, M. Hadfield, Failure modes of ceramic rolling elements with surface crack defects. Wear 256 (2004) 208-219.

[5] W. Karaszewski, Surface cracks of ceramic rolling elements, Acta Mechanica et Automatica, vol. 3, nr 1(7), 2009, 68-70.

[6] M.H. Aliabadi, D. P. Rooke, Numerical Fracture mechanics, Computational Mechanics Publications, Southampton and Kluwer Academic Publisher, Dordrecht, 1991.

[7] T.A. Cruse, Boundary Element Analysis in Computational Fracture mechanics, Kluwer Academic Publishers, 1989.

[8] A. Portela, M.H. Aliabadi, D.P. Rooke, The dual boundary element method: effective implementation for crack problems, International Journal for Numerical Methods in Engineering, 33 (1991) 1269-1287.

[9] Information on http://www.beasy.com

[10] Y. Wang, M. Hadfield, The influence influence of ring crack location on the rolling contact fatigue failure of lubricated silicon nitride: fracture mechanics analysis, Wear 243 (2000), 167-174.

[11] I. Bar-On, J.T. Beals, Fatigue 90: proceedings of 4th International Conference on Fatigue and fatigue Thresholds, vol. 2, Edgbaston, UK, 1990, 793-798.

[12] C. Cirel, T.A. Stolarski, Factors affecting the wear of ceramic balls, Wear 172 (1994) 191-196.

[13] M. Hadfield, T.A. Stolarski, R.T. Cundill, S. Horton, Failure modes of ceramic elements with ring crack defects, Tribology International 26 (1993) 157-164.

[14] Y. Wang, M. Hadfield, Failure modes of ceramic rolling elements with surface crack defects, Wear, 256 (2004) 208-219

[15] W. Karaszewski, Influence of ring cracks on fatigue in silicon nitride, Tribologia, R. 35 vol. 3 (2004) 159-168

[16] W. Karaszewski, Limitation of fatigue life of ceramic couples due to surface ring cracks, Tribologia, R. 40 vol. 2 (2009), 29-38.

Key Engineering Materials Vol. 490 (2012) pp 226-236

doi:10.4028/www.scientific.net/KEM.490.226

Numerical and experimental analysis of stress and strains in flat ends of high-pressure vessels

Bogdan Szybiński[1, a], Paweł Romanowicz[2,b] Andrzej P. Zieliński[3,c]

[1,2,3]Institute of Machine Design, Cracow University of Technology, Warszawska 24, 31-155 Kraków, Poland

[a]boszyb@mech.pk.edu.pl (corresponding author), [b]promek@mech.pk.edu.pl, [c]apz@mech.pk.edu.pl

Keywords: Pressure vessels, flat ends, stress relief groove, finite element method, experimental analysis.

Abstract. Application of flat welded ends with stress relief grooves in high-pressure vessels is a common alternative to use of dished vessel ends. It is well established and follows calculation rules given in codes: EN-12952-3 [1], EN-13445-3 [2], or in ASME code [3]. However the calculation rules do not give any definite answer what should be the choice of parameters defining a circular stress relief groove, for example, position of the groove and its radius. Usually the choice of them relies on engineering practice. The present paper clearly shows the influence of this choice on stress concentration in the cylinder-endplate junction area. The results of numerical study are verified in experimental investigations performed for a cylindrical high-pressure vessel.

Introduction

Flat ends welded to cylindrical tubes or vessels are well known alternative to commonly used ellipsoidal ends of vessels. They are mainly used as endplates in high-pressure containers or boilers and for blind tubes in piping systems. Several, typical design solutions are presented in codes [1, 2, 3]. In all admitted solutions the abrupt change of the structure shape in this junction is encountered. It means, that the joint between the vessel and the endplate is a source of relatively high stress concentration. This considerably reduces the maximum value of the applied pressure and inevitably leads to presence of plastic deformations in the junction neighborhood. Application of the stress relief groove helps to minimize that local effort. It is suggested by the codes in a circular form. Thus, at least two parameters are needed to describe the groove shape and position: the radius of the groove and location of its center. Several papers [4, 5, 6] considered various aspects of calculations of flat ends with circular stress relief grooves. The performed numerical calculations mainly apply the finite element modeling, which provides easy manipulations with geometrical parameters and introduction of complex material behavior including plasticity or creep phenomenon. However, they do not optimize the groove proportions.

The calculations performed for the flat end parameters give a set of certain admissible pairs for radius r_{ik} and the location of circular arc center (the more convenient for manipulation is the minimum thickness of endplate e_{h1} used alternatively as a second parameter, see Figure 1). These pairs usually form a triangular or quadrilateral area when plotted in (r_{ik}, e_{h1}) coordinate system. There is no clear evidence how to choose the best combination of the stress relief groove parameters, which provide the minimum stress concentration. The finite element modeling aided with optimization technique can help to set their best combination.

The paper belongs to a series of publications concerning the high-pressure vessels. The authors and their team investigated shakedown phenomenon [7], transient mechanical and thermal effects [8] optimization in low-cycle loading conditions [9] and high-pressure sealing [10].

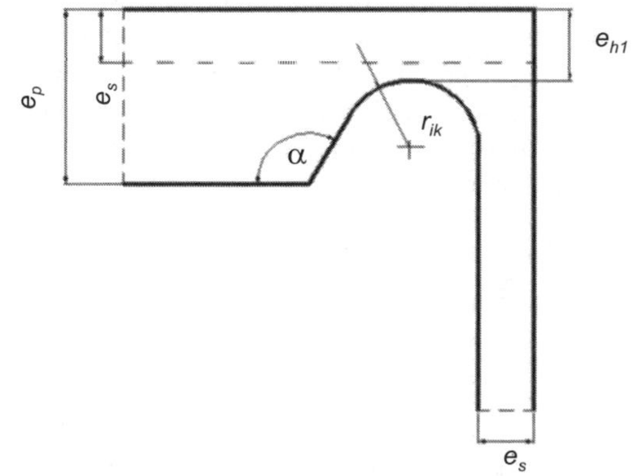

Figure 1 Circular stress relief groove in flat end plate and its parameters

Numerical analysis and optimization of stress relief groove parameters

The cylindrical tube with nominal size ϕ406.4×20 was chosen for calculation and experimental purposes. The length of the tube was set to 1200 *mm*, which is long enough to eliminate interference of the boundary effects. The opposite ends of the cylinder were made as flat endplates with two different sets of stress relief groove parameters. The tube and both endplates were made of the low alloy constructional structural steel (16Mo3), which is typical in pressure vessel applications. In the tube certificate, the tension properties were attached by the manufacturer: the yield stress was stated as $R_e = 308$ *MPa* and the ultimate stress $R_m = 514$ *MPa*. In the case of endplates, special samples for tension tests were prepared (14 samples). From the test results, the values for the yield stress R_e and the ultimate stress R_m were assumed. The tests performed for the endplate material help to establish the maximum (material 3W) and minimum (material 9P) strength properties, which were used in the numerical calculations. On the base of real stress-strain curve the material data approximation was set. For the both mentioned above samples - 9P and 3W, multilinear approximation was used, and kinematic hardening in the plastic zone was assumed. The detailed stress-strain curves are shown in Figure 2.

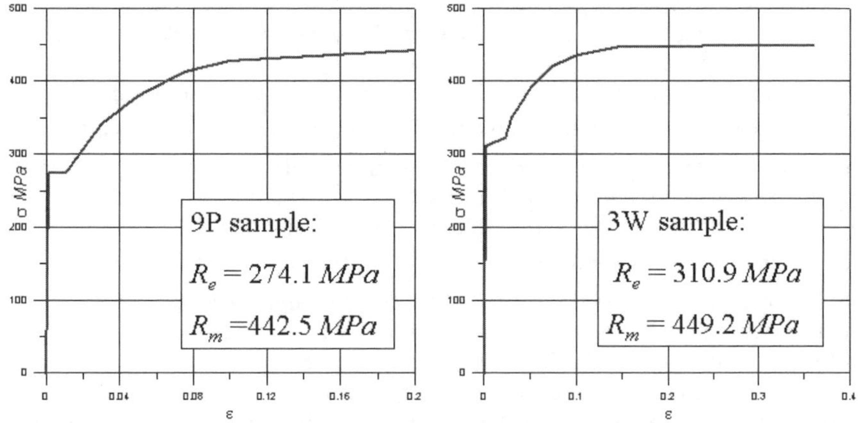

Figure 2 Stress - strain curves for endplate material (samples 9P and 3W)

In order to set the value of internal pressure p_{int} it was decided to use code EN 12952-3 dedicated for calculations of unfired pressure boilers (there are certain differences in calculations of vessels or boilers in codes [1] and [2], which influence a little the value of the nominal applied internal pressure). On the base of detailed measurement of the real tube thickness and real diameter, the internal pressure was set to p_{int} = 22 *MPa*. In the next step, the endplate thickness was designated as e_p = 61 *mm*. Then, the range of admissible values for stress relief groove parameters was set. The use of conditions given in code [2] resulted in the triangular area ABC (see Figure 3). In the above calculations no deviation of dimensions (suggested in the standards) were taken into account. It was due to the full knowledge of all the real model dimensions and no need for corrosion protection (only temporary use of the test vessel was assumed).

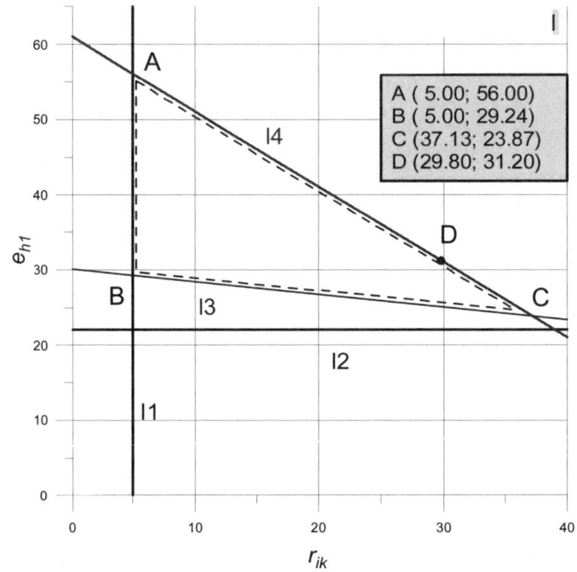

Figure 3 ABC area for admissible values of stress relief groove parameters

Next, the finite element analysis was prepared [13]. In order to maximally reduce the calculation time the axisymmetrical model was used, which enable to proceed with 1/2 of the symmetrical axial cross-section of the vessel. In the vicinity of the tube endplate junction and also along the circumference of the groove relatively dense mesh of finite elements was applied in order to keep the approximation error in energy below 2% (introduced for linear elastic solution), which was acceptable from the engineering point of view. In the final calculations the elastic-plastic material behavior was assumed (see Figure 2).

Due to existence of the stress concentration effect, the zones of plastic deformations inside the vessel were expected. The maximum equivalent plastic strain $\varepsilon_{eqv\_pl}^{\max}$ observed both in the endplate and in tube was chosen as a state parameter. Two independent control parameters influencing the solution were considered: the groove radius and the minimum thickness of the endplate. Therefore, a simple search method over the admissible area was used to find the optimal values of control parameters providing the minimum value of $\varepsilon_{eqv\_pl}^{\max}$ in both parts of the boiler. As it was expected, the values of this parameter varied significantly in the investigated area.

Figure 4 presents the shapes of studied stress relief grooves and differences in control values obtained for the vertices of admissible ABC triangle (presented results refer to the sample material 9P). The solution obtained for point C can be regarded as a several times better than solutions obtained for points A and B. Moreover, the optimal point D, which gave the minimum value of the state parameter is located nearby the corner C. This observation was confirmed by numerical results obtained for several vessel models with different dimensions and for different material combinations (endplate/cylindrical part), however, a formal prove of this phenomenon was not yet established. The results in Figure 4 and, further, Figure 5 were obtained for the nominal operating pressure (p_{int} = 22.0 *MPa*), which corresponded to the Standard calculations [1, 2] for the given material of the experimental vessel and its geometry.

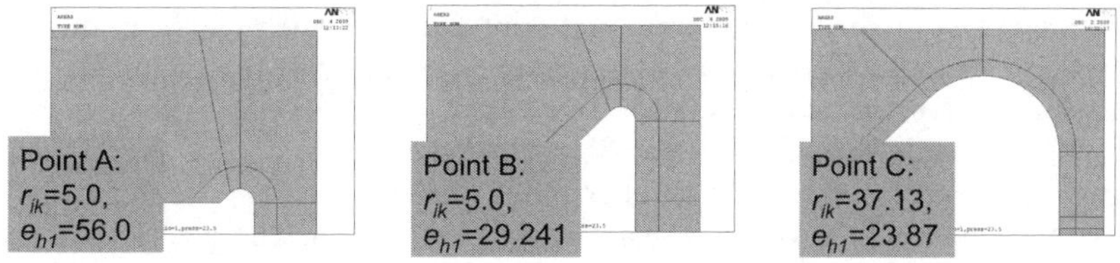

Point	Flat endplate		Cylindrical tube	
	$\sigma_{eqv}^{max}\ MPa$	$\varepsilon_{eqv\_pl}^{max} \times 10^5$	$\sigma_{eqv}^{max}\ MPa$	$\varepsilon_{eqv\_pl}^{max} \times 10^5$
A (5.0, 56.0)	281.65	559.97	312.97	163.45
B (5.0, 29.241)	278.15	615.83	201.11	0.0
C (37.13, 23.87)	277.73	151.20	278.80	0.0
D (29.80, 31.20)	276.91	142.16*	280.49	0.0

Figure 4 Comparison of results for corners of the admissible area ABC and optimal point D

The detailed results of the performed numerical investigations are presented in Figure 5, which shows the topographic map of distribution of the maximum equivalent plastic strain $\varepsilon_{eqv\_pl}^{\max}$ over the whole admissible domain ABC. As it can be observed, the plastic strains were not fully eliminated even for the optimum point D. The full elimination of plastic deformations is only possible by respective reduction of the internal pressure for the existing vessel geometry.

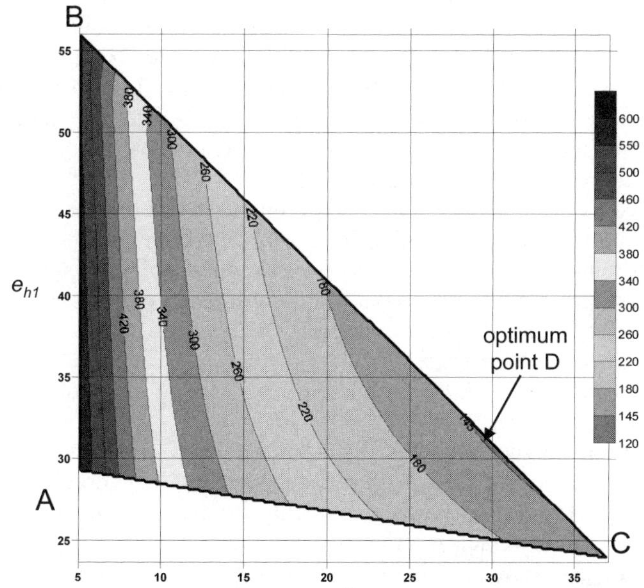

Figure 5 Distribution of maximum equivalent plastic strain $\varepsilon_{eqv\_pl}^{\max}$ over ABC admissible domain

The both cited codes [1, 2] suggest application of the circular shapes for stress relief grooves, these shapes are not optimal in many structural applications, which has been illustrated in papers [11, 12]. Certain calculations performed for pressure vessels with flat endplates with curved, but non-circular shapes give further reduction of the stress concentration. They can be permitted for application within so-called Design by Analysis formulae suggested in the above mentioned codes.

The results of numerical study for circular relief grooves were considered as a basis for final preparation of the experimental vessel. It was decided that the one endplate would have the optimal groove parameters (point D), while the other would follow non-optimal point B. Figure 6 shows the final set of dimensions of both model endplates.

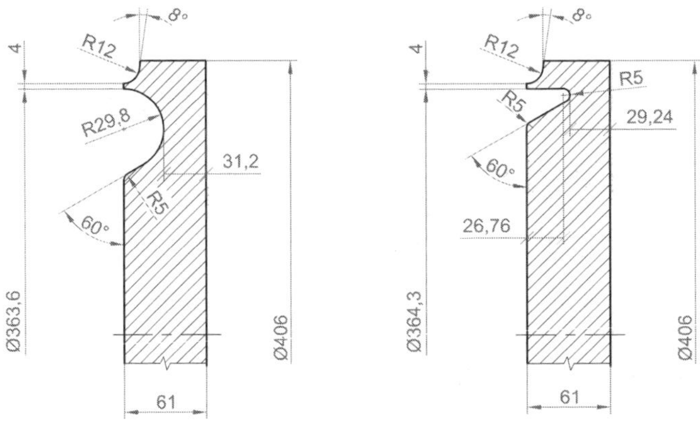

Figure 6 Final dimensions of endplates and stress relief grooves parameters; point D (a), and B (b) in Fig. 5

Experimental verification of numerical results

The main purpose of the experimental study was to confirm the existence of plastic deformation zones in the investigated vessel. The most suitable for this was the use of strain gauges placed in the areas of maximum strain values. As it can be seen in Figure 7 the zone of maximum effort was located in the vicinity of the stress relief groove. To find the zone, exemplary results given the for non-optimal groove configuration, material 9P, internal pressure $p_{int} = 33.0\ MPa$ ($1.5 \times p_O$), the von Mises equivalent stress and plastic strains were considered. It should be mentioned that the internal pressure was imposed by means of a pumping system supplying water to the vessel, which followed the standard test procedures described in codes [1, 2].

In such tests the access to the interior of the experimental vessel is very limited and, to perform the measurement, requires application of special sets of strain gauges, cables and protecting insulation. Also controlling of the leak-tightness for the cables going outside the vessel becomes a problem. To investigate a possibility of estimation of the internal obviously the largest effects from the external measurements, the detailed numerical analysis of strains in the vessel has been performed. It has been done comparatively for two material models (elastic and elastic-plastic) showing visible discrepancy between the strains on outer skin of the structure, in particular in the groove zone.

This difference became higher and higher for increasing internal pressure starting from the nominal operating pressure ($p_{int} = 22.0\ MPa$) up to the pressure of hydraulic test ($p_{int} = 35.0\ MPa$). Therefore, it was proposed to control the development of plastic effects by means of strain gauges placed on the outer skin of the vessel. In order to minimize the installation errors it was propose to use the sets of strain gauges shown in Figure 8. Obviously, before setting the gauges on the vessel outer skin the weld surface was polished in order to minimize the influence of notches introduced by the welding process.

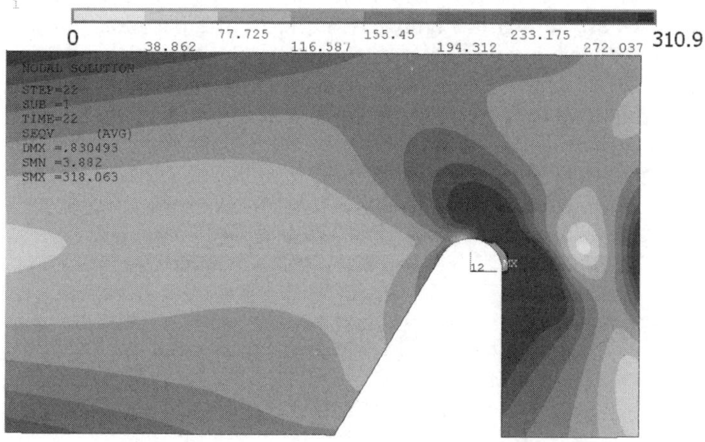

(a)

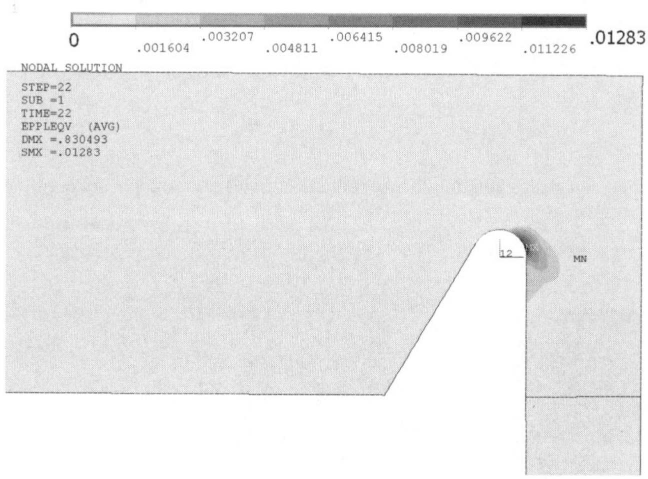

(b)

Figure 7 Distribution of the equivalent von Mises stress (a) and plastic strains (b) in vicinity of the stress relief groove, non-optimal configuration (point B Figures 4, 5) , material 9P and internal pressure p_{int} = 33.0 *MPa*

The first two sets of the gauges were placed in vicinity of the opposite ends of the vessel top. They were used only to measure the axial (larger) strains. The next two sets were placed at the bottom part and consisted of gauges, which measured alternatively strains in axial and hoop direction. In Figure 8 the schematic drawing of strain gauge sets location for optimal parameters of stress relief groove is shown.

According to suggestions given by the numerical calculations, the location of measuring sets on the opposite vessel ends were slightly shifted to observe possibly exactly the distribution of the extreme strains. Figure 9 shows the experimental vessel with strain gauges mounted to it, the detail is shown in photography (a), while the (b) photo shows more general view of the vessel. Figure 10 presents the general 3D finite element model of the experimental vessel, which was finally used to obtain the numerical results. In the model, also the dead weight of the vessel and the load of the water weight were taken into account. In that figure the position of strain gauges is marked and also its numbering is included.

The test have been made in two experimental sessions. The values of pressure for the consecutive load cycles are shown in Figure 11 in a form of a bar plot. For all the pressure values the measurement results have been registered. Subsequently, the numerical calculations were later performed for identical schedule of imposed pressures. At both ends, in the area of stress relief grooves, plastic deformations appeared for the pressure exceeding 15 *MPa*.

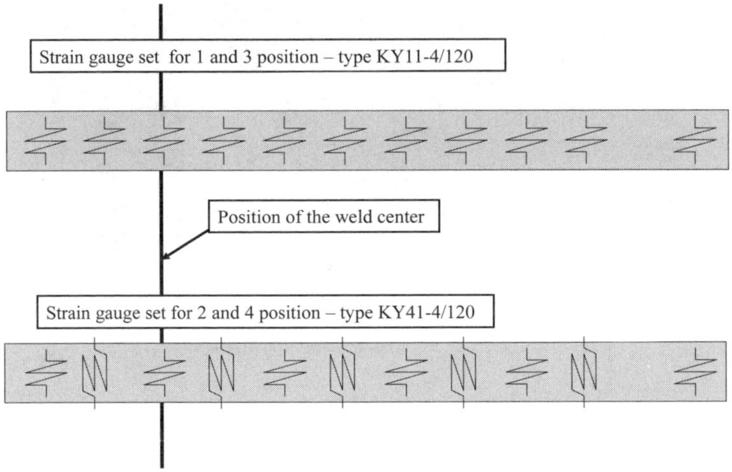

Figure 8 General view of strain gauge sets for measurement of axial and axial + perpendicular (hoop) strains [14]

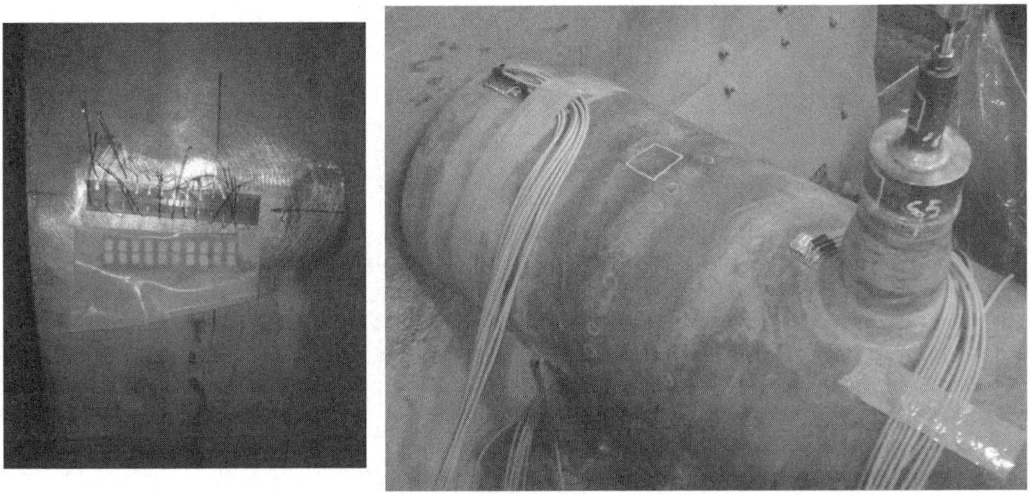

(a) (b)

Figure 9 Experimental vessel, (a) detail of strain gauge set, (b) general view of the vessel

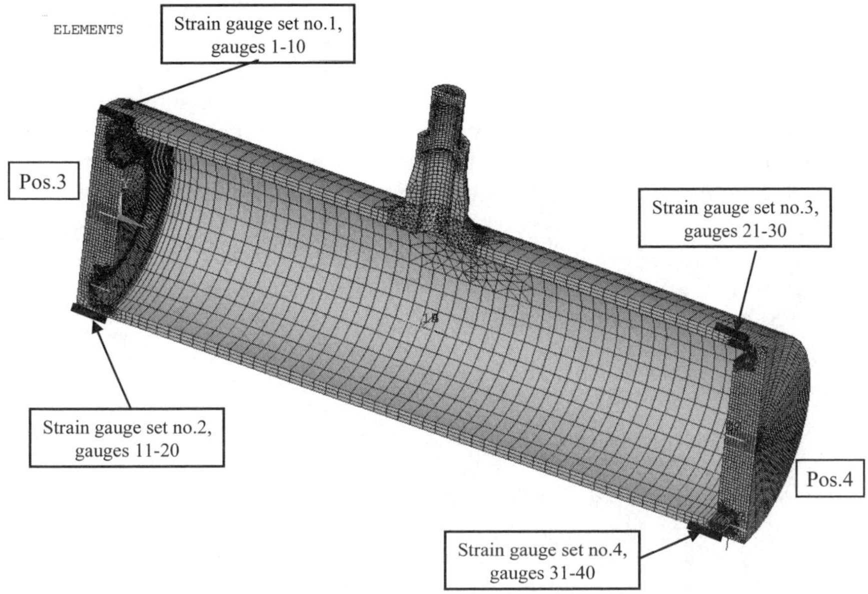

Figure 10 Finite element model of 1/2 of vessel

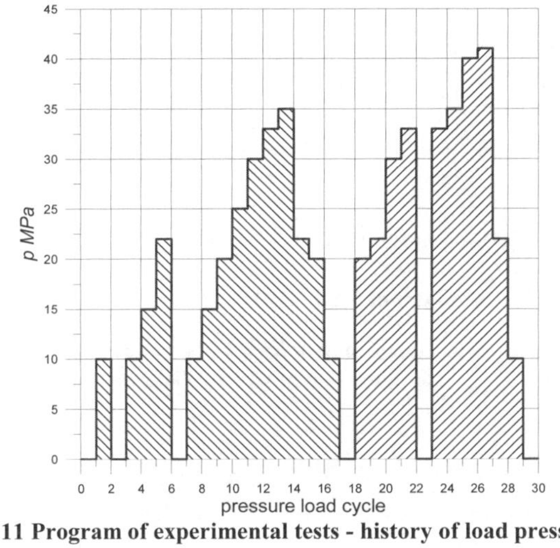

Figure 11 Program of experimental tests - history of load pressure

The history of the load in numerical simulation has been followed in the same way as in experiment. For all consecutive measurements plots with comparison of numerical and experimental results have been made. In the present paper only two exemplary plots for rather higher values of the internal pressure are presented. In Figure 12 a distribution of the axial strain along the control path AOB for the optimal stress relief groove configuration is shown. These results refer to load cycle 14, when the internal pressure was set to $p_{int} = 35.0$ *MPa*, the load cycle is marked on the pressure load history as a black solid bar. As it can be seen, the experimental and numerical results remain in acceptable, good agreement.

In Figure 13 the respective results for non-optimal configuration for load cycle 22 ($p_{int} = 33.0\ MPa$) are shown. In this case in the load history the full unloading step preceded the studied load case, which ended the first session of measurement. Therefore, due to plastic deformations in the structure, certain residual strains were registered for control path. Then, in the second session, the structure was loaded and again a reasonably good agreement of both studied results was observed. The above given comparison of strain distributions confirmed in general the proposed idea of the experiment. On both plots (12 and 13) values for maximum elastic strains ($\varepsilon_{elastic}$) are given for both material approximations and placed in the plot as horizontal lines. This serves for the comparison of the calculated strains and the measured values within the admissible strain idea. It should be underlined, the preliminary welding stresses have also been observed in the experimental process. It will be presented in the next paper of the series.

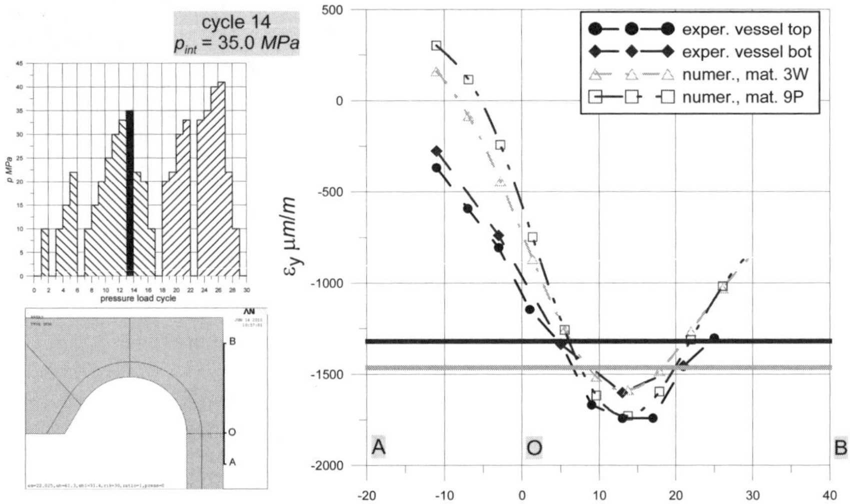

$\varepsilon_{elastic} = 1305\ \mu m/m$ (mat. 9P)

$\varepsilon_{elastic}$ 3W = 1480 $\mu m/m$ (mat. 3W)

Figure 12 Distribution of axial strains along control path AOB, numerical vs. experimental results for optimal configuration of stress relief groove

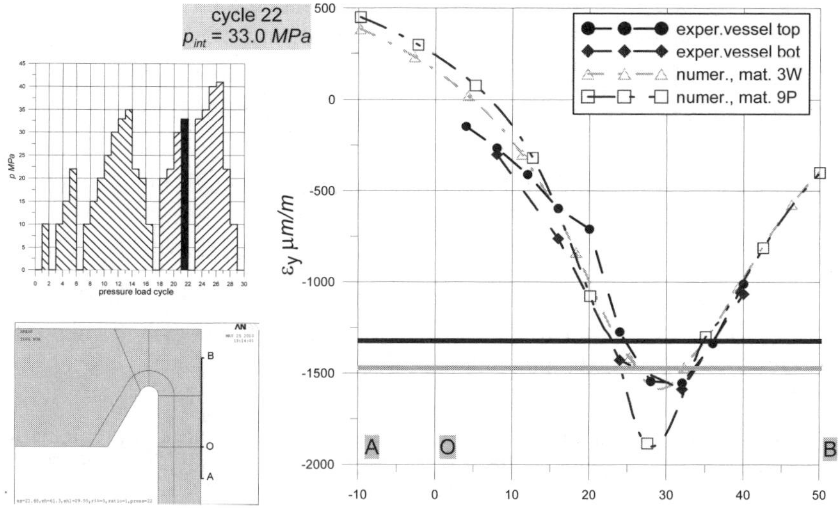

Figure 13 Distribution of axial strains along control path AOB, numerical vs. experimental results for non-optimal configuration of stress relief groove

Conclusions

In the paper, the numerical study of stress and strains distribution made for a high-pressure vessel with flat endplates and stress relief grooves was presented and supplemented by the experimental investigations. The longitudinal (largest) strains measured in vicinity of the junction between the endplate and the cylindrical part were the main studied results. The experimental results remain in good agreement with the results of finite element approximation. The observation have confirmed the proposed idea of the experiment, development of plastic zones inside the experimental vessel and determination of the residual strains observed in strain gauges after termination of the experiment (internal pressure set to zero). The investigation in the paper concentrated only on geometry of the stress relief grooves suggested by European and American Standards. However, if the family of the curves is extended to an arbitrary form, the observed local effort can further be diminished [12].

Acknowledgement

Authors would like to express their gratitude to Mr. W. Rożnowski and Mr. H. Kozioł from RAFAKO Company and the RAFAKO itself for preparation of design documentation, manufacturing of the experimental vessel and for facilitation of experimental measurement made on stand prepared in company site.

References

[1] British-Adopted European Standard EN-12952-3:2001, *Water-tube boilers and auxiliary installations. Design and calculation of pressure parts*, May-2002.

[2] European Standard EN-13445-3:2202, *Unfired pressure vessels – Part 3: Design*, May-2002.

[3] Companion Guide to the ASME Boiler and Pressure Vessel Code, K.R. Rao (Ed.), ASME Press New York, 2006.

[4] R. Preiss, Stress concentration factors of flat end to cylindrical shell connection with a fillet or stress relive groove subject to internal pressure, Int. J. Press. Vessel and Piping, 73 (1997) 183-190.

[5] R. Preiss, F. Rauscher, D. Vazda, J.L. Zeman, The flat end to cylindrical shell connection – limit load and creep design, Int. J. Press. Vessel and Piping, 75 (1998) 715–726.

[6] T. Vilhelmsen, Reference stress solutions of flat end to cylindrical shell connection and comparison with design stresses predicted by codes, Int. J. Press. Vessel and Piping, 77 (2000) 35–39.

[7] S. Łaczek, J. Ryś, A.P. Zieliński, Load capacity of thick-walled cylinder with radial hole, Int. J. Press. Vessel and Piping, 87 (2010) 433-439.

[8] G. Widłak, Local shakedown analysis of thick-walled reactor subject to mechanical and thermal loads,PhD Thesis (supervisor A.P. Zieliński), Cracow Univ. of Technology, 2010.

[9] M. Mrzygłód, A.P. Zieliński, Low-cycle fatigue constraints in two-stage structural optimization of a nozzle to pressure vessel connection, T. Burczynski, J. Periaux (Eds.), CIMNE Barcelona, 2009.

[10] B. Szybiński, A. Trojnacki, Experimental verification of strength calculations of high-pressure metal B-type sealing (in Polish), submitted to Mechanical Review (2010).

[11] P. Pedersen, Suggested benchmarks for shape optimisation for minimum stress concentration, Struct. Multidsciplinary Optim. 35 (2008), 273-283.

[12] B. Szybiński, A. Wróblewski, Parametric optimisation of stress relief groove shape in flat ends of boilers, submitted to Journal of Strain Analysis for Eng. Design (2011).

[13] ANSYS, ver.12.0 for research and educational purposes. Swanson Analysis Systems, 2010.

[14] Information on http://www.hbm.com.

Key Engineering Materials Vol. 490 (2012) pp 237-246
doi:10.4028/www.scientific.net/KEM.490.237

Numerical model of bevel gears cutting by Duplex Helical Method

Jadwiga Pisula [1, a], Mieczysław Płocica[2,b]

[1]Rzeszow University of Technology, Poland

[2] Rzeszow University of Technology, Poland

[a]jpisula@prz.edu.pl, [b]mplocica@prz.edu.pl

Keywords: bevel gears, duplex helical method, bevel gear cutting.

Abstract. In the paper it was presented a mathematical model of tooth generating of spiral bevel pinion and gear. Geometry as well as kinematic functions of the 116 machine tool were defined and mathematical model of the technological gear was built, which allows for the simulation of tooth generating.

Introduction

In the process of the new gears production there occurs a necessity to verify the correctness of the constructional conceptions and technological calculations. Virtual generation of gear units allows to specify (at the calculations stage) the correct settings of the machine tool and, as a consequence, obtain the proper tooth bearing. In the traditional approach the proper adjustment of a machine setting requires performing the technological calculations, cutting the pinion and the gear on the machine tool and introducing corrections after examination of the tooth bearing on the control machine. The possibility of numerical modelling allows for elaboration and verification of the various conceptions of a virtual teeth cutting and evaluation of the convergence of the performed simulations and actual sample results.

In the article the Gleason method called SGDH (*Spiral Generated Duplex Helical*) is presented [3]. This method is used for cutting spiral bevel gears with small modules. The teeth of every gear unit are cut to final state in one operation, because the two-sided front cutter-heads are used for milling. The gear is cut with rolling motion, whereas the pinion is made with the helical rolling motion, which is used for liquidation of the tooth bearing bias which is considered as unwanted phenomenon. The SGDH method is practically used for transmissions greater than 1:2,5. Its basic assumption is equalization of the gear and pinion finishing times. The method is particularly of the industrial importance due to the short finishing time of the units in the gear and connected with it is economic attractiveness.

Model of tool surface

Because of geometric complexity of the considered teeth cutting method, the tool representation was assumed as its working surface, obtained by the rotation of the cutting edges around the axis of a cutter head [5]. Additionally, such an assumption is correct due to the discrete way of the cutting that causes the serration of the tooth flank surface in case of the discontinuous cutting edge. Two-sided cutter head contains two groups of blades, internal and external ones, placed coaxially on the circles that create the two sides of action (Fig. 1). Cutters internal cutting edges with an angle of α_{WP} creates the acting surface for generation of the convex side of a tooth while the external edges with α_{WK} – the surface generating the concave side.

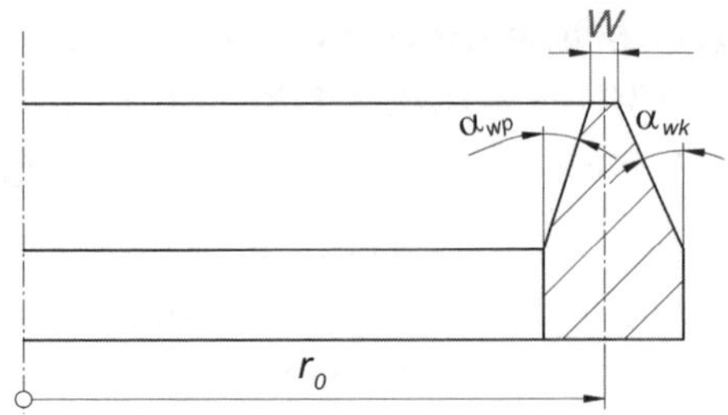

Fig. 1. Axial section of a tool model – geometry in a ALTERNATE toolhead type

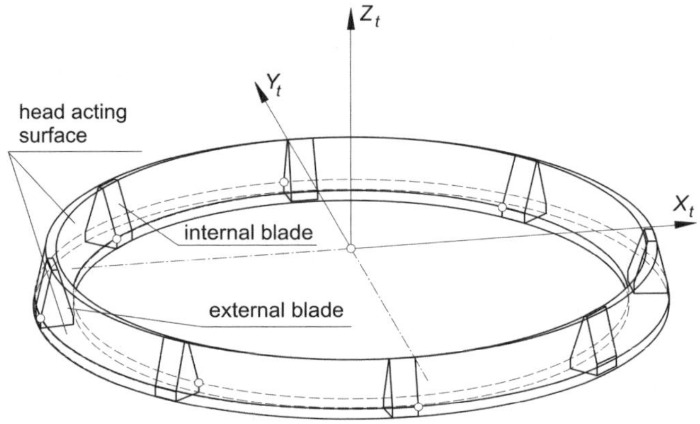

Fig. 2. Parametric model of the two-sided cutter head acting surface

Figure 2 presents the solid model of the tool acting surface, created on the basis of cutter blades geometry. This model has the angles α_{WK} and α_{WP} in its axial section which is related to the two-sided head cutters' angles.

The machine tool is modelled as the set of rectangular right-handed coordinate systems which contains the geometry of the machine main modules. Figures 4 and 6 show the machine tool model in the initial positions, accordingly to the gear and pinion cutting.

Gear generating system

In the process of the gear teeth cutting the generating system is the bevel technological gear. It consists of: cradle with the cutter head as a generating gear and the gear blank. While cutting the gear with the SGDH method the cutter head axis is parallel to the cradle's axis and the plane of the head cutters' apices – perpendicular. Thus the generating gear is a ring gear

Cutting with the two-sided cutter causes the generation of space with the constant width of the bottom, equal to the width of the blades apices spacing. In order to reach the full depth of the tooth space in the whole face width it is necessary to provide the tangency of the workpiece end's taper with the head cutters apices plane. In order to fulfil these conditions the fixed headstock's torsion angle of a workpiece has to be equal to the root angle of a workpiece, that means $\delta_{m2} = \delta_{f2}$. Toolhead's coordinates are determined on the basis of the assumed geometry of the workpiece and the chosen cutter. Since the cutter model is represented by its acting surface, the rotation of the S_t system around the Z_t axis is not introduced [4, 7].

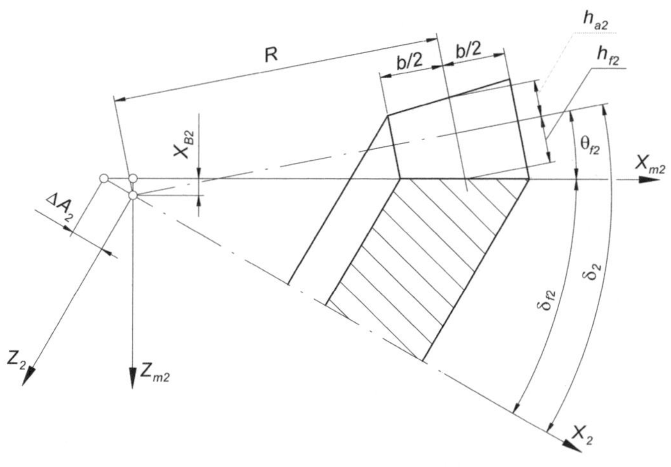

Fig. 3. Axial section of a gear

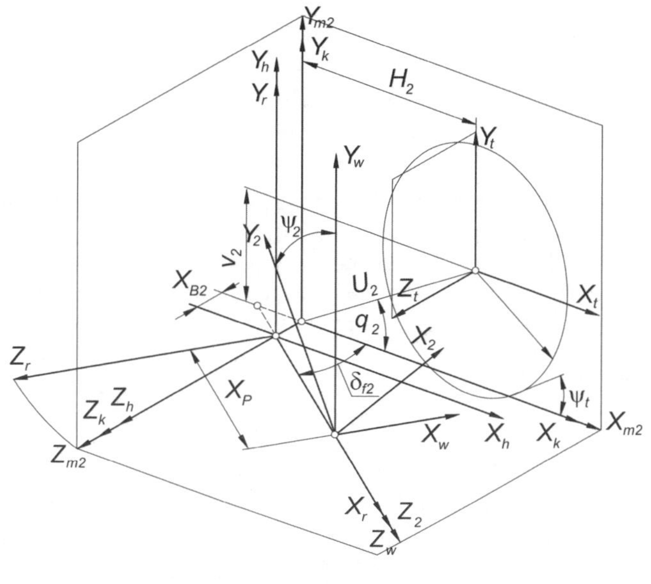

Fig. 4. Generating system of the gear

Pinion generating system

Due to simultaneous cutting of the both sides of the tooth space (pinion and gear), the bearing contact bias occurs after assembling of a transmission made in that way. Such an effect is undesirable since it causes loud work of the transmission and non-uniform transfer of the motion which leads to the faster fatigue wear. In order to remove the bias the helical motion is applied during pinion finishing. It is realized by the axial offset of the fixed headstock of the workpiece connected with the generating gear rotation.

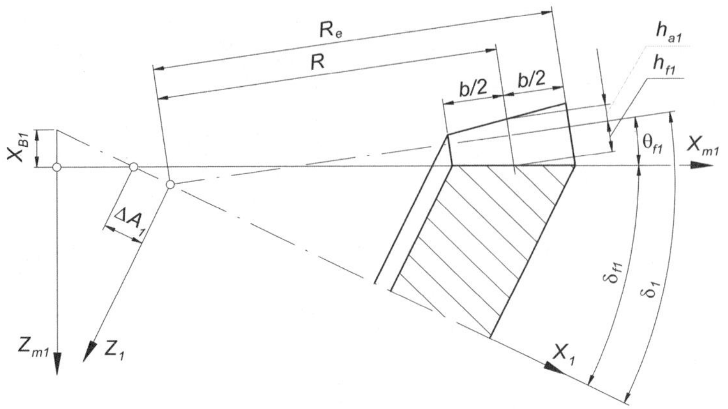

Fig. 5. Axial section of a pinion

The generating system of the pinion is the technological hypoid gear obtained thanks to the hypoid offset of the workpiece's axis by the E value with respect to the cradle's axis. The cradle with inclined toolhead (*tilt*) creates the bevel generating gear. For the pinion cutting the head's axis inclination is applied in order to compensate the difference between the pressure angle and tool profile so in the initial position the S_c system is rotated with respect to the S_d system by the angle of j around X_d axis.

The head's axis inclination angle (in degrees and minutes) and the value of the hypoid technological offset E is determined on the basis of the gear geometry analysis and calculation cards. Positive values mean down-shift by the pinion with left inclination line direction of a tooth or up-shift by the pinion with right inclination line direction of a tooth. The negative value means up-shift by the pinion with left inclination line direction of a tooth or down-shift by the pinion with right inclination line direction of a tooth.

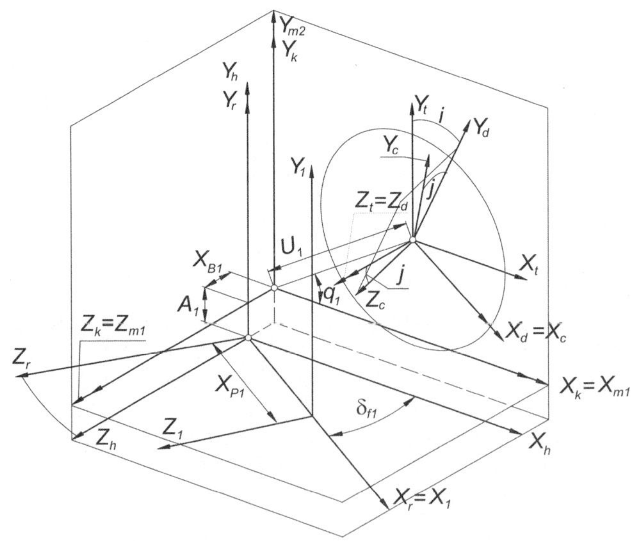

Fig. 6. Generating system of the pinion – initial position

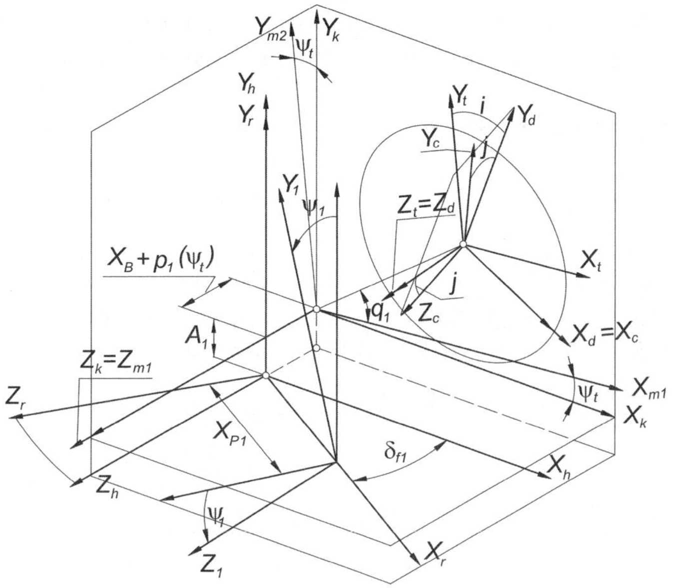

Fig. 7. Generating system of the pinion – cutting with helical motion

Mathematical model of tooth flank surface

Mathematical notation of the tooth flank is apparent from the equation of the surface of action of the tool, kinematics, and accepted treatment technological system. The following discussion presents the side of the tooth surface obtained with the use of a technological system with a bevel generating gear - this is the case more generally in comparison with a ring generating gear. While processing the envelope, the equation of the tooth flank, which is the bounding surface of the utility, is determined from the system of equations [1, 2, 6]. This system includes the equation of the family of tool surfaces and the equation of meshing, resulting from the method for determining the kinematic envelope:

$$\begin{cases} \boldsymbol{r}_I\left(s_t,\theta_t,\psi_t\right) \\ \boldsymbol{n}_I\cdot\boldsymbol{v}_I^{tI}\left(s_t,\theta_t,\psi_t\right)=0 \end{cases} \tag{1}$$

where: $\boldsymbol{r}_I\left(s_t,\theta_t,\psi_t\right)$ - determining the vector function of the family tool surfaces system bounded with treating pinion (S_1), $\boldsymbol{n}_I$ - the unit normal vector defined in S_1 , $\boldsymbol{v}_I^{tI}\left(s_t,\theta_t,\psi_t\right)$ - the relative velocity vector defined in S_1.

Based on the defined technological model, the family of tool surfaces is determined as follows:

$$\boldsymbol{r}_I\left(s_t,\theta_t,\psi_t\right)=\boldsymbol{M}_{It}\left(\psi_t\right)\cdot\boldsymbol{r}_t\left(s_t,\theta_t\right) \tag{2}$$

where: $\boldsymbol{r}_t\left(s_t,\theta_t\right)$ - vector equation of the tool surface referred to the system associated with the tool, S_t, s_t,θ_t - curvilinear coordinates surface form, $\boldsymbol{M}_{It}\left(\psi_t\right)$ - the transformation matrix being the product of a transformation matrix representing the rotations and translations of homogeneous coordinate systems included in the technological gear model ψ_t - parameter of motion (in this case - angle of the cradle rotation).

Vector equation of the tool surface as a function of curvilinear coordinates s_t, θ_t shows the relationship (3), which involves the processing of the active side of the tooth.

$$\boldsymbol{r}_t(s_t, \theta_t) = \begin{bmatrix} \cos\theta_t (r_{wk} + s_t \sin\alpha_{wk}) \\ \sin\theta_t (r_{wk} + s_t \sin\alpha_{wk}) \\ -s_t \cos\alpha_{wk} \end{bmatrix} \tag{3}$$

where: α_{wk} - the angle of the external blades, r_{wk} - the radius of the cutterhead.

Based on the model of technological gear it is designated a family of the tool surfaces according to equation (2), for which it determines conversion matrix equation (4).

$$\boldsymbol{M}_{1t}(\psi_t) = \boldsymbol{M}_{1w}(\psi_1(\psi_t)) \cdot \boldsymbol{M}_{wr} \cdot \boldsymbol{M}_{rh} \cdot \boldsymbol{M}_{hm} \cdot \boldsymbol{M}_{mk}(\psi_t) \cdot \boldsymbol{M}_{kc} \cdot \boldsymbol{M}_{cd} \cdot \boldsymbol{M}_{dg} \tag{4}$$

where: $\boldsymbol{M}_{1t}(\psi_t)$ - conversion matrix, $\boldsymbol{M}_{ij}$ - the elementary transformation matrices representing the rotations and translations of homogeneous coordinate systems, ij - subscript indicating the direction of the transformation from system S_j to system S_i.

Technological gear model is also used to determine (in the chosen system, for example (S_1 or S_m) a unit normal vector and relative velocity. The normal vector to the surface of the tool sets at any of the predefined layouts. In this way we can get the components of meshing equation. Solving equations (1) by eliminating one of the variables such as: s_t from the meshing equation and then substituting into the equation of the family of tool surfaces $\mathbf{r}_1(s_t, \theta_t, \psi_t)$ we can obtain the tooth flank surface equation in two-parametric form (5).

$$\boldsymbol{r}_1(\theta_t, \psi_t) = \boldsymbol{r}_1(s_t(\theta_t, \psi_t), \theta_t, \psi_t) \tag{5}$$

where: $\boldsymbol{r}_1(\theta_t, \psi_t)$ - the equation of the pinion tooth surfaces in the two-parametric form, $s_t(\theta_t, \psi_t)$ - the variable s_t in the function of other parameters.

Model of technological gear is designed to create the flank surface of the gear and pinion teeth, which will be used for the analysis of meshing for constructional spiral bevel gear. In order to obtain a tooth surface as the family of tool surfaces it should be designated in the system rigidly bounded with cutting pinion S_1, which represents the envelope to family of surfaces of the cut gear tooth $\Sigma^{(1)}$. Family of tool surfaces is shown in the following way:

$$\boldsymbol{r}_1^{(t)} = \boldsymbol{L}_{1\leftarrow w}(\psi_1(\psi_t)) \cdot \boldsymbol{L}_{w\leftarrow r} \cdot (\boldsymbol{L}_{r\leftarrow h} \cdot (-\boldsymbol{T}_{h\leftarrow m} + \boldsymbol{L}_{m\leftarrow k}(\psi_t) \cdot (\boldsymbol{L}_{c\leftarrow d} \cdot \boldsymbol{L}_{d\leftarrow t} \cdot \boldsymbol{r}_t^{(t)} - \boldsymbol{T}_{k\leftarrow c})) - \boldsymbol{T}_{r\leftarrow h}) \tag{6}$$

In this equation the vector record $\boldsymbol{r}_1^{(t)}$, $\boldsymbol{r}_m^{(t)}$, $\boldsymbol{r}_g^{(t)}$, concerns a family tool surfaces $\Sigma^{(t)}$ (as evidenced by the superscript (t)), set respectively in the cutting gear system S_1, the basic system (stationary) and the tool system. Other signs are bounded with the transformations of coordinate systems used in the model of technological gear. The applied rotation and translation matrixes of the coordinate system are as follows:

$$\boldsymbol{L}_{d\leftarrow g} = \begin{pmatrix} \cos i & 0 & \sin i \\ 0 & 1 & 0 \\ -\sin i & 0 & \cos i \end{pmatrix}, \tag{7}$$

$$\mathbf{L}_{c\leftarrow d} = \begin{pmatrix} cos\, j & sin\, j & 0 \\ -sin\, j & cos\, j & 0 \\ 0 & 0 & 1 \end{pmatrix}, \tag{8}$$

$$\mathbf{T}_{k\leftarrow c} = \begin{pmatrix} -U \cdot cos\, q \\ -U \cdot sin\, q \\ 0 \end{pmatrix}, \tag{9}$$

$$\mathbf{L}_{m\leftarrow k}(\psi_t) = \begin{pmatrix} cos\,\psi_t & -sin\,\psi_t & 0 \\ sin\,\psi_t & cos\,\psi_t & 0 \\ 0 & 0 & 1 \end{pmatrix}, \tag{10}$$

$$\mathbf{T}_{h\leftarrow m} = \begin{pmatrix} 0 \\ -A_1 \\ X_{B1} + p_1(\psi_t) \end{pmatrix}, \tag{11}$$

$$\mathbf{L}_{r\leftarrow h}(\delta_{f1}) = \begin{pmatrix} cos\,\delta_{f1} & 0 & sin\,\delta_{f1} \\ 0 & 1 & 0 \\ -sin\,\delta_{f1} & 0 & cos\,\delta_{f1} \end{pmatrix}, \tag{12}$$

$$\mathbf{T}_{r\leftarrow h} = \begin{pmatrix} X_{P1} \\ 0 \\ 0 \end{pmatrix}, \tag{13}$$

$$\mathbf{L}_{w\leftarrow r} = \begin{pmatrix} 0 & 0 & -1 \\ 0 & 1 & 0 \\ 1 & 0 & 0 \end{pmatrix}, \tag{14}$$

$$\mathbf{L}_{1\leftarrow w}(\psi_1(\psi_t)) = \begin{pmatrix} cos(\psi_1(\psi_t)) & sin(\psi_1(\psi_t)) & 0 \\ -sin(\psi_1(\psi_t)) & cos(\psi_1(\psi_t)) & 0 \\ 0 & 0 & 1 \end{pmatrix}. \tag{15}$$

Results

Numerical simulation of the SGDH method was performed for the transmission gears which geometrical characteristics are included in table 1. The angles were given in degrees and the linear values in millimetres. Geometry of tools and the machine set parameters (tables 2, 3) were determined basing on the teeth shaping geometry analysis and as a result solved technological calculations.

Table 1. Transmission gear geometric data

z_1	Number of pinion teeth	9
z_2	Number of gear teeth	44
L_{zz}	Hand of pinion spiral	left
β	Spiral angle	35°0' 0"
Σ	Shaft angle	90°
α	Average blade angle	20°
b	Face width	39mm
d_{2e}	Outside diameter	259,85mm

Table 2. Cutter specification

d_0	Cutter diameter	191,77 [mm]	190,500 [mm]
W_2	Point width	2,016 [mm]	3,302 [mm]
R_{02}	Blade edge radius	0,635 [mm]	1,016 [mm]
α_{WK}	Outside blade angle	20 [°]	20 [°]
α_{WP}	Inside blade angle	20 [°]	20 [°]

Table 3. Basic machine-tool settings for pinion and gear cutting

q	Setting angle head	58°14' 0"	53°8' 0"
U	Radial setting head	89,672 [mm]	97,509[mm]
E	Blank offset	2,219 [mm]	0 [mm]
δ_m	Machine root angle	10°17' 0"	75°33' 0"
X_P	Machine center to back	-1,667 [mm]	0 [mm]
X_B	Sliding base	-4,521 [mm]	-1.771 [mm]
i	tilt angle	2°24' 0"	0 [°]
j	swivel	22°4' 0"	0 [°]
p_d	Helical motion in 20 deg	0,02 [mm]	0 [mm]
i_{odt}	Rolling motion ratio	0,436922	0,899604

As a result of the numerical simulation the virtual models of the gear and pinion were obtained (Fig. 8, 9) which flank tooth surfaces are the mappings of the family of tool surfaces. The presented models are the basis for the creation of the constructional transmission gear meshing simulation and for obtaining the tooth bearing.

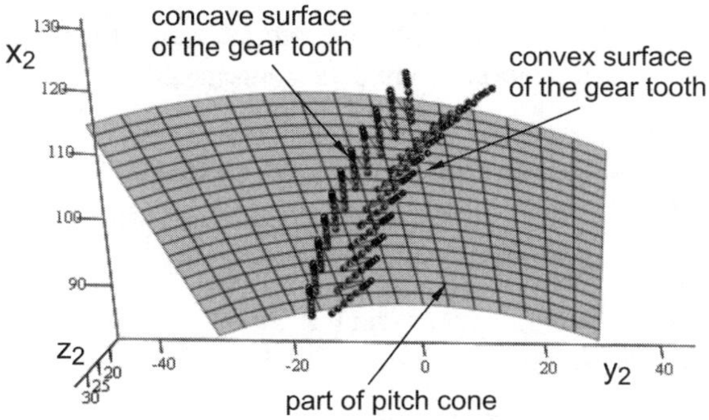

Fig. 8. Flank tooth surfaces of the gear

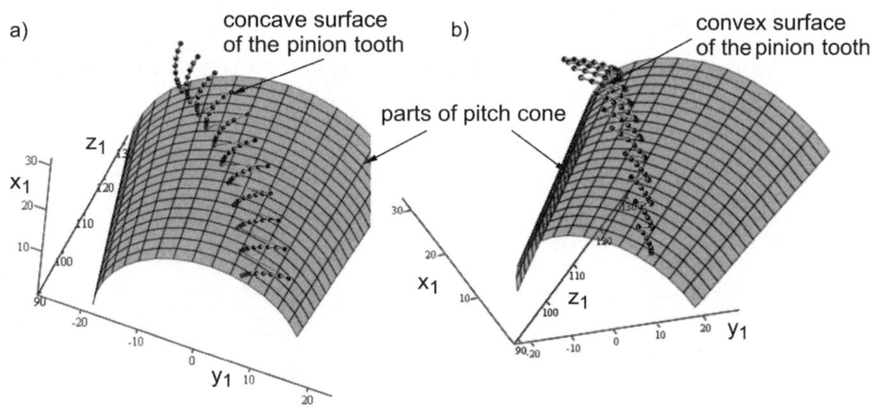

Fig. 9. Flank tooth surfaces of the pinion

Summary

Methods of numerical modeling of cutting, which was presented above, is of considerable practical importance. We can speed up the process thanks to its introduction into the production of the new transmission, obtaining at the design stage of technology, appropriate landscaping the lateral surfaces of the teeth, which will trace the desired mesh. Through numerical simulation, we can quickly determine appropriate values for machine settings and check the effect of treatment, setting the received transmission units in constructional gear pair.

Financial support of Structural Funds in the Operational Programme - Innovative Economy (IE OP) financed from the European Regional Development Fund - Project No POIG.0101.02-00-015/08 is gratefully acknowledged.

References

[1] L. Chung-Yun, T. Chung-Biau, F. Zhang-Hua: Computer-aided manufacturing of spiral bevel and hypoid gears by applying optimization techniques. Journal of Materials Processing Technology, 114 (2001), 22-35.

[2] G. Figliolini, J. Angeles: Algorithms for Involute and Octoidal Bevel-Gear Generation. ASME Journal of Mechanical Design, 127 (2005), 664-672.

[3] Gleason Works: Generated Spiral Bevel Gears Duplex Helical Method. Rochester, 1965.

[4] F.L. Litvin, A. Fuentes.: Gear Geometry and Applied Theory, second edition, Cambridge University Press, New York, 2004.

[5] A. Marciniec: Synthesis and analysis of meshing for spiral bevel gears, Publishing House of the Rzeszow University of Technology, Rzeszow, 2002.

[6] W. Pei-Yu, F. Zhang-Hua: Mathematical Model of Face-Milling Spiral Bevel Gear with Modified Radial Motion (MRM) Correction. Mathematical and Computer Modelling, 41 (2005), 1307-1323

[7] H. J. Stadtfeld: Advanced Bevel Gear Technology. The Gleason Works, Rochester, 2000.

Key Engineering Materials Vol. 490 (2012) pp 247-256
© (2012) Trans Tech Publications, Switzerland
doi:10.4028/www.scientific.net/KEM.490.247

Problems of Analytical Determination of Journal Bearing Bush Fatigue Strength Estimates

Rafał Gawarkiewicz[1, a], Jan Sikora[1,b], Bogusław Siwek[1,c]

[1]Gdańsk University of Technology, Poland

[a]gawar@pg.gda.pl, [b]jsikora@pg.gda.pl, [c]bosiw@pg.gda.pl

Keywords: journal bearings, fatigue strength estimates, analytical methods, slide bearing research

Abstract. Problems connected with determination of stress distribution in sliding layer of thin-walled bearing bushes, investigated in bearing fatigue test rigs, have been presented. Using an example of plain bearings tested in the fatigue machine SMOK (built at the Gdańsk University of Technology) problems with obtaining a convergence of iterative procedure for determining the fatigue strength estimators of bearing alloy surface layer are analyzed. Calculations consisting in successive iterations of pressure distribution in oil film and corresponding elastic deformation of the bearing bush and housing were based on finite-element method and utilized ANSYS program. The objective of calculation was the estimation of values of radial, tangential, axial and reduced stresses in surface layer of the bearing bush corresponding to experimentally determined load magnitudes leading to fatigue cracks initiation. The convergence of analytical procedure has been achieved by the use of a relaxation method.

Introduction

Fatigue failures in lining of bearing bushes are one of the forms of destructive damages in technical devices in which cyclic loadings occur. The fatigue phenomena begin at surface cracks axially oriented, which is generally explained as the effect of tangential stresses [1,2]. The magnitudes of load leading to bearing fatigue for a given load pattern can only be determined experimentally whereas the estimate of corresponding critical stress may be evaluated analytically [3].

Fatigue strength of a sliding bearing depends on many factors such as mechanical properties of bearing materials, structure of lining and design features of the whole bearing unit. The last factor is often underestimated. Contemporary computational procedures based on the finite-element method enable modelling a real shape of the bearing and calculate the stresses in its lining with needed accuracy. Calculations consist in successive iterations of pressure distribution in hydrodynamic oil film and corresponding elastic deformations and stresses in bearing bush and its housing. In this article problems with obtaining convergence of such an iterative computational procedure are analyzed using as an example the bearing investigated in the fatigue testing rig SMOK [4]. The testing machine was built at the Gdańsk University of Technology. An objective of the calculation was estimation of values of radial, tangential, axial and reduced stresses in the surface layer of thin-walled bearing bushes under cyclic load which led to fatigue cracks during the tests in the machine.

Bearing Test Machine SMOK

A bearing fatigue test rig SMOK (Fig. 1) at the GUT tribological laboratory produces the cyclic load due to a short eccentric movement of the test journal causing a pressure built-up in the hydraulic cylinder. A pattern of generated load and an example of bearing damaged in fatigue test are shown in Fig. 2. The magnitude of the load is controlled by the valve and orifice system. The bearing is centrally located in the head of the rig. Due to this, despite deflection of the shaft under load, symmetry of the oil film in the axial cross-section of the bearing is maintained.

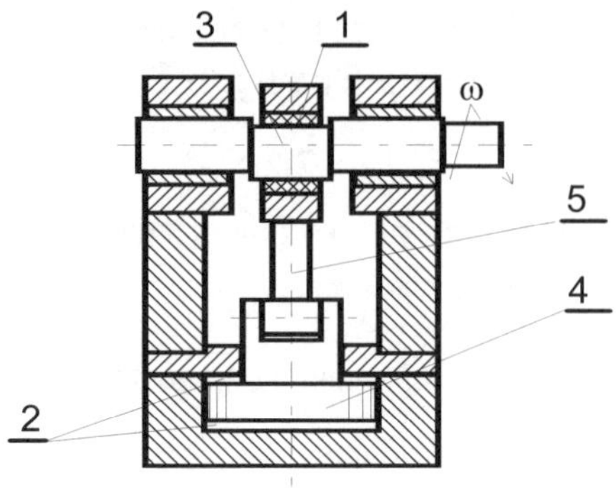

Fig. 1. Principle of operation of plain bearing fatigue test rig SMOK; 1 – tested bearing, 2 - oil chambers, 3 - eccentric journal of the shaft, 4 - piston, 5 - connecting rod

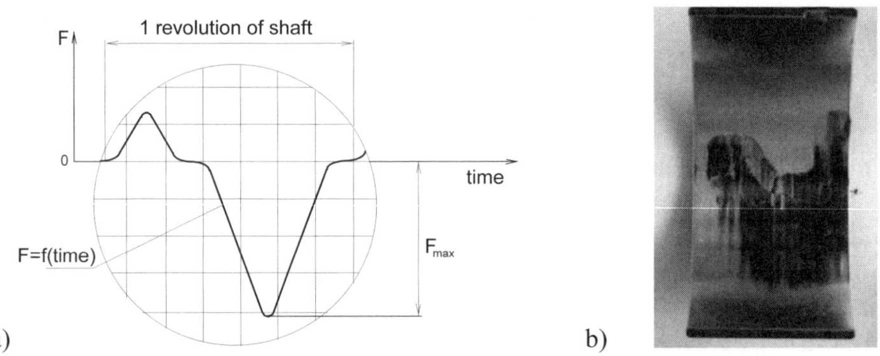

Fig. 2. Pattern of loading in test machine SMOK (a) and view of bearing bush surface destroyed in fatigue test after $3{,}6\times10^6$ load cycles (b)

The surface layer of the bush is loaded with varying oil pressure that fluctuates between zero and maximum value. The value of peak hydrodynamic pressure depends on test parameters and design features of the bearing.

The bearing machine SMOK meets requirements of ISO 7905-1 standard [3] in relation to fatigue testing of slide journal bearings. Full information about operating characteristics of the machine is given in [4].

Testing procedure for the test rig should ensure evaluation of the endurance limit stresses for a specified number of load cycles – a basis of the fatigue test. In case of a standard test in the SMOK machine it is equal to $3{,}6\times10^6$ cycles. Due to the scatter of test results and stochastic nature of the fatigue limit, it is recommended that the results should be evaluated using a statistical method. It has been proved that the most efficient sequential procedure that can be used in plain bearing fatigue research is the two-point strategy [5]. Metallographic examination of damaged specimens may provide important information about the crack growth mechanism, corrosive attack and diffusion resulting from thermal effects.

Object of Analysis

Shape and nominal dimensions of half-bearings tested in the SMOK test rig are shown in Fig. 3. Bimetallic half-bushes consisting of steel shell and lining made of bearing alloy are usually investigated. The bush is fixed in the steel housing (Fig. 4) with special stiffness profile. The housing profile has been properly selected to maintain, under working conditions, steady clearance between the shaft and bush surfaces in loaded zone. It ensures almost parallel arrangement of journal and bush generating lines and parabolic shape of oil pressure distribution in axial direction. That simplifies an analytical model of the bearing during calculation. Moreover, owing to the flexibility of bush and housing edges the fatigue cracks are initiated in central zone of the bearing bush where the maximum oil pressure occurs.

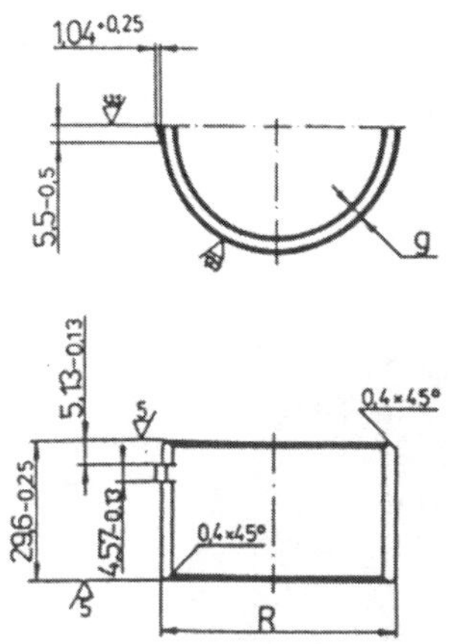

Fig. 3. Standard specimen tested in the SMOK machine

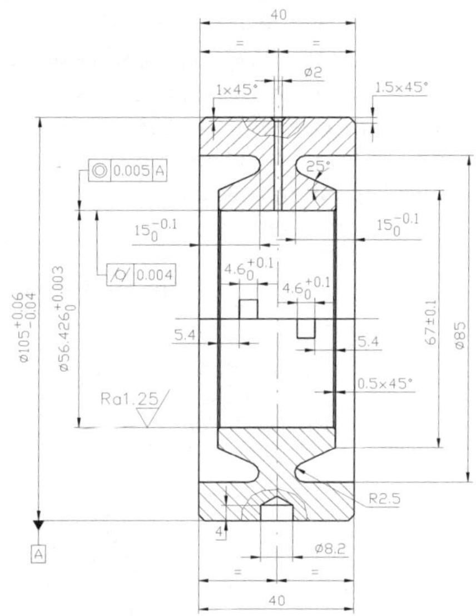

Fig. 4. Design scheme of elastic housing in the SMOK machine

Geometry of the bearing and fatigue test parameters used for analysis and stress calculation procedure are specified in Table 1.

Table 1. Bearing data admitted in calculation procedure

Feature or parameter	Value
rotational speed	3000 [rpm]
journal diameter	52.7 [mm]
width of the bush	29.6 [mm]
total wall thickness	1,82 [mm]
relative bush width	0.56
bearing alloy thickness	0.25 [mm]
bearing radial clearance ratio	0.001
housing dimensions [mm]	acc. to Fig. 4
Young's modulus of steel shell	210 [GPa]
Poisson's number of steel	0.3
Young's modulus of bearing alloy	90 [GPa]
Poisson's number of bearing alloy	0.34
lubricating oil viscosity	0.015 [Ns/m^2]
maximum specific load	36.4 [MPa]
resultant hydrodynamic force	56.5 [kN]

Numerical Model of the Object

For structural modelling of the bearing unit the finite-element technique has been applied. Eight-nodal solid finite elements of three degrees of freedom in each node were used for modelling the bimetallic bearing bush and steel housing. For simplicity, the bearing steel shell and the housing were assumed to be one piece. To increase computation accuracy the mesh of the elements in bearing alloy layer was substantially refined.

It should be noted that both the bearing geometry and the loading are symmetric across the axial direction of the bearing towards the bearing cross-section central plane. Therefore, only half of the unit has to be modeled. In the result the finite-element representation of the object may be shown like the image in Figure 4. This figure also shows the coordinate system and the example of oil hydrodynamic pressure imposed on the bearing surface.

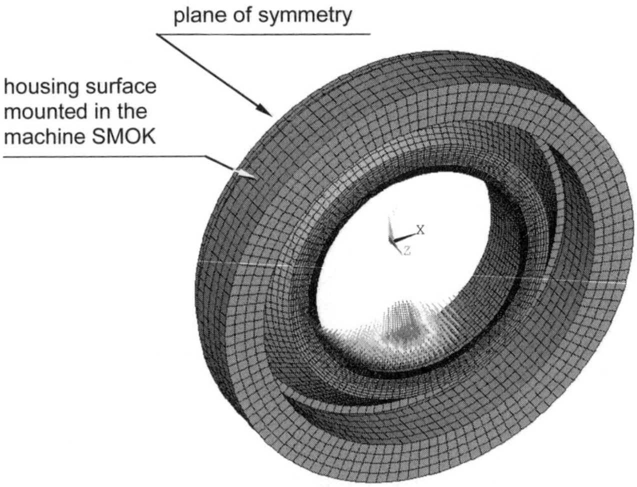

Fig. 4. Finite-element representation of bearing bush and housing and the example of oil hydrodynamic pressure imposed on the bearing surface

Numerical Calculation

As it was said earlier the goal of the calculation is an estimation of stress values in bearing alloy layer responsible for fatigue damage initiations. The analysis is carried out in two stages repeated in an iterative procedure. The first one includes the determination of oil hydrodynamic pressure distribution versus angular and axial position on bearing alloy surface for a given bearing clearance and expected value of hydrodynamic force. The determined oil pressure distribution constitutes the input data to the analysis in the second stage where the calculations of bush and housing deformation and stress distribution in bearing alloy are performed by means of finite-element method using the computer program ANSYS [6].

A significant difficulty in analytical calculation of pertinent critical values of stresses was the finding that for large deformations of bush and elastic housing the displacement of the journal centre corresponds to the eccentricity ratio greater than 1. It means the substantial change in geometry of oil gap and pressure distribution in oil film and, in consequence, a significant change in loading that decides stress values in bearing surface layer. So, after determining the bearing surface deformation a new oil pressure distribution should be calculated for which the next oil gap deformation is estimated and so on - up to achieving a convergence of the procedure. Fig. 5 shows the detailed flow chart of the calculation method.

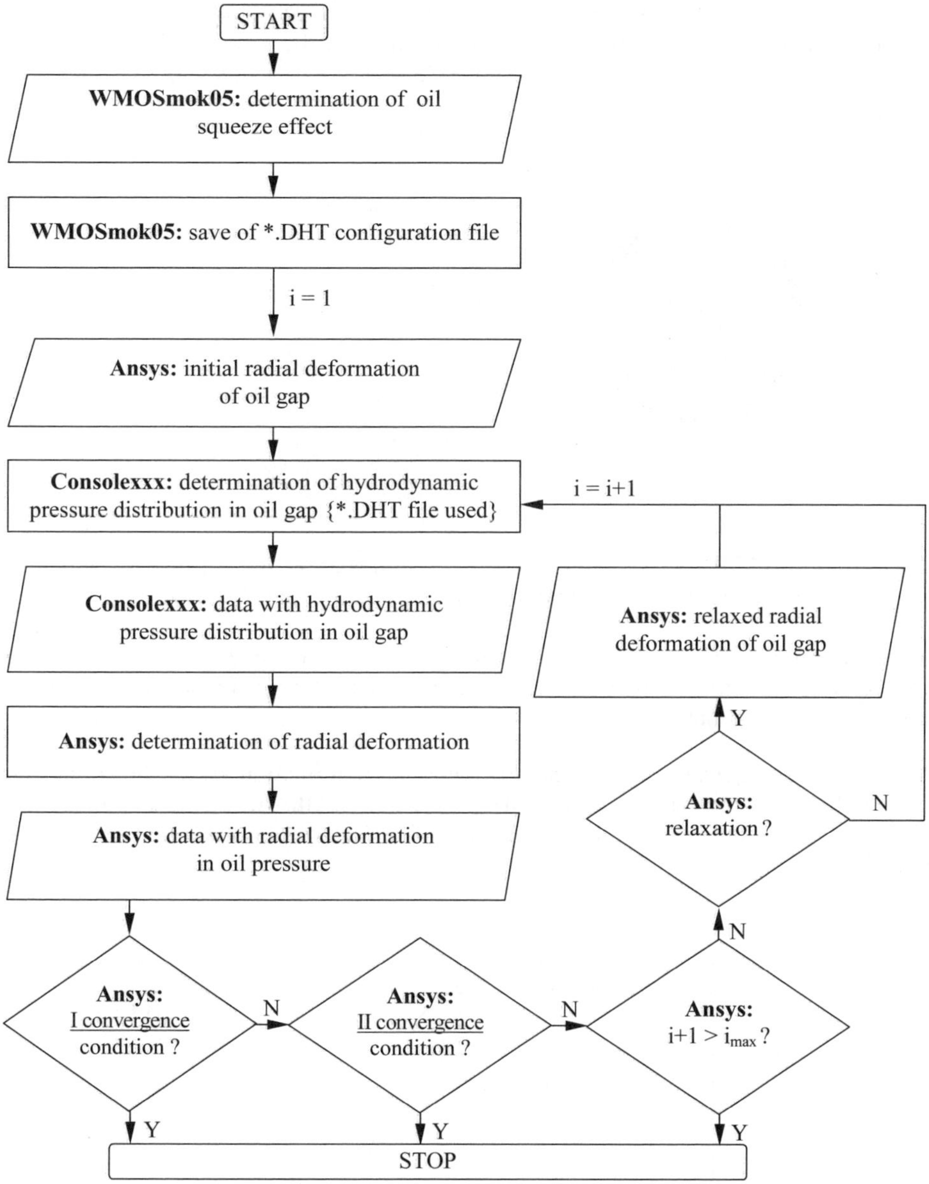

Fig. 5. Flow chart for estimation of final oil pressure distribution and bearing surface layer deformation

Calculation Algorithm. Before starting the essential iterative loop (Fig. 5), a journal centre movement path and an oil pressure distribution in rigid bearing bush fixed in rigid housing is determined using a program WMOSmok05 (Fig. 6). The journal movement is accompanied by lubricant squeeze effect. The pertinent data are written into configuration file (*.DHT) and employed by the second program Consolexxx (both programs were developed by B.J. Siwek). This program is employed in computational loop for determination of the oil hydrodynamic pressure distribution in the defined and deformed oil gap. It should be noted that the Consolexxx enables performing the calculation also for the journal eccentricity ratio greater than 1. The data set describing the oil pressure distribution on the bearing bush surface is collected by ANSYS program that determines the radial displacement in each surface node of the bearing FEM model. The output data from ANSYS are the input data for program Consolexxx – in this way the calculation cycle closes itself.

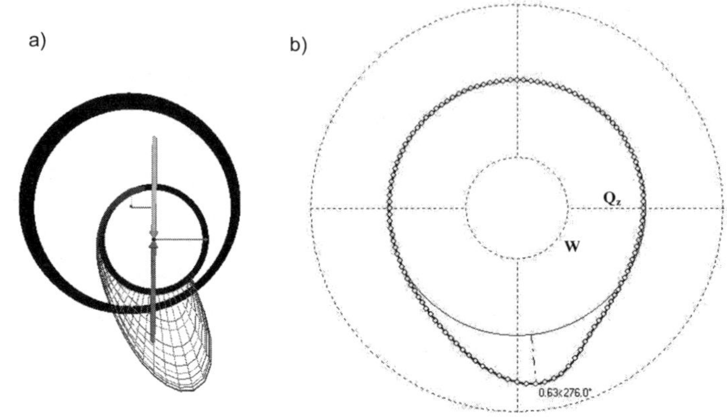

Fig. 6. Distribution of oil hydrodynamic pressure in rigid bearing and corresponding deformation of elastic bush at the moment of maximum load (Fig. 2a) in analyzed bearing (Table 1)

The iterative process of data exchange between programs is automatic. It comes to an end when convergence conditions are fulfilled or, in the case of lack of convergence, after performing the assumed number of iterations.

Conditions of Convergence. After calculating the radial deformation for all the surface nodes the change in radial displacement at each node in relation to the result of previous iteration is determined. As an example, Fig. 7 shows the view of surface transformation in selected (n) cross-section of the bearing for two successive (i and $i+1$) iterations.

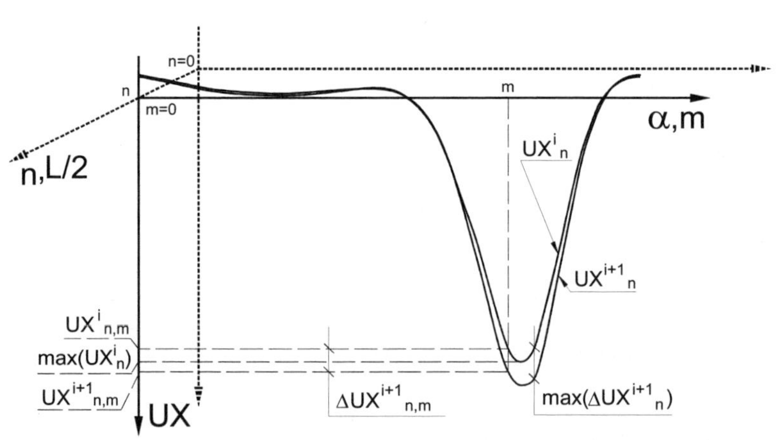

Fig. 7. Division of the bush and housing into finite-elements and bearing slide surface deformation at central cross-section in two successive iterations

(Denotation: α – angular position of the given node in circumferential direction,
m – node number in the circumferential division of the bearing model,
n – node number in the surface axial division of the bearing model,
UX – node displacement in the radial direction,
L/2 – half of the bearing width)

Variation of deformation at each node is analyzed:

$$\mathbf{\Delta UX^{i+1}{}_{n,m} = \left|UX^{i+1}{}_{n,m} - UX^{i}{}_{n,m}\right|} . \tag{1}$$

Two conditions are checked (fulfilling any one brings the calculation to an end):
Convergence I condition (checked at each node):

$$\mathbf{\Delta UX^{i+1}{}_{n,m} \le \Delta UX_{dop} \cdot UX^{i}{}_{n,m}} , \tag{2}$$

where $\mathbf{\Delta UX_{dop} = 1\%}$.

Convergence II condition (it concerns the nodes that do not meet the I condition):

$$\max(\Delta UX^{i+1}{}_{n,m(if:\Delta UX^{i+1}{}_{n,m} > \Delta UX_{dop} \cdot UX^{i+1}{}_{n,m})} \le \Delta UX_{filtr} \cdot \max\left|UX^{i}{}_{n,m}\right| , \tag{3}$$

where $\Delta UX_{filtr} = 1\%$.

The convergence II condition allows to disregard the nodes in which the elastic deformation are small but the convergence I condition is not fulfilled because their relative displacement ΔUX^{i+1} related to the deformation UX^{i} at previous stage, that was also small, gives the great relative value. Hence the condition II enables to eliminate the influence of deformation of peripheral nodes outside the zone of high hydrodynamic pressure.

Relaxation. An attempt of calculation of the resultant radial deformation in bearing lining in a standard way showed that the process is unstable and the attainment of the convergence is impossible. For high loads typical for bearing fatigue investigations it became evident that the elastic surface deformation at the beginning leads to unimodal pressure distribution area (Fig. 6) whereas in the next iteration the hydrodynamic pressure is divided into two regions (two-modal area) at both sides of the deformation zone of the previous step (Fig. 8). In the next step two-modal deformation area appears. This process either repeats itself alternately (Fig. 9) or leads to so great oil gap deformation that it is not possible to determine the oil pressure distribution that counterbalances the assumed load.

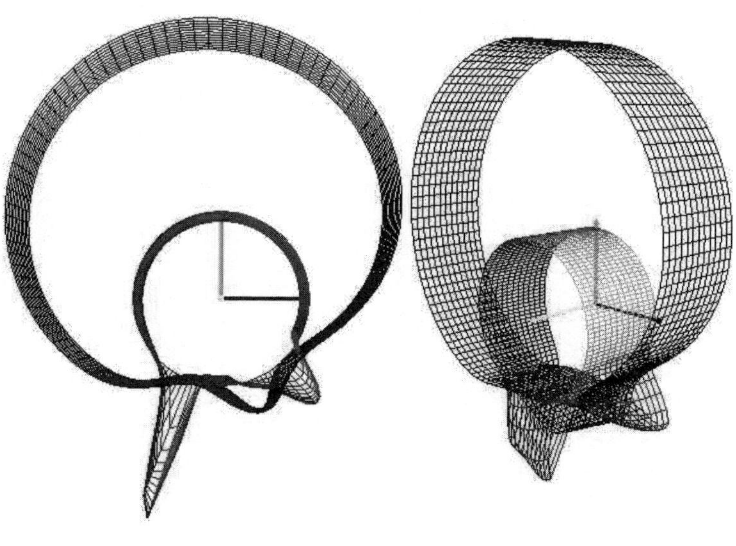

Fig. 8. Distribution of oil hydrodynamic pressure in the deformed oil gap after 8th iteration performed in the standard way

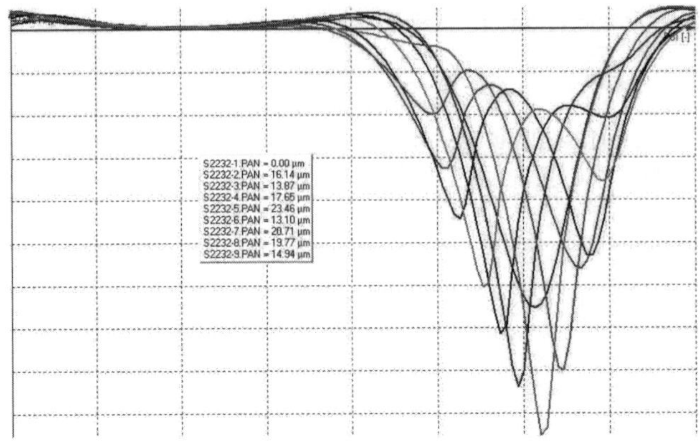

Fig. 9. Deformation of bearing slide surface at central cross-section in successive iterations of computation in the standard way (lack of the convergence)

Several methods for obtaining the convergence of calculation procedure have been tried. They proved that the problem can be solved with using the relaxation method (described by J. Kiciński in [7]). It consists in decreasing the value of radial displacements $(UX^{i}_{n,m})_{relx}$ before sending the data to the program that determines the pressure distribution. So the deformations computed with ANSYS are modified in accordance with the relationship:

$$\left(\mathbf{UX^{i}}_{\mathbf{n,m}}\right)_{\mathbf{relx}} = (\mathbf{1} - \xi) \cdot \mathbf{UX^{i-1}}_{\mathbf{n,m}} + \xi \cdot \mathbf{UX^{i}}_{\mathbf{n,m}}, \tag{4}$$

where $\xi = <0.2 \ldots 1>$.

Selected Results of Calculation

The final calculations have been performed using the relaxation method. The convergence was attained – iteration loop came to an end after 17th step (meeting the convergence II condition). The stresses were determined in both bearing central cross-section and the bush lining.

To illustrate changes in deformation shape in successive iterations the view of surface profiles in bearing central cross-section is presented in Fig. 10.

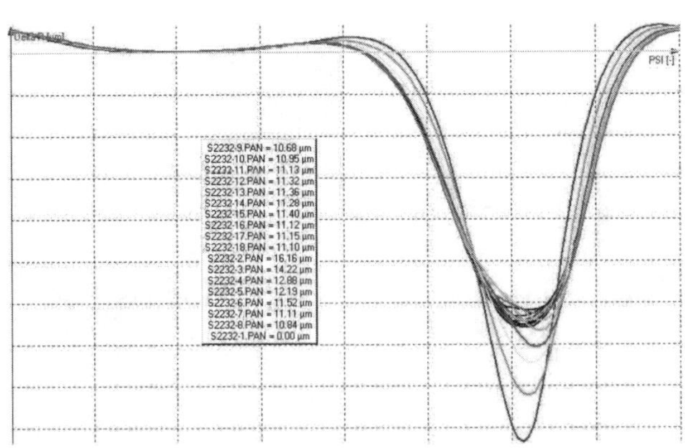

Fig. 10. Deformation of bearing slide surface at central cross-section in successive iterations of computation with using the relaxation method (convergent procedure)

Fig. 11 shows the hydrodynamic pressure distribution in oil film corresponding to bearing load capacity of 56 kN after the last iteration. This figure also shows the shape of deformed bush surface and values of the bearing main operating parameters.

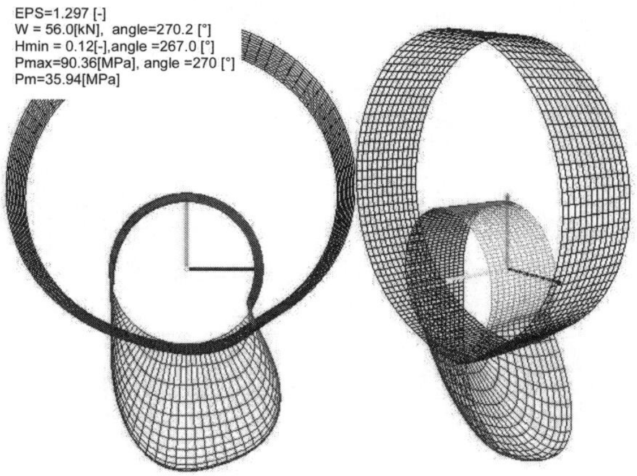

Fig. 11. Distribution of oil hydrodynamic pressure in the deformed oil gap for the load of 56 kN after 17th iteration with using the relaxation method
(Denotation: *EPS* – journal eccentricity ratio,
W – bearing load carrying capacity and the angle of the resultant force,
H_{min} – oil gap relative minimum thickness and its position angle,
Pmax – maximum hydrodynamic pressure and its position angle,
Pm – mean hydrodynamic pressure)

Results of stress calculation for pressure distribution corresponding to the bearing hydrodynamic force of 56 kN are presented, in graphical form, in Fig. 12a – for the bush and housing, and in Fig. 12b – for the bearing alloy surface layer. The figures also show the pressure distribution imposed on bearing surface. In the figures the upper scale concerns the oil pressure values, the lower scale – the reduced stresses.

Analogous data concerning the tangential stresses in surface layer of the bearing are presented in Fig. 13.

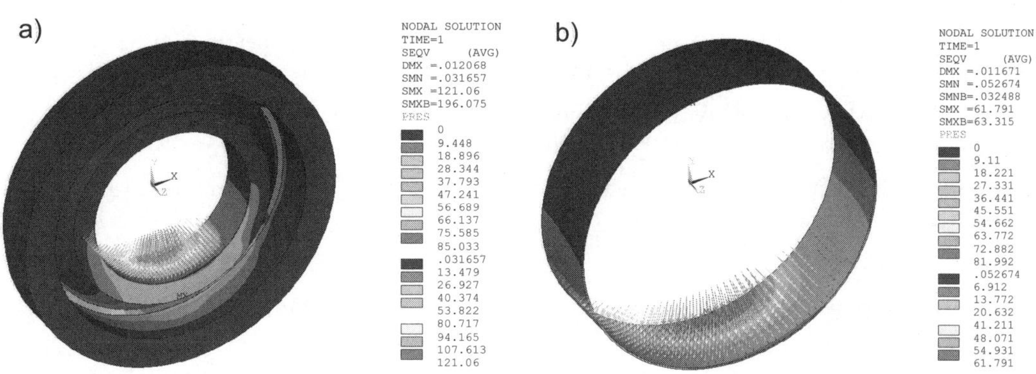

Fig. 12. Reduced stress in [MPa] according to Huber's hypothesis and oil pressure distribution (*PRES*) for the bearing capacity of 56 kN (after 17th iteration) for: a) the bush and housing, b) the surface layer of bearing alloy

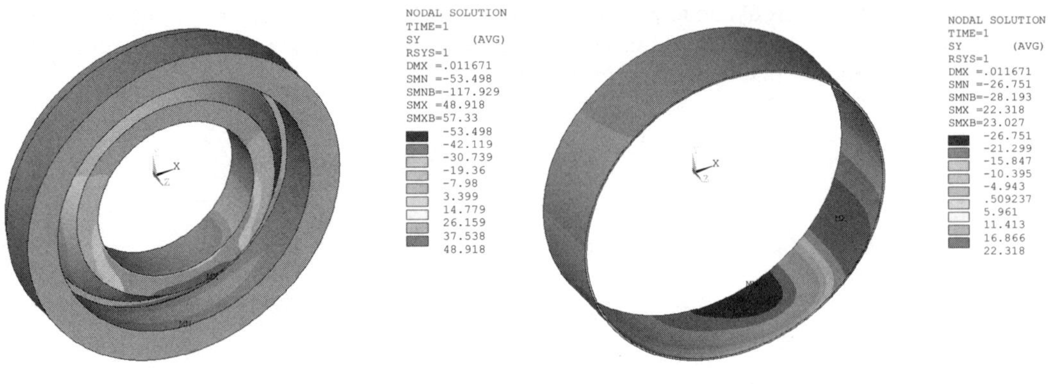

Fig. 13. Tangential stress in [MPa] for the bearing capacity of 56 kN (after 17th iteration) for: a) the bush and housing, b) the surface layer of bearing alloy

Summary

Results presented in the article proved that plain bearing element flexibility, which modifies considerably oil gap shape in the high pressure zone compared to that of a rigid bearing, may cause a significant change in the estimation of critical stress values leading to fatigue cracks in bearing alloy layer during laboratory tests. Without elasto-hydrodynamic analysis the load distribution on the component will be inaccurate, resulting in inaccurate stress calculation. The drawbacks of the standard method of iterative EHD calculation consist in difficulties with convergence of the procedure and may be successfully solved by using relaxation approach presented in the work.

This work has been financed from budget sources on science in 2011 year as a research project (grant) No N N504 340136.

References

[1] O.R. Lang, Surface Fatigue of Plain Bearings, Wear 1977, Vol. 43, 25-33.

[2] F.A. Martin, D.R. Garner, D.R. Adams, Hydrodynamic Aspects of Fatigue in Plain Journal Bearings, ASME, Journal of Lub. Technology 1981, Vol. 103, 150-161.

[3] ISO 7905-1:1995, Plain bearings - Bearing fatigue - Part 1: Plain bearings in test rigs and in applications under conditions of hydrodynamic lubrication.

[4] J. Sikora, Methodology of investigation of plan journal bearing fatigue strength, Gdańsk University of Technology, Gdańsk 1996, Mechanika Nr 534, (in Polish).

[5] J. Sikora, Research Problems of Fatigue Strength of Bearing Slide Layers, Gdańsk University of Technology, Gdańsk 2009 (in Polish).

[6] www.ansys.com

[7] J. Kiciński, Theory and investigation of hydrodynamic plan journal bearings, Rotating Machinery, t. 15, PAN, Wrocław-Warszawa-Kraków 1994 (in Polish).

Key Engineering Materials Vol. 490 (2012) pp 257-264

doi:10.4028/www.scientific.net/KEM.490.257

Radial stiffness of thrust ball bearings

Andrzej Raczynski[1,a], Jaroslaw Kaczor[1,b]

[1]Technical University of Lodz, Department of Heating and Ventilation Engineering,

al. Politechniki 6, 90-924 Lodz, Poland

[a]andrzej.raczynski@p.lodz.pl, [b]jaroslaw.kaczor@p.lodz.pl

Keywords: thrust ball bearing, radial displacement, stiffness.

Abstract. The paper presents the unusual question of determining the dependency between the radial strength applied to the thrust ball bearing against the radial displacement of one ring to another. Contrary to appearances, the shift may occur in a standard bearing system and consequently it may lead to its premature wear and tear. The article depicts a method of determining this dependency (called 'radial stiffness') and shows the examples of calculation results obtained through a special computer program.

Introduction

Thrust ball bearings are designed to carry loads directed axially, that is, along the axis of the shaft. According to textbooks, it is forbidden to load these bearings with any force that is perpendicular to the axis of the shaft [1] due to the ease of "exploding" (leading out) of one of the rings from its proper position with respect to the other one. The radial ball bearing may be installed in the vicinity of the thrust bearing (as shown in Figure 1) to take the load of radial forces. Another possibility is to install a radial displacement limiting pin to prevent the mentioned "explosion" of the thrust bearing.

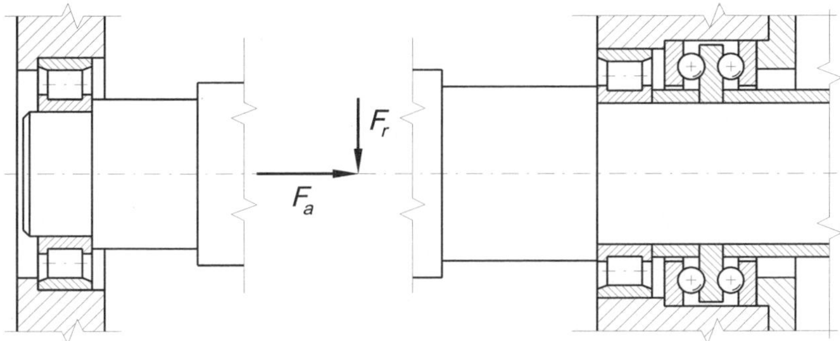

Fig. 1. Bearing support with two bearings (radial and thrust double direction) in a common unit (right)

However, despite the measures taken, in the technical reality, there occurs a radial load in the thrust bearing. This happens, for example, when in the accompanying radial bearing, there is significant inner clearance or when it has been installed loosely in the body. This situation is illustrated in Figure 2. Even a small axial force F_a is trying to set the thrust bearing rings centrally to one another. If the thrust bearing is centered in the base with the radial bearing and the radial force F_r is relatively small, the axial force F_a forces setting the shaft in the axis of the whole in the body without the radial bearing interaction. The larger the internal clearance in the radial bearing section, the more radial force it takes for the radial bearing to cooperate in transferring the radial load. The radial bearing will transfer the radial force until the clearance is fully eliminated.

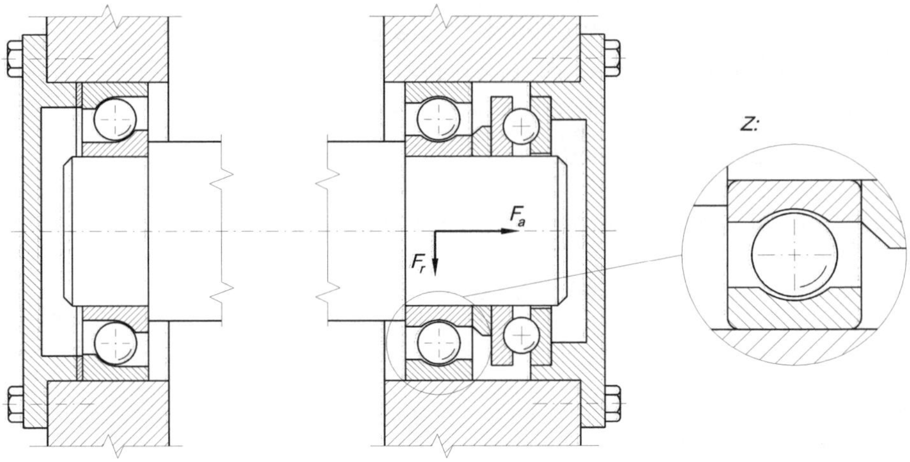

Fig. 2. In the right support of a radial-thrust bearing unit; clearance is visible in radial bearing when zoomed (Z)

Another case of an axial bearing load with radial force is illustrated in Figure 3. It depicts the so-called "toe" bearing that involves taking over the vertical forces (mainly gravity) by the built-in thrust bearing at the bottom. The radial bearing has not been installed next to it. With the dominance of vertical loads, and only accidental and low horizontal (radical) load, such a system can work safely. But even small radial forces that have an impact on the thrust bearing to increase its wear and tear and cause higher or lower eccentricity of the shaft axis against the lower bearing rim.

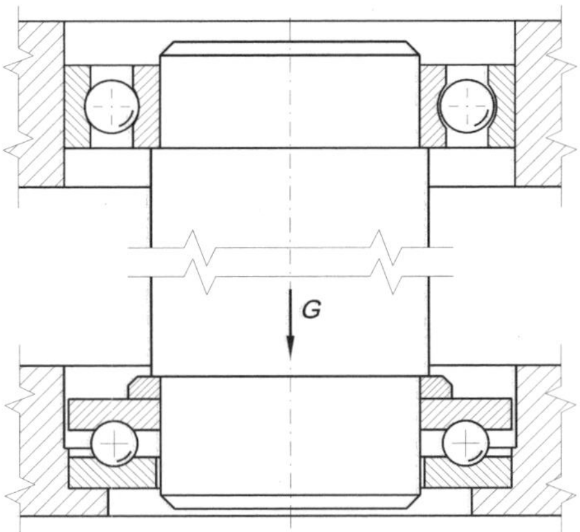

Fig. 3. In the bottom - "toe" bearing without radial bearing

These examples show that, in certain situations, the thrust bearing can be loaded with a radial force. In the first example, the value of this radial force is dependent on the clearance in the radial bearing next to it and directly from the radial displacement between the rings. In the second example, it may be interesting and important to know, from a practical point of view, what radial misalignment (eccentricity) will be produced in the thrust bearing when subjected to a given radial force. As one can see, regardless of the design, it is crucial to determine the dependency between the radial force and radial displacement of one ring against another the other in the thrust bearing. In other words, it is about determining the radial stiffness of the thrust bearing.

This question is not straightforward. In what axial load conditions the radial stiffness is to be determined still remains an open question. It turns out that there may be two opposing models of these conditions. The first model is illustrated in Figure 3. In this case, the axial bearing load is equal to G force regardless of the radial displacement. This follows from the fact that there is no other bearing that is setting the shaft in the vertical direction. If there exists a tendency to raise the upper ring by a few or several micrometers, as a result of radial displacement in the thrust bearing (ΔH dimensions in Figure 4), there will be nothing to oppose the tendency and it will not change the axial load.

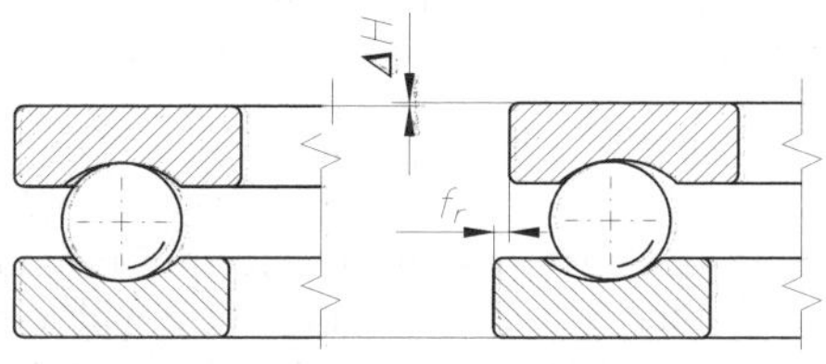

Fig. 4. Increasing of the distance between the rings of the thrust bearing through radial displacement

The second model of the conditions is illustrated in Figure 5. In this case (double direction bearing) the distance between the outer rings of the bearing is rigidly fixed. If the central ring is forced to move in a radial direction, the balls will be fiercely pushed in the races and, as a consequence, the axial force will increase.

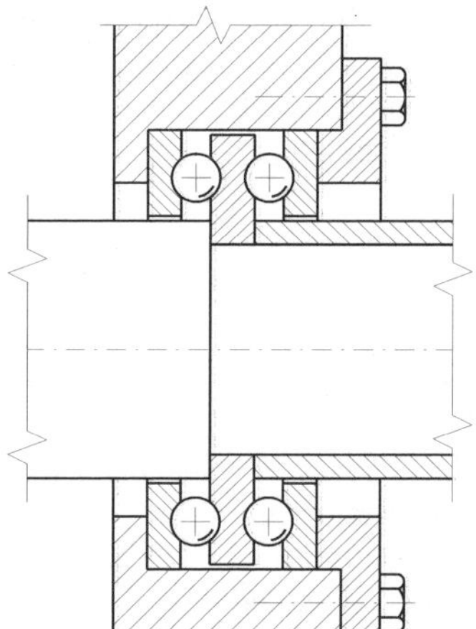

Fig. 5. Double direction thrust bearing - an example of a case, in which it is impossible for the rings to come apart

Thus, the bearing stiffness should be examined in two model cases:
1) assuming a constant axial force,
2) assuming a constant distance between the rings.

These are extreme cases. In the most common bearing systems, in which the thrust bearing is forced in one direction and another bearing (radial or axial) is forced in the other direction (as in Figure 1), there is a long chain of susceptible components and, consequently, neither the force nor the distance between the rings is a constant. In this case, the axial force increase that occurs as a result of radial displacement in the thrust bearing will be less than the above described in the second extreme case.

The calculation model presented below was used to determine the thrust bearing stiffness.

Physicomathematical model

A single-row thrust bearing made from standard bearing steel is a subject under consideration. The following simplistic assumptions were made:
1. Axes of the two rings of each bearing are always parallel to each other.
2. Errors in the shape of balls and races are not considered.
3. The rings remain in an unchanged (ideal) shape of free surfaces regardless of the radial and axial loads.
4. The only deformed areas are those of contact between the balls and the races of the rings.
5. Deformations are considered statically, following the theory of Hertz.
6. The axial (thrust) load is applied exactly in the axis of the bearing.

Calculation method

The force in the roller bearing is transferred from one ring to another by means of balls. The close-up of race of one ring to another δ, defined in the place where the ball is located, equals the sum of contact deformations of both the race and the ball in their contact with both races. According to Harris [2], the pressing force causing a determined contact of the rings is reflected in the below equation:

$$Q = \xi \cdot E' \cdot D_k^{0,5} \cdot \delta^{1,5}, \tag{1}$$

where: D_k – the ball diameter, ξ – non-dimensional parameter of Hertz's contact, dependent on the ratio of race radius R_b to the diameter of the ball bearing D_k, according to [2] or [3], δ – the sum of the standard deformation in the abutment of the balls and rings (the close-up of the outer and inner ring races measured in the direction indicated by the line of action of the balls), E '– Young's substitute module, determined by the formula:

$$E' = \frac{E}{\left(1 - \nu^2\right)}. \tag{2}$$

where: E – Young's module, ν- Poisson ratio.
For the bearing steel, E '= 227.1 GPa was applied.

The axial force F_a on the bearing causes elastic bends of balls and the race. With a centric positioning of one ring of the bearing against the other, the bearing load forces Q of the balls are identical and can easily be calculated by dividing the force F_a by the number of balls. Transforming the formula (1), the following equation is obtained:

$$\delta = 2{,}6865 \cdot 10^{-4} \cdot Q^{\frac{2}{3}} \cdot D_k^{-\frac{1}{3}} \cdot \xi^{-\frac{2}{3}}, \tag{3}$$

which allows for an initial determination of the close-up of bearing rings loaded only with an axial force F_a:

$$f_a = 2{,}6865 \cdot 10^{-4} \cdot \left(\frac{F_a}{Z}\right)^{\frac{2}{3}} \cdot D_k^{-\frac{1}{3}} \cdot \xi^{-\frac{2}{3}}, \tag{4}$$

where: f_a – the close-up of rings, Z – number of balls.

When the radial displacement is introduced into considerations, it interferes with the distribution of axial forces on individual balls. Some of the balls (mostly the ones located on the direction of the displacement) will be subjected to an increased load because, in the places where they are located, the races are the closest to each other, as shown in Figure 4.

It is not possible to determine the radial displacement using the knowledge of the radial and axial forces because you cannot predict the number of the balls loaded upon at the same time, nor the dependency between the radial and the axial bend. The only way is to assume the value of the axial f_a and radial f_r displacement, the calculation of the close-up value of δ for each ball, the definition of force Q on each ball and then designating the axial and radial forces as the sum of views of all the forces Q in the appropriate directions.

Calculations of contact deformations on the basis of displacement f_r and f_a were made according to method as in [4]. The procedure in relation to the axial ball bearing is illustrated in Figure 6. The Figure 6a shows the ball in the unloaded state. The S_n dimension - the distance between the centers of race sections in this condition equals to:

$$s_n = 2 \cdot R_b - D_k, \tag{5}$$

where: R_b – radius of the ring race section.

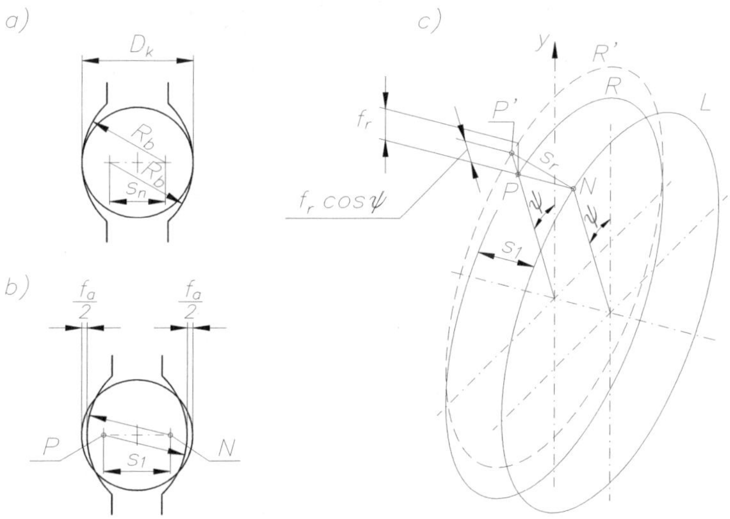

Fig. 6. The illustration of the method for calculating the contact deformation

Figure 6b shows the same ball in the bearing loaded with an axial force, but without radial displacement. The close-up of the rings resulting from the axial force f_a causes an increase in the distance between the centers of the race sections to the value of s_1:

$$s_1 = s_n + \frac{f_a}{2} + \frac{f_a}{2} = s_n + f_a. \tag{6}$$

The location of these centers is marked with points P (center of the right ring section) and N (center of the left ring section).

Figure 6c shows the effect of radial displacement in a bearing in the direction y. The envelope R is the circumferential axis of the torus forming the right ring race, and the envelope L is the circumferential axis of the torus forming the left ring race. The points P and N are found on these envelopes. Let us consider the ball placed in any radial area from the axis y at an angle ψ. Now let us assume that the ring moves in the right direction y about the distance f_r. This is illustrated by a shift of the envelope R about such distance to R' (the envelope is indicated by a broken line). Thus, the center of the section of the right in the considered area moves from point P to point P'. The distance between points P and P' is the product of displacement and the cosine of the angle ψ, which defines the location of the considered ball. The distance between the points P' and N is defined as:

$$s_r = \sqrt{(s_n + f_a)^2 + (f_r \cdot \cos\psi)^2}\,. \tag{7}$$

The difference in distance s_r and s_n is a summary of contact deformation of the ball and both races:

$$\delta = s_r - s_n\,. \tag{8}$$

Knowledge of the summary allows the designation of the contact deformation of normal force Q operating between the ball and the races of the bearing rings according to formula (1). One should only consider that in certain positions of the balls, the calculations may lead to a negative value of δ strain. In such cases, a zero force value Q is applied.
The Q forces operate along the drawn line through the points N and P' (it is a so-called line of action of the ball). Due to the need for adding up the projections of these forces in the axial direction and in the direction of the y-axis, it is necessary to know the operating angle of each ball. In Figure 6, it is marked as a PNP'angle and as α in Figure 7. The angle is different for the balls placed in different places, in accordance with the dependency:

$$\alpha = \text{arc tg}\frac{f_r \cdot \cos\psi}{s_1}\,. \tag{9}$$

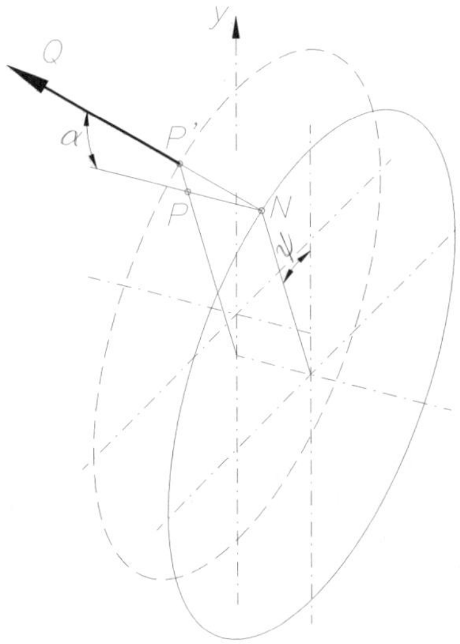

Fig. 7. The location of the normal force operating between the ball and the race

Considering the angles of the direction of the Q vectors, one can specify the total of projections of all the Q forces in the axial direction and in the y direction, which is synonymous with designating of external forces (axial and radial) corresponding to a given displacement f_a and f_r:

$$F_a = \Sigma\,(Q \cdot \cos\alpha), \tag{10}$$
$$F_r = \Sigma\,(Q \cdot \sin\alpha \cdot \cos\psi)\,. \tag{11}$$

Based on the presented dependencies, a computer program was developed, which enables one to determine the force of the radial bearing, depending on the radial displacement of the rings. The software can make calculations based on two model cases discussed in the introduction. One can make calculations following the assumption that the axial force remains unchanged with the increase of radial displacement or following the assumption of an unchanged distance between the rings when the radial displacement increases. The software was also equipped with a control function, which enables monitoring whether the radial displacement is not absurdly high, that is, whether it does not make the balls move out of their races (the "explosion" of the bearing). The boundary shift was adopted on the basis of the assumption that the theoretical contact point of the ball with the race must be found within the race area and the shortening ("cutting") of the ellipse of the contact cannot be greater than 15% over the full length of the ellipse. The 15% limit is equivalent to keeping 70% of the length of semi-axis of the ellipse of contact, which according to [5], reduces the amount of contact stress to the level of the stresses in the middle of the ellipse of contact p_H. Such limit is illustrated in Figure 8.

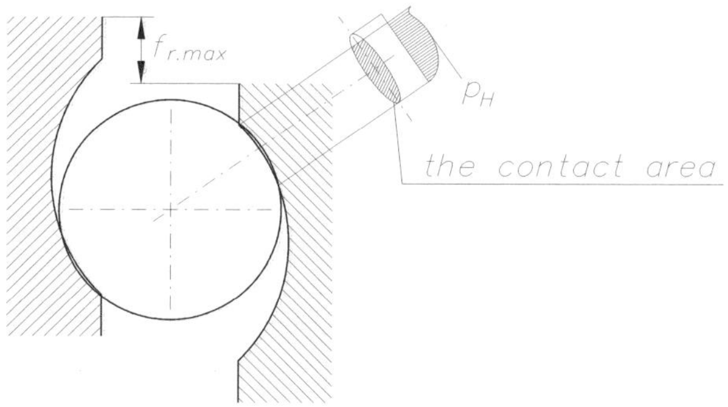

Fig. 8. The illustration of the boundary displacement of the bearing rings

Calculation results

Sample calculations were performed for a single-row axial ball bearing with the following dimensions:
- ball diameter: 12.7 mm,
- number of balls: 8,
- race radius: 6.73 mm,
- race depth: 1.25 mm,
- axial load: 48 kN.

According to the above findings, the calculations were made for two cases. As a result of the calculations, the dependence of radial force from forced radial displacement was obtained as illustrated in Figure 9.

The results show that, in a situation when the distance between the bearing rings remains unchanged, the radial force is greater than in the second case. This follows from the fact that maintaining an unchanging distance causes greater ball pressing into the rings. As established in the introduction, these two cases under consideration are extreme for technical reality. It can be concluded that, in practice, the radial force at a given displacement equals the value found between one and another line in the presented graph.

The resulting lines are curved. Therefore, one cannot specify a single value of radial stiffness of the axial ball bearing, understood as the ratio of the increase of force displacement F_r to the increase of displacement. The stiffness may be identified in the compartments, but it is not useful. In practice, it is sufficient that the radial force is dependent on the radial displacement, as shown in Figure 9.

The available calculating software can be used with any size of thrust ball bearing.

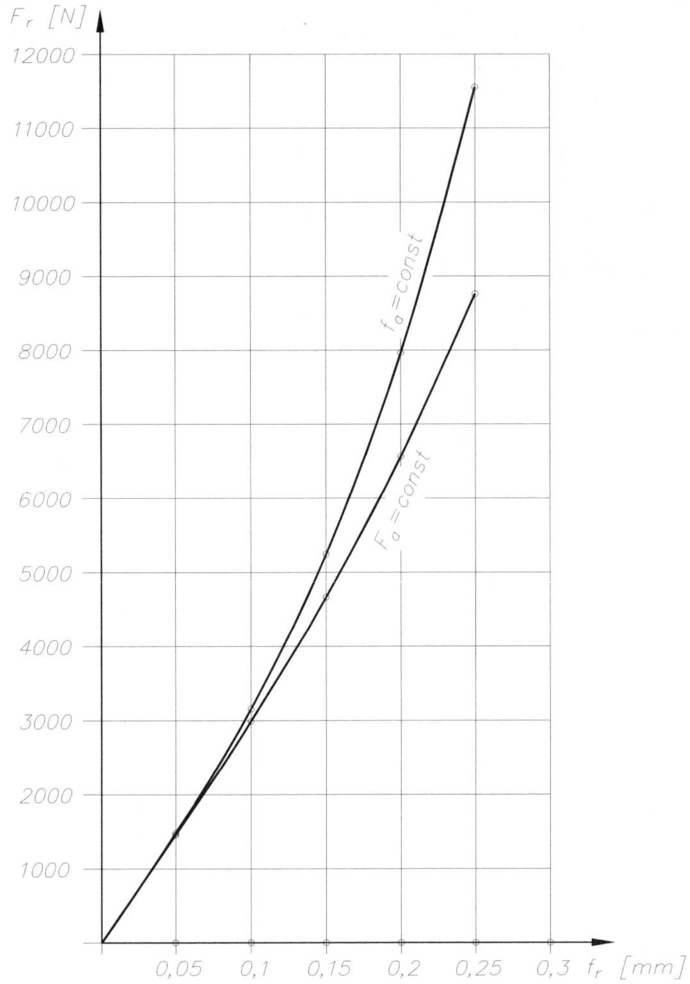

Fig. 9. The radial force dependence to radial displacement

References

[1] H. Krzemiński-Freda, Rolling Bearings, PWN, Warsaw 1989.

[2] T.A. Harris, Rolling Bearing Analysis, NewYork-London-Sydney 1984.

[3] S. Andreason, On load distribution in rolling bearing with special reference to the influence of bearing misalignment, Chalmers University, Göteborg 1973.

[4] J. Kaczor, The Methodology of Calculations of Rolling Bearing Durability of Supporting Shaft, Technical University of Lodz, Lodz 2007.

[5] H. Krzemiński-Freda, The dependence of roller-race contact durability and capacity on pressure distribution along generatrix, The Archive of Mechanical Engineering 1993,T XL, z.1-2.

Key Engineering Materials Vol. 490 (2012) pp 265-272
doi:10.4028/www.scientific.net/KEM.490.265

Relieving the Measuring System of the Trailer for Tyre/Road Rolling Resistance Measurements from the Inertia Force

Ryszard Wozniak[1a], Stanislaw Taryma[1b] and Grzegorz Ronowski[1c]

[1] Gdansk University of Technology, Faculty of Mechanical Engineering, ul. Narutowicza 11/12, 80-233 Gdansk, Poland

[a]rwozniak@pg.gda.pl, [b]staryma@pg.gda.pl, [c]gronowsk@pg.gda.pl

Keywords: Road Measurement, Tyre, Inertia Force, Vibration

Abstract. In the article the principle of measuring of tyre/road rolling resistance tyres in road conditions applied in a special measuring trailer is presented. This trailer was built in Faculty of Mechanical Engineering at Gdansk University of Technology. The construction of two special devices used in this trailer was discussed. These devices eliminate the influence of some chosen confounding measurements factors such as: the inertia force acting on the measuring arm with the wheel and the tested tyre and vibrations of the measuring arm. The operation of these devices was proven in practice. The conclusions arising from the use of these systems are described.

Introduction

European Parliament voted to introduce new rules on tyre labelling [1]. Since November 2012 all new tyres sold in Europe will have to be marked in terms of external rolling noise, wet grip and fuel efficiency. System is going to remind the system used to determine household appliances. The aim of introducing the new system is to enable buyers to compare different tyres and make a better choice of them. It is expected that introducing a new system of tyre labelling will reduce vehicle maintenance costs and will decrease emissions of environmentally harmful substances [2].

Fuel efficiency is directly related to tyre/road rolling resistance [3]. Reduction of tyre/road rolling resistance reduces fuel consumption and emissions of toxic substances contained in exhaust gases. Therefore it is important to have ability of measuring tyre/road rolling resistance in real road conditions.

Such road tests can be performed using a special measurement trailer. The trailer of this type was built in Faculty of Mechanical Engineering at Gdansk University of Technology. The results of measurements performed using this trailer were presented among others in [4, 5]. The principle of measuring tyre/road rolling resistance tyres using this trailer is shown in Fig. 1. A measure of the rolling resistance force F_t is the deflection angle Θ of the measuring arm on which the wheel with the tested tyre is assembled. The coefficient of tyre/road rolling resistance C_{RR} is determined. It is the ratio of the rolling resistance force F_t to the normal surface reaction force Z acting on the tested tyre [6] (see Fig. 1). During measurements of the tyre/road rolling resistance coefficient several confounding factors disturb the tests. In this article two of them are mentioned. The first one is the inertia force disturbance. The second one are the vibrations of the measuring arm.

Elimination of the Inertia Force Disturbance

If during the measurements the longitudinal acceleration or deceleration is acting on the measuring trailer the measuring arm with the tested tyre is deflected additionally by the inertia force. This additional deflection increases the measurement error.

In the measurement trailer the special design solution was put into practice to eliminate the influence of the inertia force. Schematic drawing of the device is shown in Fig. 1, the photo in Fig. 2 and the operating description is mentioned further.

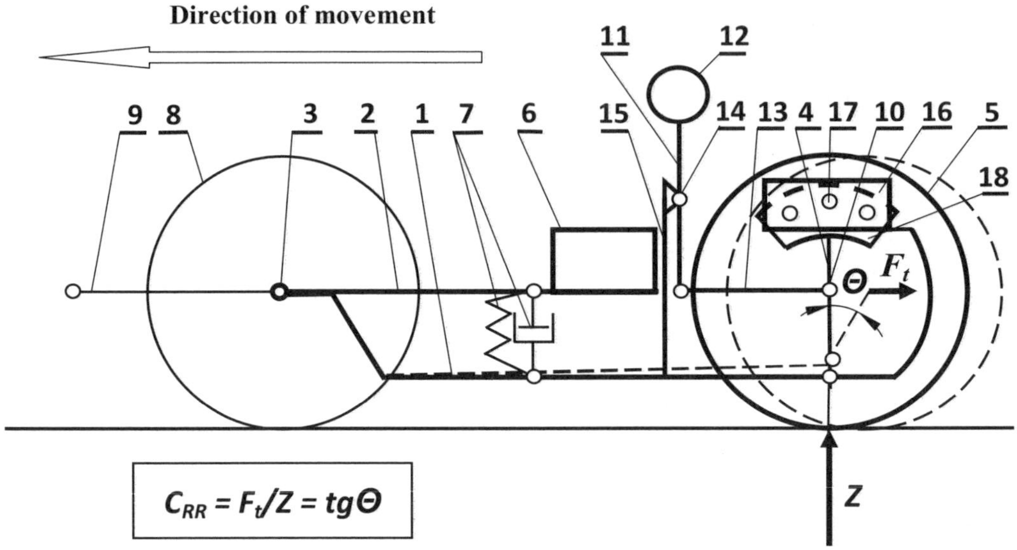

Figure 1. Scheme of the measuring trailer with the device for relieving the measurement arm from the inertia force:
1 - frame of the trailer, 2 - swing arm with weights, 3 - axis of rotation of the frame and swing arm with weights, 4 - measuring arm, 5 - tested tyre, 6 - pod with weights, 7 - elastic-damping column, 8 - front wheel, 9 - trailer hitch, 10 - axis of rotation of the tested tyre, 11 - vertical lever, 12 - correction mass, 13 - connector, 14 - axis of rotation of the vertical lever, 15 - bracket, 16 - electrodynamic brake casing, 17 - electromagnets, 18 - sector of the copper disc.

Figure 2. The device for relieving the measurement arm from the inertia force mounted in the measurement trailer (on the right side).

To the axis 10 of the tested tyre 5, (which is the center of mass of the measuring arm 4) the force is applied with direction and value such as the inertia force operating at the same moment, but with the opposite sense. Because of this it results in relief of the measuring arm with the tested tyre from the inertia force. The advantage of this solution is that it can compensate the disturbance influence of longitudinal acceleration or deceleration on results of the tyre rolling resistance coefficient measurements. The axis of the tested tyre 10 is jointedly connected by the connector 13 to the vertical lever 11.

In the middle of its length the vertical lever 11 is assembled jointedly by the bracket 15 to the frame of the trailer 1. At the upper end of the lever 11 the correction mass 12 is fitted. This mass ($m_{correction}$) is equal to the sum of the masses of the tested tyre m_{tyre} and measuring arm $m_{measuring\ arm}$ ($m_{correction} = m_{tyre} + m_{measuring\ arm}$). If during the measurements the longitudinal acceleration or deceleration a is acting on the trailer the inertia force $(m_{tyre} + m_{measuring\ arm})\cdot a$ is acting on the measuring arm 4 with the tested tyre 5 in the axis of rotation of the tested tyre 10.

On the correction mass 12 an inertia force $m_{correction.}\cdot a$ acts as well. Taking it's value into account it is equal to the force $(m_{tyre} + m_{measuring\ arm})\cdot a$. By the vertical lever 11 and the connector 13 this force is transferred to the axle of the wheel with the tested tyre as the force with the opposite sense. These forces: $m_{correction.}\cdot a = (m_{tyre} + m_{measuring\ arm})\cdot a$ neutralize each other. Because of this it is possible to relieve the measuring arm 4 from the inertia force during the rolling resistance coefficient measurements. In these considerations the vertical lever 11 and the connector 13 are treated as weightless. During the measurements the angle Θ of the measuring arm becomes a few degrees only. For such small angles of deflection in these reflections the influence of the moments of inertia caused by the rotation have been omitted.

The following figures (Fig. 3, Fig. 4 and Fig. 5) show the results of experimental selection of mass $m_{correction}$ to obtain minimal fluctuations of the angle Θ of the measuring arm during acceleration and deceleration of the measuring trailer.

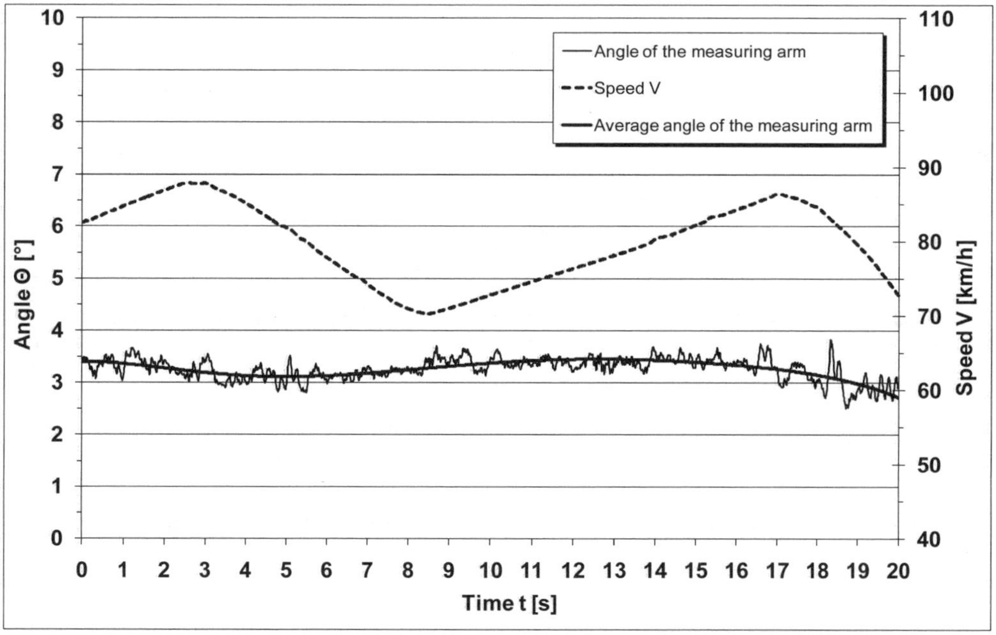

Figure 3. Fluctuations of the angle Θ of the measuring arm during acceleration and deceleration of the measuring trailer for mass $m_{correction} = 5$ kg (mass too big).

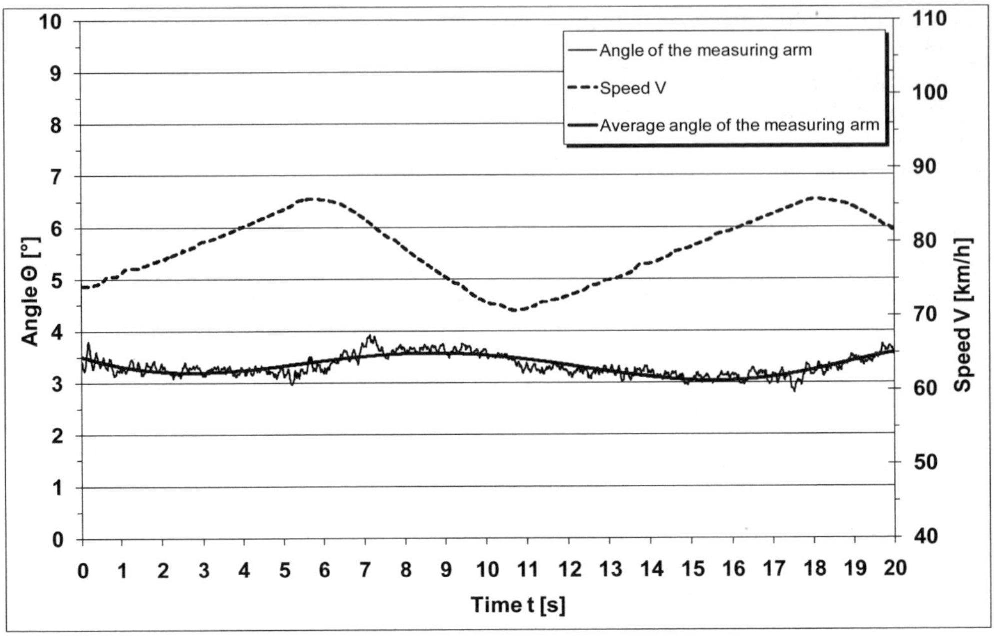

Figure 4. Fluctuations of the angle Θ of the measuring arm during acceleration and deceleration of the measuring trailer for mass $m_{correction}$ = 3 kg (mass too small).

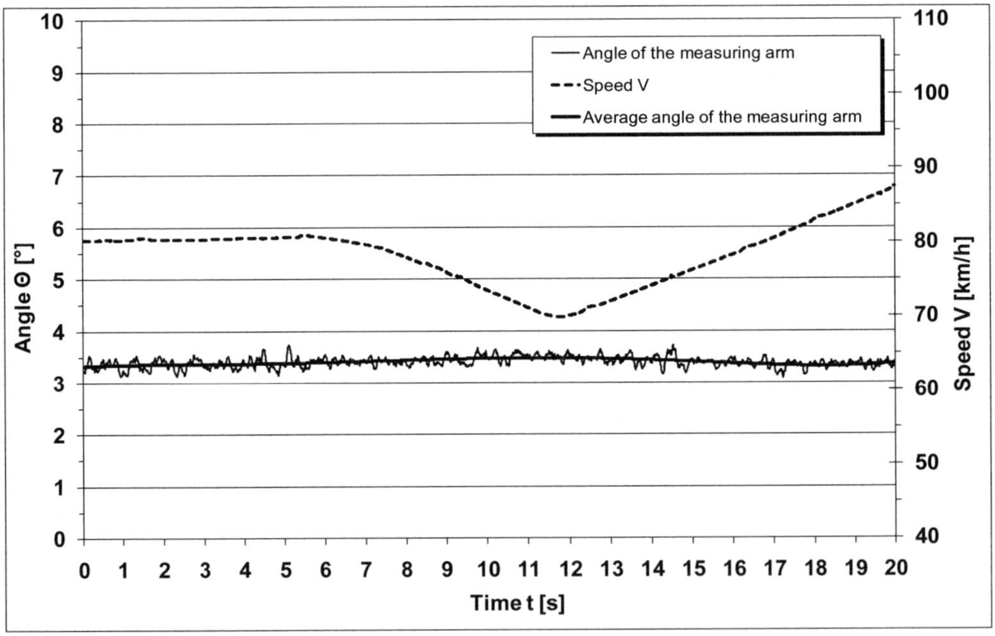

Figure 5. Fluctuations of the angle Θ of the measuring arm during acceleration and deceleration of the measuring trailer for mass $m_{correction}$ = 4 kg (correct mass).

The traffic measurements show that the device described above also eliminates the disturbance effects of the longitudinal slope of the road.

This strictly mechanical device has been reported as the patent application.

Dumping Vibrations of the Measuring Arm by the Electrodynamic Brake or the Magnetic Brake

During measurements forced vibrations of the measuring arm were damped by the electrodynamic brake mounted in the measuring trailer (and Fig. 6).

Figure 6. The electrodynamic brake.

The maximum deflections of the measuring arm are limited by the rubber bumpers. The electrodynamic brake has been built using the phenomenon of interaction between the permanent magnetic field and an electrical conductor moving in this field. The sector of the copper disc 18 (see Fig. 1) with a thickness of 2 mm was applied as the electrical conductor. The electromagnets 17 were the source of electromagnetic field. They were placed in the casing 16. This casing was assembled to the frame 1 of the trailer. The sector of the copper disc 18 was mounted on the measuring arm 4 and was moving with it together.

The electrodynamic brake was dumping vibrations very well. Fig. 7 shows the undamped forced vibrations of the measuring arm (the electrodynamic brake switched off).

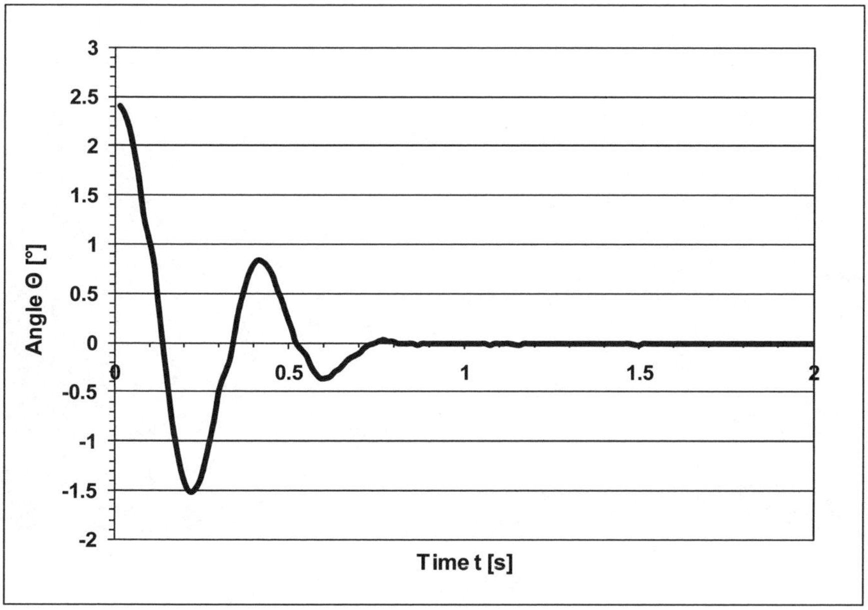

Figure 7. The dependence of the measuring arm deflection angle Θ as function of time for data obtained from measurements for the electrodynamic brake switched off.

In Fig. 8 we can see the damped forced vibrations of the measuring arm (the electrodynamic brake switched on).

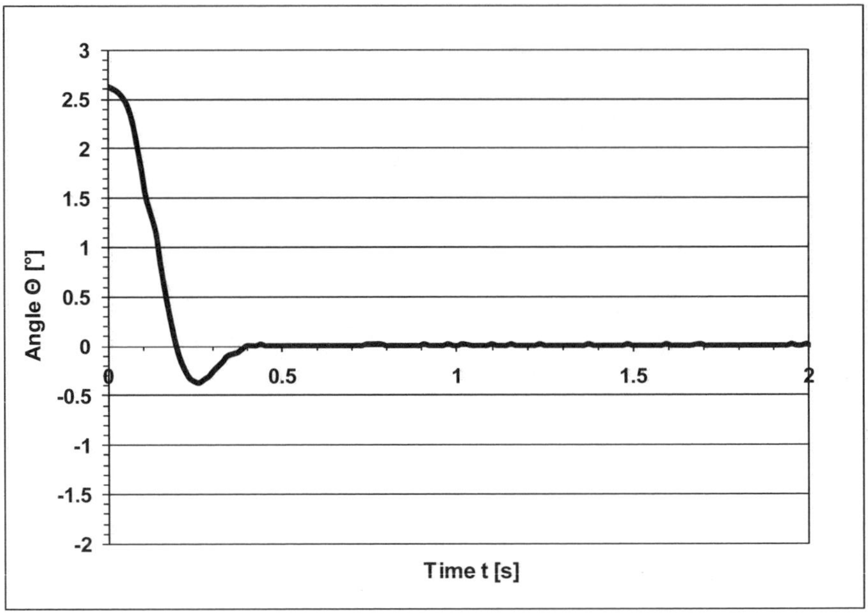

Figure 8. The dependence of the measuring arm deflection angle Θ as function of time for data obtained from measurements for the electrodynamic brake switched on.

Unfortunately the inclusion of the electrodynamic brake resulted in a large energy consumption which could not be compensated by either large-capacity battery nor the constant recharging from the alternator of the towing vehicle. Conducting long-time road measurements were therefore impossible. That is why it was decided to replace the electrodynamic brake by the magnetic brake with the neodymium (permanent) magnets which do not require power supply (Fig. 9).

Figure 9. The magnetic brake.

To be sure that the magnetic brake will work well the tyre/road rolling resistance coefficient measurements were done on 18 different road surfaces (one by one) for SRTT tyre (UNIROYAL P225/60R16 97S M+S with tread TIGER PAW) at 80 km/h. The internal pressure was fixed at 210 kPa and the vertical load Z was fixed at 4002 N. The results (not filtered) are shown in Fig. 10 and in Fig. 11.

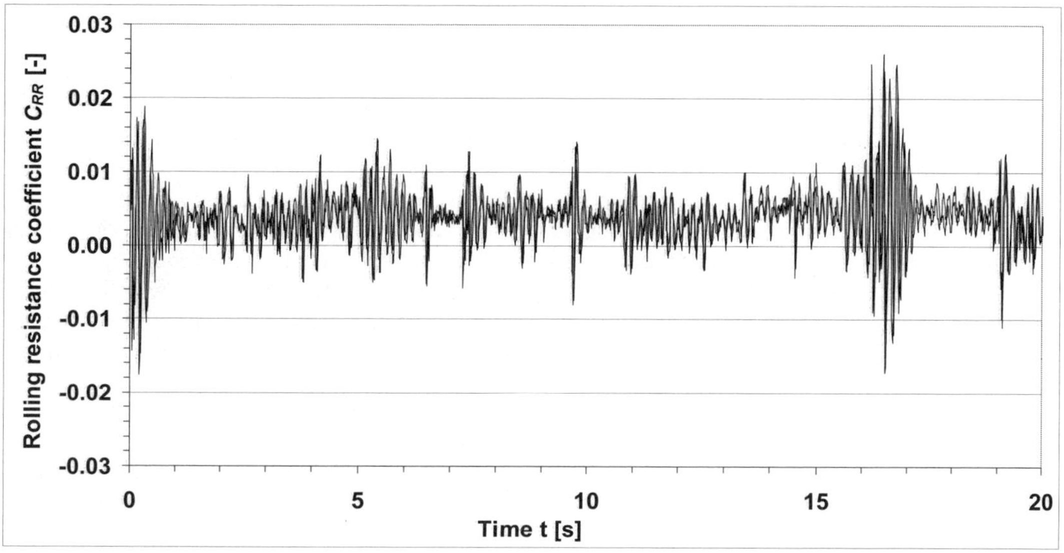

Figure 10. Tyre/road rolling resistance coefficient of SRTT tyre as a function of time on 18 different road surfaces (one by one) for the magnetic brake taken out from the measuring trailer.

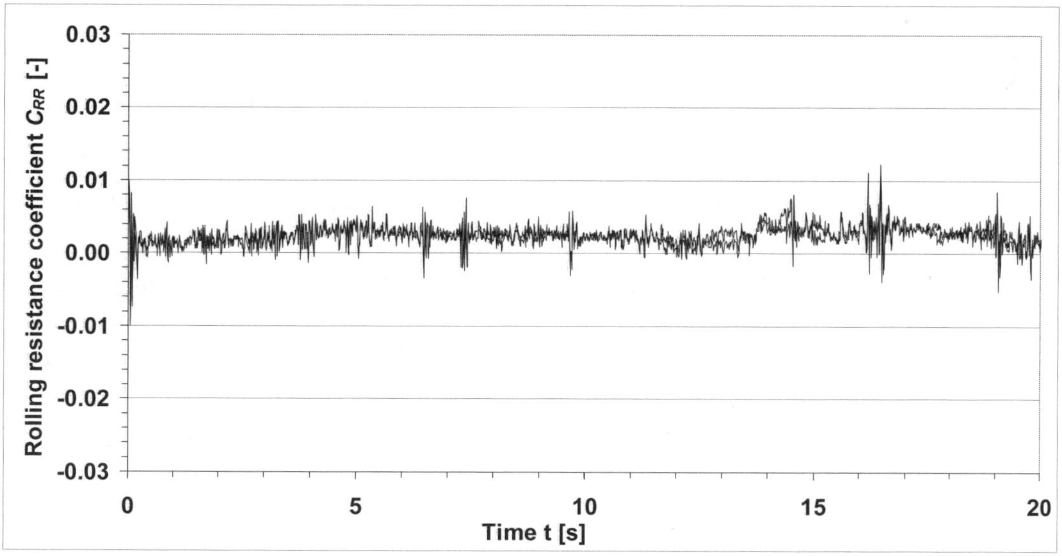

Figure 11. Tyre/road rolling resistance coefficient of SRTT tyre as a function of time on the same 18 different road surfaces (one by one) for the magnetic brake in operation.

The magnetic brake was dumping the vibrations very well. The phases of the vibrations were exactly the same in both cases (see Fig. 10 and Fig. 11). The amplitude of vibrations of the measuring arm decreased very much when the magnetic brake was in operation. This device seems to be very useful during the tyre/road rolling resistance coefficient measurements.

Conclusions

Compensation system used in the trailer effectively relieves the measuring arm with the wheel and the tested tyre from the inertia force disturbances. Measured fluctuations of the deflection angle Θ of the measuring arm caused by the inertia force were minimal. Correction mass $m_{correction}$ was chosen in an experimental way.

The electrodynamic brake was dumping forced vibrations of the measuring arm very well. Unfortunately the electrodynamic brake was characterized by a large energy consumption which could not be compensated by either the large-capacity battery nor the constant recharging from the towing vehicle alternator in traffic conditions. Because of this the period of time which could be applied for road measurements was too short and could not be acceptable. Therefore it was decided to exchange the electrodynamic brake for an equally effective brake with the neodymium (permanent) magnets which do not require external power supply.

Elimination of the inertia force disturbance can be done applying additional mathematical formula in the algorithm for determining of the tyre/road rolling resistance coefficient. There exists another way. It makes use of eliminating the source of disturbances by the special devices described by the authors. The method of eliminating the sources of disturbances in authors' opinion is much more effective.

The most reliable tyre/road rolling resistance measurements can be done only in real tyre maintenance conditions. It means that they should be done on real road using special equipment such as a measuring trailer.

References

[1] Regulation (EC) No .../2009 of the European Parliament and of the Council of on the Labelling of Tyres with Respect to Fuel Efficiency and Other Essential Parameters, Brussels. 14639/09 2009.

[2] Z. Chlopek, Natural Environment Protection, Communication Publishing House, Warsaw, 2002.

[3] J. R. Bang, Influence of Tire Rolling Resistance on Fuel Consumption, Bang Consult, Sandnes, 1999.

[4] S. Taryma, R. Wozniak, Tyre/Road Rolling Resistance Measurements in Road Conditions, Asphalt Pavements. 2 ($2008_{(14)}$) 6-9.

[5] J. A. Ejsmont, S. Taryma, P. Mioduszewski, R. Wozniak, G. Ronowski, M. Cwikla-Duda, M. Wisiorek, Rolling Resistance Results - Silvia Project Report, Gdansk, 2005.

[6] K. S. Clark, Mechanics of Pneumatic Tires, Washington, 1978.

Key Engineering Materials Vol. 490 (2012) pp 273-281
© (2012) Trans Tech Publications, Switzerland
doi:10.4028/www.scientific.net/KEM.490.273

Selected Properties of Bearings with Helical Groove on the Journal

SEP Jaroslaw

Rzeszow University of Technology
Faculty of Mechanical Engineering and Aeronautics, W. Pola 2
35-959 Rzeszow, Poland
jsztmiop@prz.edu.pl

Key words: slide bearing, wear, oil, contamination, load carrying capacity, helical groove, oil flow rate.

Abstract. The paper presents results of tests concerning slide bearing with helical groove on the journal selected properties. It has been stated that during lubrication with contaminated oil helical groove results in wear decrease. On the other hand while lubricating agent was clean oil helical groove results in load carrying capacity decrease and oil flow rate increase. The proposition of non-dimensional index describing the groove and helical line is also presented in the paper. This index is correlated with bearing characteristics and its wear

Introduction

Hard particles in the oil are one of the reasons of premature failures of slide bearings. Contaminants appeared in oil in production process but during operating their number considerably increases [1]. Sizes of physical oil contaminants can be as large as 200 μm [2], they are greater than the minimum thickness of oil film. Classical slide bearing, consisting of a soft bearing sleeve mating with a hard shaft is exposed to damage in case of hard particles presence in oil.

The first analysis of the effect of presence of abrasive particles in oil on the tribological wear of the sliding elements began more than 90 years ago [3] and a more systematic analysis started in 1950s [4, 5]. The problem of hard contaminants presence in the lubricant on wear processes is being investigated at present, see for example [6-11]. Results of these investigations indicate that hard particles contained in the lubricating oil intensify wear processes. Abrasive particles can be embedded into co-acting surfaces, rolled between them or disintegrated.

Change of the bearing journal design is one of the possibilities of destructive hard particles affect decrease. It was found [12] that a two-component surface layer manufacturing on the bearing journal caused 5 times smaller bearing sensitivity to destructive contaminants action. A two-component surface layer was created by copper located in helical groove in the base material (steel). This layer was manufactured also in order to create triangular groove in materials contact zone. During the shaft rotational movement the helical groove intensified axial oil flow [13] and was a place of contaminants location [14]. Therefore the smaller number of abrasive particles were supplied to the friction contact zone and as the consequence wear decreased.

These interesting from practical and scientific points of view results prove necessity of further research of bearings with helical grooves on journal. The aim of this paper is presentation of results of the experimental investigations concerning the effect of the groove geometry on the bearing load capacity and oil flow rate and wear in presence of abrasive particles in the lubricating fluid. The attempt was additionally done in order to determine index describing the groove and helical line; this index will be correlated with bearing characteristics and its wear.

Bearing with journal having helical groove

Figure 1 presents scheme of bearing with journal having helical groove.

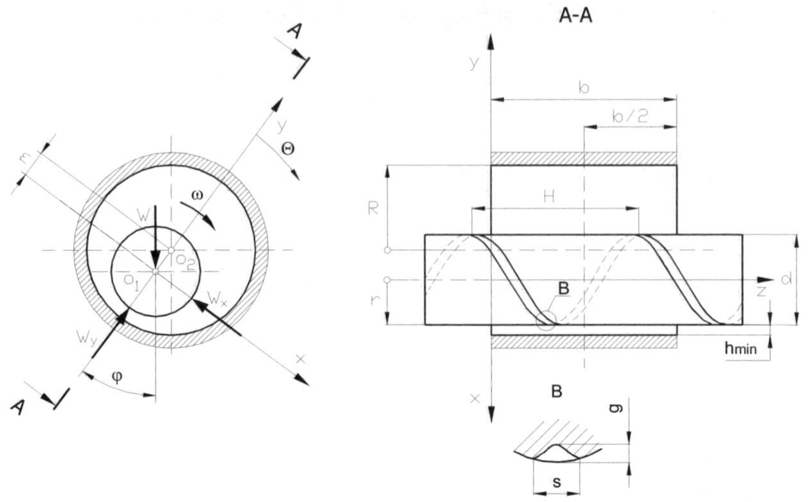

Fig. 1. Bearing with journal having helical groove: R – sleeve radius, d(r)– journal diameter (radius), b – bearing length, H – spiral lead of the groove, s – groove width, g – groove depth, m – journal eccentricity, W – load carrying capacity, φ - attitude angle, Θ - coordinate in circumferential direction as measured from the point where bearing interspace thickness is maximum, ω - journal angular velocity, h_{min}- minimum of oil clearance thickness.

Helical groove of journal bearing surface causes substantial changes of an oil clearance geometry. For hydrodynamic slide bearings it has effect on conditions of the oil flow and then bearing characteristics. The oil clearance geometry changes particularly in a place of the smallest thickness; this change affects considerably oil flow. Therefore non-dimensional index describing the oil clearance was proposed; this index was then compared with results of experimental investigations.

Index I_1 characterizes the oil clearance in place of its smallest thickness. It can be described as the ratio of grooves cross-sectional area to area of rectangle of sides of the bearing width b and the minimum oil clearance thickness. Therefore for triangular shape of the groove:

$$I_1 = \frac{\frac{s \cdot g}{2} \cdot \frac{b}{H}}{h_{min} \cdot b} = \frac{s \cdot g}{2 \cdot H \cdot h_{min}} \qquad (1)$$

This index will be combined with other indices describing changes in load carrying capacity, oil flow rate and wear.

Methodology of testing

Experimental investigations of slide bearings with the helical groove on journal were conducted on two testers. The first of them was used in order to analyse the grooves effect on load carrying capacity and oil flow rate. These tests were carried out using clean oil under condition of the fluid lubrication. However the second test rig was lubricated with oil contaminated with hard particles. Under this condition the wear resistance tests were carried out.

In order to compare universality of the proposed index, two kinds of tests were done with applications of different lubricating oils, test parameters and bearing geometries.

Investigation of load carrying capacity and oil flow rate

Experimental tests of the effect of helical groove on journal on the load carrying capacity and the oil flow rate, when clean oil was used, were conducted on the tribological tester, which scheme is shown in Figure 2. These tests were supplement to wear resistance tests, since the groove presence on the journal can cause decrease in the load carrying capacity and also increase in the oil flow rate.

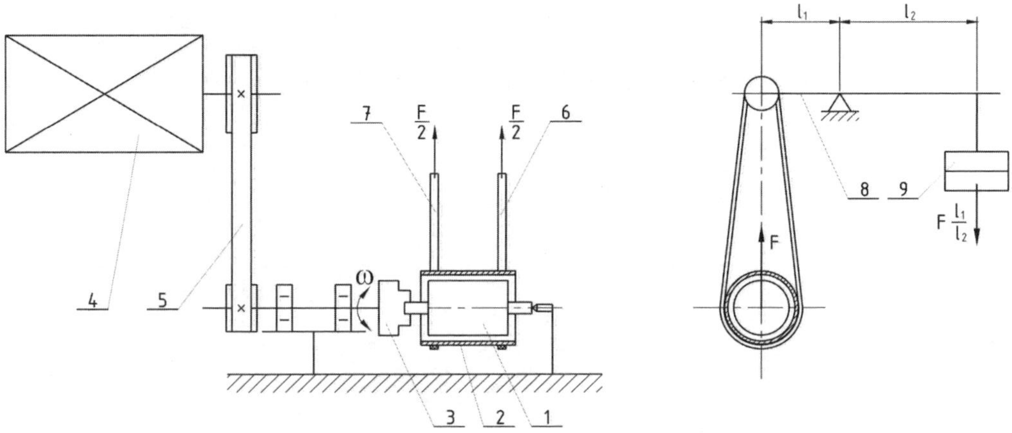

Fig. 2. Diagram of test stand for load capacity investigations: 1- bearing journal, 2- bearing sleeve, 3- holder, 4- electric motor, 5- belt transmission, 6,7 – flexible connectors, 8 – lever, 9- load

The journal bearing (1) was assumed as element of fixed axis, however the bearing sleeve (2) could displace in the oil clearance range. The bearing journal was driven by the electric motor (4) through the belt transmission (5) and shaft with the holder (3). The electric motor collaborated with inverter which allowed to a stepless control of journal bearing rotational speed. Work [15] presents the detailed description of this rig. The investigations were carried out under fluid friction condition. Bearing having smooth journal and two variants of bearings having journal with helical groove were investigated and the dependences between load carrying capacity and relative eccentricity for each bearing were obtained. After loading and thermal stabilization the bearing sleeve displacement was measured, on the basis of it the relative eccentricity and then I_1 index were calculated. This index was combined with the other index I_2 describing decrease in load carrying capacity caused by groove on the bearing journal presence. This index is the ratio of load carrying capacities of the bearing with groove (L_g) and the bearing with smooth journal (L_s).

$$I_2 = \frac{L_g}{L_s} \tag{2}$$

In addition the measurement of the oil flow rate was made. In order to assess the effect of the groove presence on the oil flow rate the I_1 index was also combined with I_3 index, characterizing change in oil flow rate caused by the groove presence. In mathematical sense this index is the ratio of oil flow rate for bearing with the groove (Q_g) and bearing with smooth journal (Q_s).

$$I_3 = \frac{Q_g}{Q_s} \tag{3}$$

There were the following test parameters:
- lubricating agent - LAN 46 oil (kinematic viscosity 46 mm^2/s at 313K, viscosity index 96),
- journal diameter - 60 mm,

- bearing length - 42 mm,
- relative clearance ratio - 0.004,
- journal angular velocity - 25 rad/s.

The wear tests:

Wear tests were carried out at Machine Design and Vehicles Department of Gdansk University of Technology on the ZAN test stand, which scheme is presented in Figure 3. The ZAN testing machine has been designed specially for the testing bearings which are lubricated with contaminated oil.

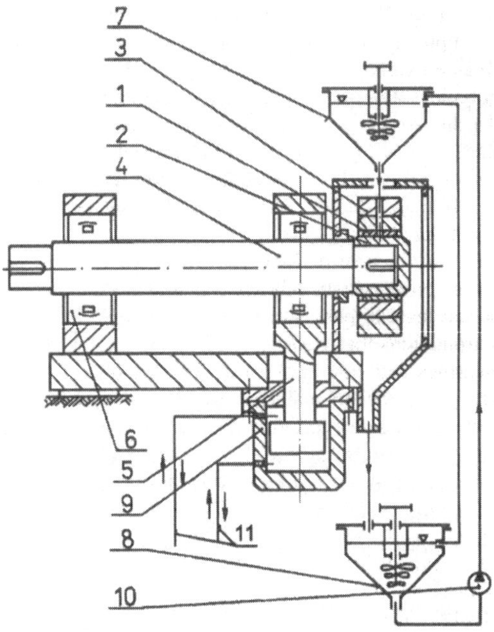

Fig. 3. Diagram of testing head in ZAN test stand: 1-test bearing, 2-replaceable sleeve, 3-housing, 4 -shaft, 5-connecting road, 6-shaft, supporting bearing, 7-upper mixer, 8-lover mixer, 9-hydraulic actuator, 10 - rotodynamic pump, 11 - supplying of hydraulic actuator.

The bearing to be tested 1, is placed in the two-part housing 3 , which is the end support for the shaft 4. Replaceable sleeve 2, i.e. placed on the shaft pin, constitutes the tested bearing journal. ZAN test stand is described more precisely in [12].

The subject of the test were six variants of slide bearing with helical groove on the journal. The wear characteristics of the conventional bearing with smooth journal has been accepted as a reference level. In both cases journals were made from heat treated 34CrNiMo6 steel (hardness 52 HRC) and they were mating with bimetal sleeves made from MB 58 material (78% Al, 20% Sn, 1% Cu, 1% Mn). The criterion of evaluation of the tested bearing was the volumetric wear of bearing journal and sleeve. Volumetric wear was determined based on profilograms of elements after tests. There were the following test parameters:
- lubricating agent - SD SAE 20W/40 oil (kinematic viscosity 15 mm^2/s at 373K, viscosity index 135),
- contaminant - hard aluminium oxide powder (concentration in oil - 0,5 g/l, average diameter of grains - 21 μm) ,
- journal diameter – 52.69 mm,
- bearing length - 30 mm,

- relative clearance ratio - 0.002,
- journal angular velocity - 62.8 rad/s,
- nominal contact pressure - 1.57 MPa,
- total time of the test - 20 h.

In order to assess the effect of the groove presence on the wear the I_1 index was also combined with I_4 index, characterizing change in wear caused by the groove presence. In mathematical sense this index is the ratio of wear for bearing with the groove (W_g) and bearing with smooth journal (W_s).

$$I_4 = \frac{W_g}{W_s} \tag{4}$$

Index I_4 was calculated separately for journals and sleeves.

In order to determine scatter of results load carrying capacity, oil flow rate and wear investigations were replicated six times for selected variants. Conventional bearing with smooth journal (the first point on diagrams) and bearing with journal having the smallest of investigated grooves (the second point on diagrams) were selected variants. Based on experimental investigations six values of I_2, I_3, I_4 coefficients were calculated. After that standard deviation and 95% confidence interval for each coefficient were calculated. Values of 95% confidence intervals were assumed for rest investigations and marked on diagrams.

Results of research and their analysis

The results of I_1 index calculation combined with the results of measurement and computations of I_2 index (describing decrease in the load carrying capacity) are shown in Figure 4.

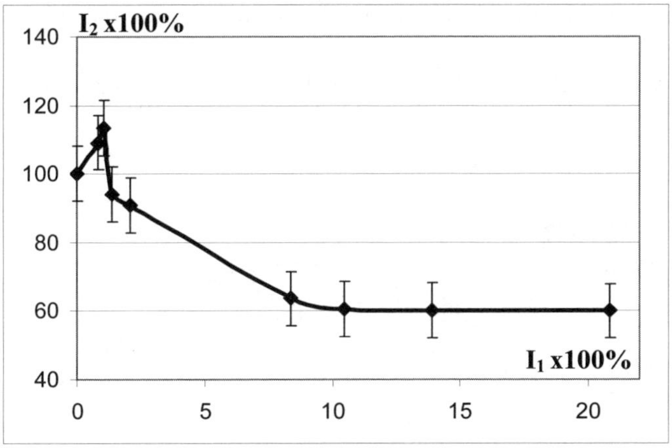

Fig. 4. The influence of journal groove cross section area (I_1) on load carrying capacity (I_2)- lubrication with clean oil.

The analysis of presented in figure 4 results of investigations indicates that the bearing load capacity depends on groove sizes and lead of screw line. The presence of grove of small dimensions (I_1 up to 0.01) caused an increase in the bearing load capacity. It is a results of flow phenomena in oil film, described in References [1,13,16]. However when I_1 index is higher than 0.01, the negative effect of the groove presence on load capacity is takes place. For its value of 0.1 the load carrying capacity of bearing with the grove is 60% of that for conventional bearing (with smooth journal). This dependence should be taken into consideration in bearing with helical groove on journal testing.

Figure 5 shows the results of research concerning the effect of groove presence on oil flow rate.

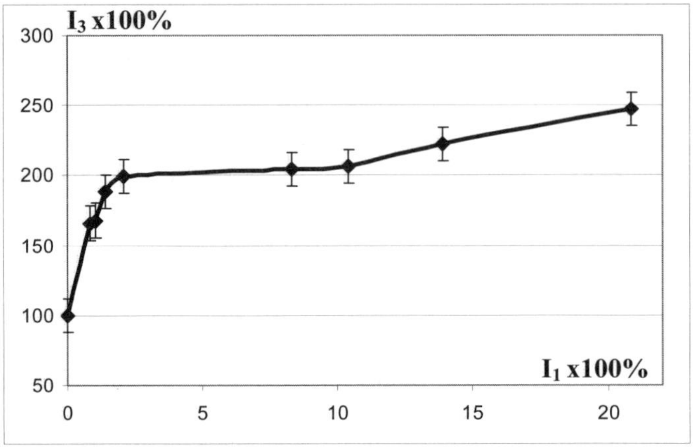

Fig. 5. The influence of journal groove cross section area (I_1) on oil flow rate (I_3)- lubrication with clean oil.

It is found after the analysis of Figure 5 that helical groove on journal presence intensifies oil flow through bearing. For the I_1 parameter of 0.02 value intensify of oil flow through bearing is twice than for conventional bearing. Intensify of oil flow through bearing is a substantial factor of possible effect on easier contaminations removal from bearing and then wear decrease. The presented results (Figs 4 and 5) pointed also that proposed parameter is correlated with changes in load carrying capacity and oil flow rates caused by groove presence.

Figures 6 and 7 present the effect of groove presence on wear under condition of lubrication with contaminated oil. They are also presented in non-dimensional form as dependence of the I_4 coefficient, determined for the journal (Fig. 6) and the sleeve (Fig. 7), on the I_1 coefficient.

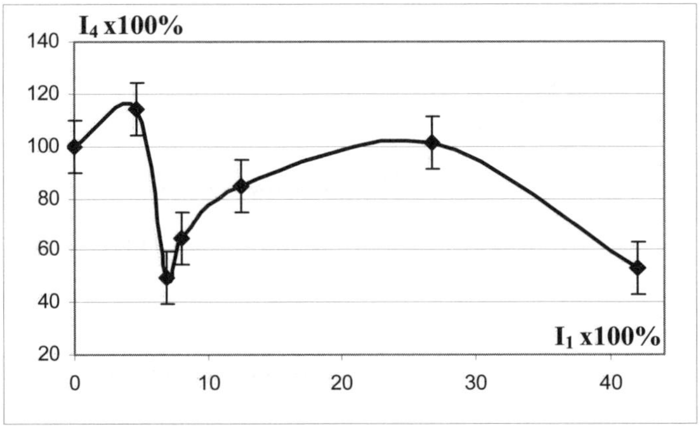

Fig. 6. The influence of journal groove cross section area (I_1) on journal wear (I_4) - lubrication with contaminated oil.

One can see after the analysis of Figs 6 and 7 that the helical groove presence on the bearing journal can substantially decrease wear in the case of hard particles presence in the lubricating oil. The journal wear depended on groove sizes and lead of screw line. The groove of too small dimensions did not assure positive results, it even caused small wear increase. When the parameter I_1 was higher than 0.05 one can see wear decrease. When this index was equaled 0.07 wear of journal with helical groove was less than 50% of wear of conventional journal.

One should remember that in this case the bearing is characterized by the smaller load carrying capacity than the bearing without the groove. It is evident that the parameter I_1 is correlated with journal wear.

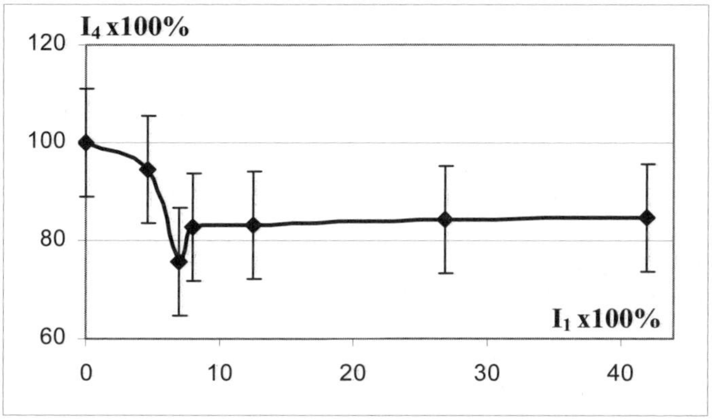

Fig. 7. The influence of journal groove cross section area (I_1) on sleeve wear (I_4) - lubrication with contaminated oil.

Wear of the bearing sleeve mating with journal with the helical grooves, under lubrication by contaminated oil, is smaller than that of conventional bearing. When this index was equaled 0.07 wear of sleeve mating with journal with helical groove was about 75% of wear of sleeve mating with conventional journal However when I_1 index is higher than 0.08 wear of sleeve stabilizes on value of about 80% of wear of the sleeve mating with the smooth journal.

Presence of the helical groove on the journal bearing intensifies axial (along z axis) oil flow through bearing. It makes easier for contaminants to move outside the bearing. In addition the groove is place for particles location. Therefore bearing sensitivity with the modified journal on destructive activity of oil contaminants is smaller than that of the classical journal. However this is accompanied with load carrying capacity decrease. It is important element limiting possible application of the analysed design. Obtained results justify further research oriented on minimization of the negative groove effect on load carrying capacity and minimization of groove dimensions ensuring wear decrease.

Conclusions

1. The helical groove on the journal bearing causes decrease of its wear in case of presence of hard contaminations in lubricating oil depending on groove dimensions and lead of screw line. In conducted research wear was maximum twice smaller than for conventional journal.

2. Journal modification caused also 25% smaller bearing sleeve wear for contaminated oil with hard particles.

3. The helical grooves presence also causes decrease in the bearing load carrying capacity (up to 40% in carried out research) and increase of oil flow rate (above twice in conducted research).

4. All obtained results of investigations are correlated with non-dimensional parameter being the ratio of grooves cross-sectional area to area of rectangle of sides of the bearing width and the minimum oil clearance thickness.

Acknowledgements

Author would like to thank Prof. J. Sikora from Gdansk Technical University for performing the experiments.

Apparatus/equipment purchased in the project no POPW.01.03.00-18-012/09 from the Structural Funds, the Development of Eastern Poland Operational Programme co-financed by the European Union, the European Regional Development Fund were used in investigations.

References

[1] J. Sep, The oil film properties in journal bearings with non-typical geometry. Rzeszow University of Technology, Rzeszow 2006 (in polish).

[2] M. Szczerek, M. Wiśniewski M.: Tribology and tribo-technique. Institute for Sustainable Technologies, Radom 2000 (in polish).

[3] S.A. McKee, Effect on abrasive in lubricant. Journal of S.A.E, 20 (1927) 3-6.

[4] A.E. Roach, Performance of oil-film bearings with abrasive-containing lubricant. Transactions ASME, 73 (1951) 677-686.

[5] H.G. Rylander, Effects of solid inclusions in sleeve-bearing oil supply. Mech. Eng., 74 (1952) 963-966.

[6] J.A Williams, A.M. Hyncica, Mechanisms of abrasive wear in lubricated contacts. Wear, 152 (1992) 57-74.

[7] V. Wikström, E. Höglund, R. Larsson R, Wear of bearing liners at low speed rotation of shafts with contaminated oil. Wear, 162-164 (1993) 996-1001.

[8] M.M. Khonsari, M.D. Pascovici, B.V. Kucinschi, On the scuffing failure of hydrodynamic bearings in the presence of an abrasive contaminant. Trans. ASME, J.Tribology., 121 (1999) 90-96.

[9] M.D. Pascovici, M.M. Khonsari, Scuffing failure of hydrodynamic bearings due to an abrasive contaminant partially penetrated in the bearing over-layer. Trans. ASME, J. Tribology, 123 (2001) 430-433.

[10] J.K. Duchowski, K.G Collins, Experimental evaluation of filtration requirements for journal bearings operating under different contaminant levels. Lubrication Engineering, 58 (2002) 34-39.

[11] M. Tomimoto, Experimental verification of a particle induced friction model in journal bearings. Wear, 254, (2003) 749-762.

[12] J. Sep, A. Kucaba-Pietal, Experimental testing of journal bearings with two-component surface layer in the presence of an oil abrasive contaminant. Wear, 249 (2001) 1090-1095.

[13] J. Sep, Journal bearing with an intensive axial oil flow- computer simulations. Scientific Problems of Machine Operation and Maintenance, 43 (2008) 7-19.

[14] J. Sep, Wear characteristics of journal bearings of two-component surface layer. Zagadnienia Eksploatacji Maszyn, 116 (1998) 571-582 (in polish).

[15] J. Sep, Journal bearing with an intensive axial oil flow- experimental investigation. Scientific Problems of Machine Operation and Maintenance, 43 (2008) 21-29.

[16] J. Sep, Three-dimensional hydrodynamic analysis of a journal bearing with a two-component surface layer. Tribology International, 38 (2005), 97-104.

Key Engineering Materials Vol. 490 (2012) pp 282-287

doi:10.4028/www.scientific.net/KEM.490.282

The Influence of Gas Mixture in the Glow-discharge Nitriding Process of Austenitic Stainless Steel on Characteristic of Nitrided Cases

Artur Sitko[1,a], Marek Szkodo[1,b], Maria Gazda[2,c]

[1]Gdańsk University of Technology, Faculty of Mechanical Engineering, Department of Materials Science and Bonding, Poland

[1]Gdańsk University of Technology, Faculty of Applied Physics and Mathematics, Department of Solid State Physics, Poland

[a]artursitko@o2.pl, [b]mszkodo@pg.gda.pl, [c]maria@mif.pg.gda.pl

Keywords: Austenitic stainless steel, nitrided case, nitriding process, nitrogen austenite, thermochemical treatment

This paper presents investigations of nitrided cases after the glow-discharge nitriding process. The nitrided cases were obtained by using a different chemical composition of gas mixture at the temperature of 450°C. The glow-discharge nitriding process was carried out on austenitic stainless steel, grade of steel X5CrNi18-10. The chemical composition and phase identification of the nitrided cases were examined by using the glow-discharge optical emission spectrometry (GD-OES) and X-ray diffractometry (XRD)

Introduction

Austenitic stainless steels characterize with excellent corrosion resistance, although low mechanical properties. In many cases that limits their exploitation in industrial applications. Moreover, austenitic stainless steel is susceptible to local corrosion in the presence of halide ions [1]. In present works, it was emphasized that the nitriding of austenitic stainless steels gives some potential abilities to obtain beneficial effect of termochemical treatment on their mechanical properties and corrosion resistance. The nitriding process below the temperature of 450°C allows to achieve above mentioned effects. The low temperature of glow-discharge nitriding process causes the sufficient decrease of chromium mobility. It allows to obtain uniform microstructure which consists of supersaturated nitrogen austenite, called γ_N. The present works prove [2-8], that phase (γ_N) has some beneficial effect on mechanical properties and corrosion resistance of these steels.

Experimental Procedures and Results

The technology of the low temperature glow-discharge nitriding process was carried out according to the parameters in table 1.

Table 1. Parameters of glow-discharge nitriding process

Number of process	Temperature [^{o}C]	Vacuum pressure [hPa]	Time [h]	Chemical composition of gas mixture [% wt.]	
				N_2	H_2
1	450	4	6	10	90
2	450	4	6	30	70
3	450	4	6	50	50
4	450	4	6	70	30

The chemical composition analysis of austenitic stainless steel was done by using glow- discharge optical emission spectrometer (RF DG-OES: *Glow Discharge Optical Emission Spectrometer).* Device RF DG-OES uses a glow-discharge (GD) supplied by radio source RF and optical emission spectrometer (OE). That solution allows to carry out the volumetric analysis of specimen. The investigations were done in vacuum chamber in argon atmosphere. The resolution of depth profile was 5 [nm]. The result of chemical analysis of the reference material was presented in fig. 1.

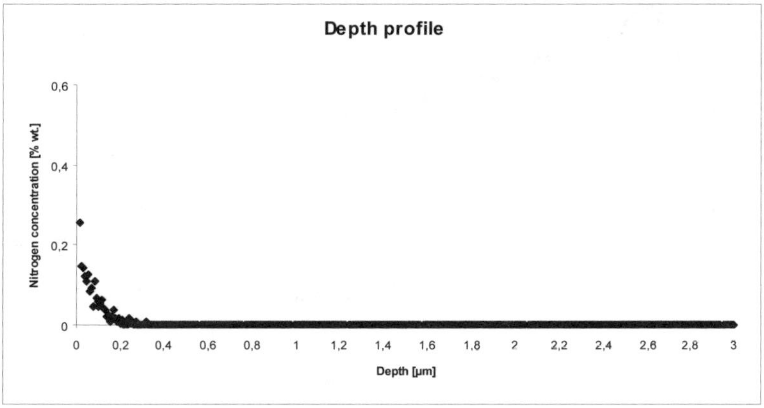

Fig. 1. Content change of nitrogen in the depth of the reference material

In fig. 2 results of chemical analysis of nitrided cases were presented.

a)

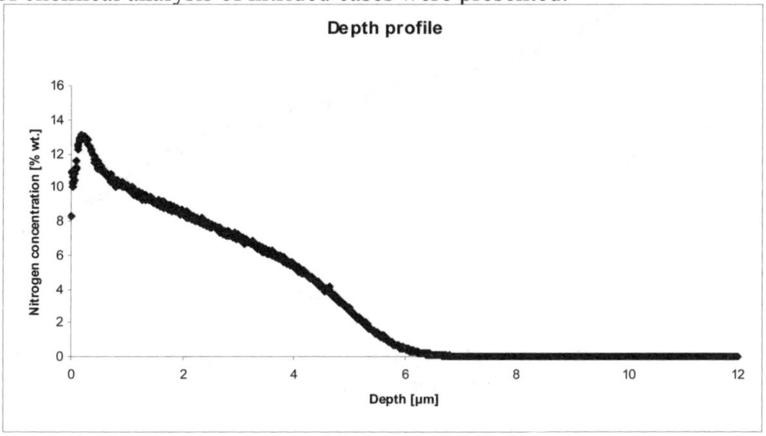

b)

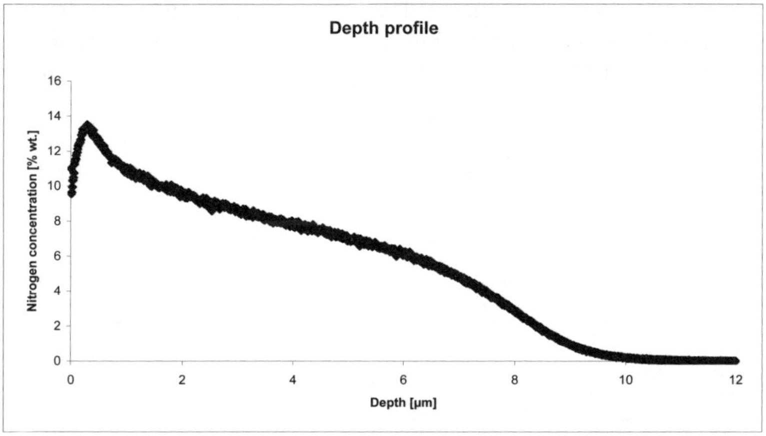

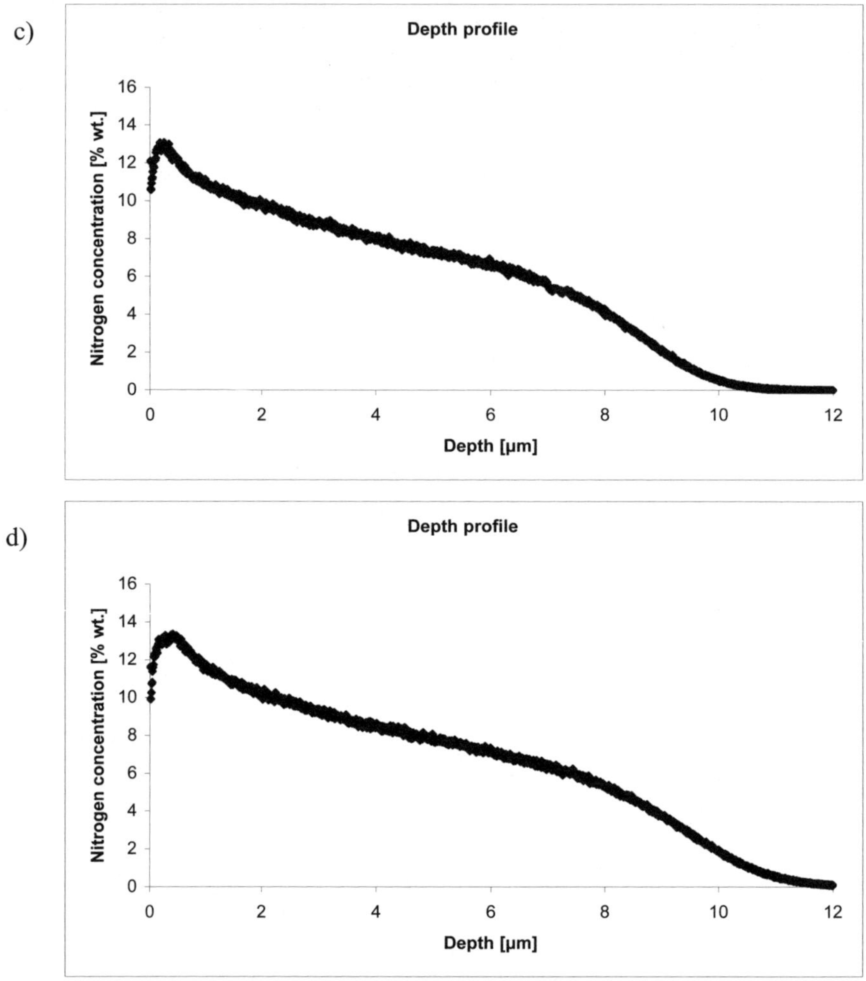

Fig. 2. Content change of nitrogen in the depth of the nitrided cases after glow-discharge nitriding process: a) 1, b) 2, c) 3, d) 4

The investigations of the thickness of nitrided cases were carried out on the basis of depth profiles. The boundary value of nitrogen was established in order to determine the thickness of diffusion layers. In accordance with the standard PN-EN 10088-1 the maximum content of nitrogen was 0,11 [% wt.].

In fig. 3, the results of the thickness of nitrided cases were presented. The diffusion layers were obtained by using the different chemical composition of gas mixture during the glow-discharge nitriding process at the same temperature of 450°C.

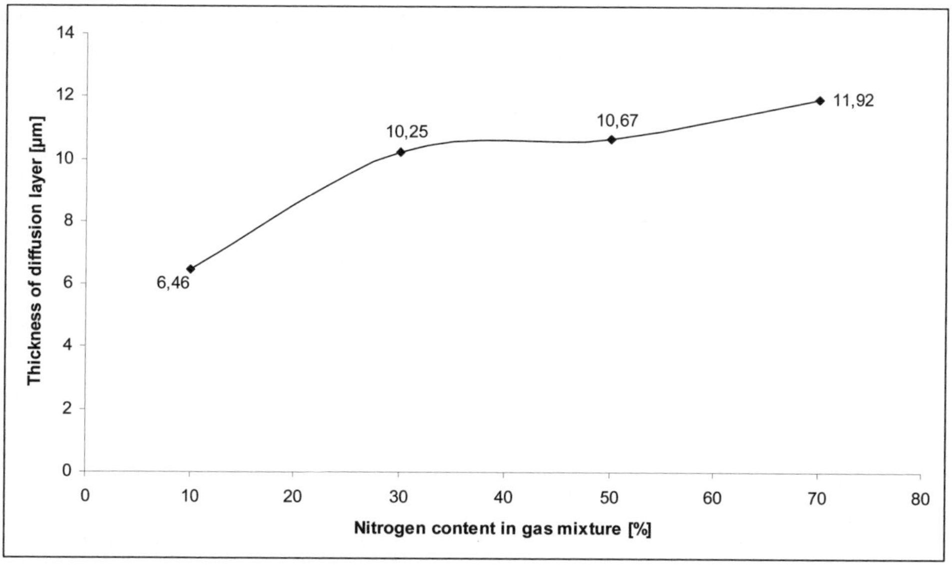

Fig. 3. Influence of gas mixture composition on diffusion layers' thickness

The X-ray diffraction was done by using X-ray diffractometer type PHILIPS X'PERT PRO with Cu Kα radiation, wavelength radiation: $\lambda_{K\alpha 1}$ = 0,154056nm, $\lambda_{K\alpha 2}$ = 0,154439nm, ratio $\lambda_{K\alpha 1}$ / $\lambda_{K\alpha 2}$ = 2. The result of phase identification of the reference material was presented in fig. 4. The microstructure of the reference material consists of austenite phase (γ) and delta ferrite phase (δ).

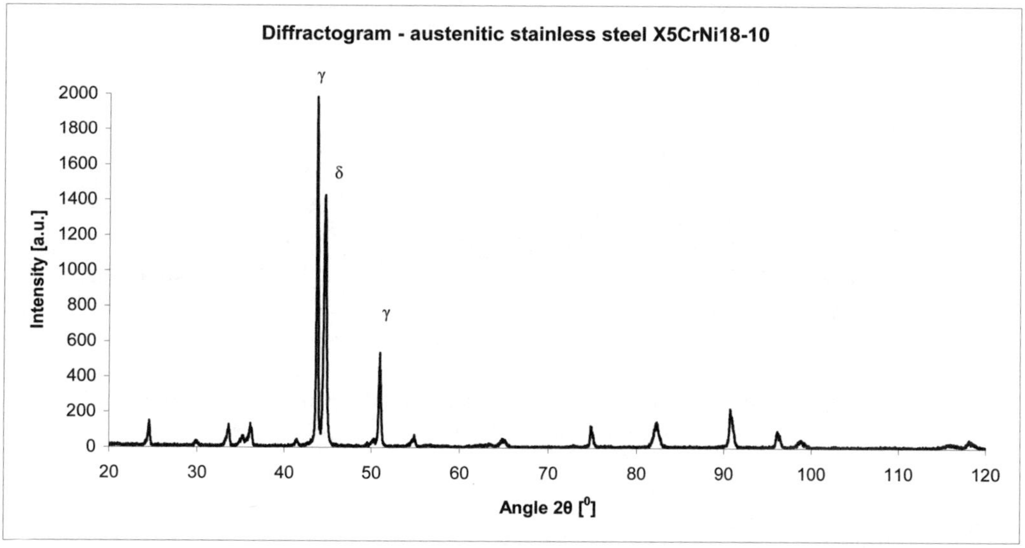

Fig. 4. Diffraction pattern of the surface of austenitic stainless steel X5CrNi18-10

In fig. 5 the results of phase identification of nitrided cases were shown. The microstructure of the diffusion layer after process no 1 consists of nitrogen austenite (γ_N) only (fig. 5a). The higher nitrogen content of gas mixture used in process no 4 has some influence on the microstructure. The microstructure consists of nitrogen austenite (γ_N) and some amount of precipitates on the surface. The results of X-ray diffraction present that the main peaks of phase γ_N in nitrided cases are shifted on the left in relation to the main peaks of phase γ which occurred in the reference material (fig. 4). It indicates that the nitrided cases have higher inner stress in compersion with the reference material.

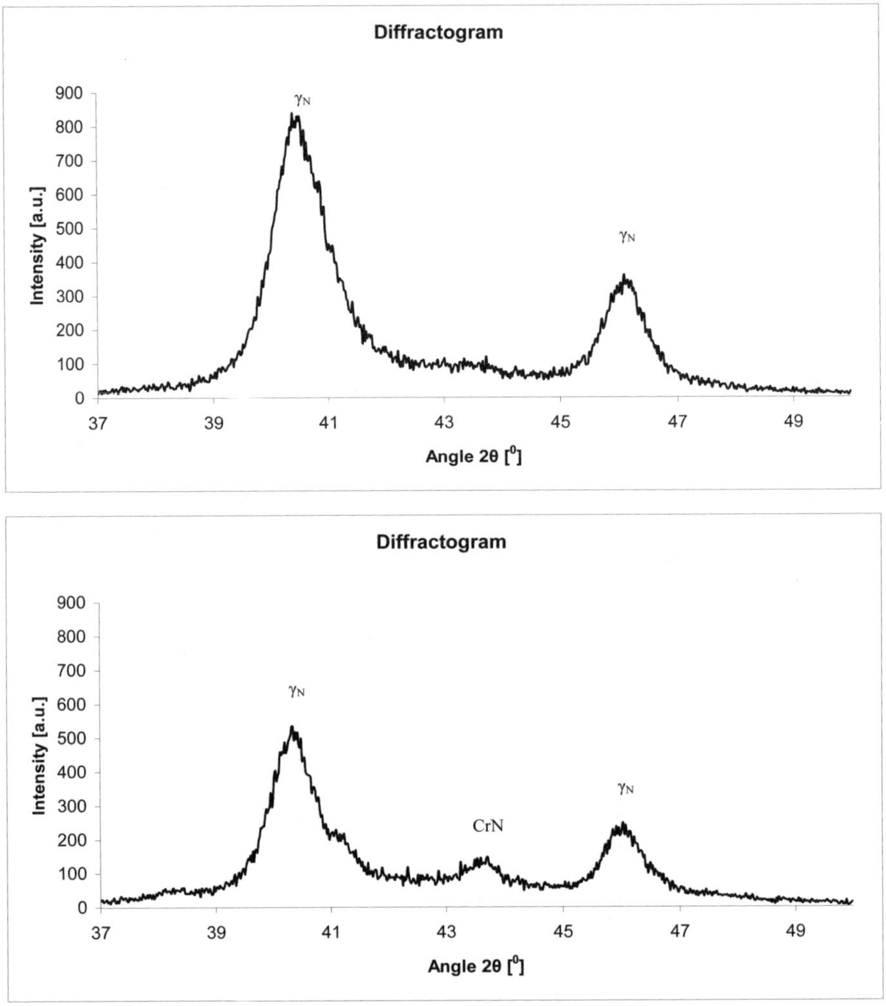

Fig. 5. Diffraction pattern of the surface of austenitic stainless steel after glow-discharge nitriding process: a) 1, b) 4

Summary

1. The nitrogen content increase in gas mixture in the glow-discharge nitriding process at the temperature 450°C causes the increase of the thickness of diffusion layers.

2. The chemical composition investigations of diffusion layers do not reveal any influence of nitrogen content in gas mixture on maximum amount of nitrogen in nitrided cases. After every process the amount of nitrogen is slightly higher than 13 [% wt.].
3. The phase investigations reveal that the chemical composition of gas mixture affect on the phase composition of diffusion layers. The small amount of chromium nitride precipitates is observed after process no 4. The same effect was observed after process no 3 in the work [9].
4. The X-ray diffraction shows that the usage of gas mixture with its composition ($10N_2$: $90H_2$) do not affect on phase composition change. The microstructure consists of supersaturated nitrogen austenite (γ_N) only. The same result was shown in the work [9] after process no 2.
5. The X-ray diffraction reveals higher inner stress in nitrided cases in comparison with the reference material.

References

[1] W. Liang, X. Xiaolei, X. Jiujun, S. Yaqin: Characteristic of low pressure plasma arc sources ion nitrided layer on austenitic stainless steel at low temperature. Thin Solid Films 391 (2001) 11-16.

[2] W Liang, S. Juncai, X. Xiaolei: Low pressure plasma arc source ion nitriding compared with glow-discharge plasma nitriding of stainless steel. Surface and Coatings Technology 145 (2001) 31-37.

[3] W. Liang, X. Xiaolei, Y. Zhiwei, H. Zukun: Low pressure plasma arc source ion nitriding of austenitic stainless steel. Surface and Coatings Technology 124 (2000) 93-96.

[4] W. Liang, X. Xiaolei, X. Bin, Y. Zhiwei, H. Zukun: Low temperature nitriding and carburizing of AISI 304 stainless steel by a low pressure plasma arc source. Surface and Coatings Technology 131 (2000) 563-567.

[5] G. A. Collins, R. Hutchings, K. T. Short, J. Tendys, X. Li, M. Samandi: Nitriding of austenitic stainless steel by plasma immersion ion implantation. Surface and Coatings Technology 74-75, Part 1, (1995) 417-424.

[6] Collins G., Hutchings R., Short K., Tendys J.: Ion-assisted surface modification by plasma immersion ion implantation. Surface and Coatings Technology, 103-104 (1998) 212-217.

[7] Gunzel R., Betzl M., Alphonsa I., Ganguly B., John P., Mukherjee S.: Plasma-source ion implantation compared with glow-discharge plasma nitriding of stainless steel. Surface and Coatings Technology 112, (1999) 307-309.

[8] Larisch B., Brusky U., Spies H.: Plasma nitriding of stainless steels at low temperatures. Surface and Coatings Technology 116-119, (1999) 205-211.

[9] Sitko A., Szkodo M., Gazda M.: The characteristic of surfach layers on austenitic stainless steel after glow-discharge nitriding process. Solid State Phenomena 165, (2010) 165-168.

Key Engineering Materials Vol. 490 (2012) pp 288-295
doi:10.4028/www.scientific.net/KEM.490.288

The Influence of Geometrical Parameters on the Friction Process in the Needle Bearing

Jerzy Nachimowicz[1,a] Robert Korbut[1,b]

[1]Faculty of Mechanical Engineering, Bialystok University of Technology, Poland

[a]nachim@pb.edu.pl, [b]holden77@wp.pl

Keywords: Needle bearing, Friction, Sliding, Wear

Abstract. The article analyses the process of friction in the needle bearing as the function of the alterations of geometrical parameters, namely the diameter of the shaft – D, the needle – d and the bearing clearance. The change in the relation between the diameters D and d results in the change of the contact area of the friction pair; in the case of two shafts with parallel axes (the needle of the bearing and the shaft pin) the contact area, if compared with the analogical process in the ball bearing, alters significantly and leads to resistance in motion. There are two types of friction in the needle bearing: the rolling friction and the sliding friction; the analysis of the movement of the bearing elements enabled the estimation of the extent to which the sliding friction matters in the overall balance of motion resistance. The article also defines the type of wear as the function of initial clearance.

Introduction

The construction of the rolling bearing was initially based on the assumption that the friction loss during the bearing work is significantly smaller than during the sliding. However, during the work of the bearing in the operating conditions there exists both the bearing and the sliding friction. Different factors result in appearance of resistance to motion while the bearing is operating:

- Hysteresis of deformations
- Interior friction in lubricant
- Sliding and microsliding caused by deformations, the geometry of contact area and the movement caused by the gyroscope moment
- Sliding between the bearing cage, the bearing elements and the bearing ring
- Sliding in the bearing seal

When bodies are deformed in the operating conditions, the phenomenon of pure bearing exists if the cooperating elements possess the same diameter, length, the properties of the material and parallel axes. Also, the roughness of the cooperating surfaces should be minimal. In such a case, when there is no lubricant, there appear only some losses caused by hysteresis of deformations [5]. This stems from the fact that the difference between the length of the contact arc and the corresponding arcs before deformations is identical for both bodies. As a result, there is no mutual sliding of the surfaces in the deformation area.

If the curvatures of the cooperating surfaces are different, in the elastic deflection the length of the contact arc for both bodies is identical, whereas before the deflection it was different. Consequently, the deformations in the contact area are accompanied by microsliding; if, however, the speed of both cooperating surfaces is identical, their mutual movement is called ‘bearing’.

In a typical situation, in the process of bearing of two bodies with different peripheral speed there occurs the bearing with sliding; such a situation is the subject of examination here. The aim is to estimate of the extent to which the sliding friction matters in the overall balance of motion resistance.

The analysis of contact areas of the needle bearing elements

In the place of contact of two elastic bodies pressed against each other with some force, some contact stresses within a certain field of mutual contact occur. They reach significant values even in the situation when the pressing force is relatively small, which, as a consequence, may lead to exceeding the acceptable limit of the material effort. This is of paramount importance during the work of needle bearings which are under considerable load. Fig. 1.b shows a situation when the axis of the needle and the axis of the shaft pin are parallel. The stresses that occur (Fig. 1.a) are evenly spread along the length of the needle, and the area of contact between the two elements equals the field of the ellipse of the length which is the same as the length of the needle and the width 2b calculated by means of Hertz's solution:

$$b = \sqrt{P\frac{D_1 \cdot D_2}{D_1 + D_2}\left(\frac{1-v_1^2}{E_1} + \frac{1-v_2^2}{E_2}\right)}, \tag{1}$$

Where: P is the strength per each unit of the rollers' length; $D1$, $D2$ are the diameters of the shaft pin and the needle respectively; $v1,v2$ are Poisson's figures; $E1,E2$ are Young's modulus. The indexes *1,2* refer to the roller *1* (the shaft pin) and *2* (the needle roller) respectively.

In the case when the axes of the touching rollers (the shaft pin and the needle) are not parallel (Fig. 1.c), the whole load and the stress connected with it concentrate in a relatively small point [3].

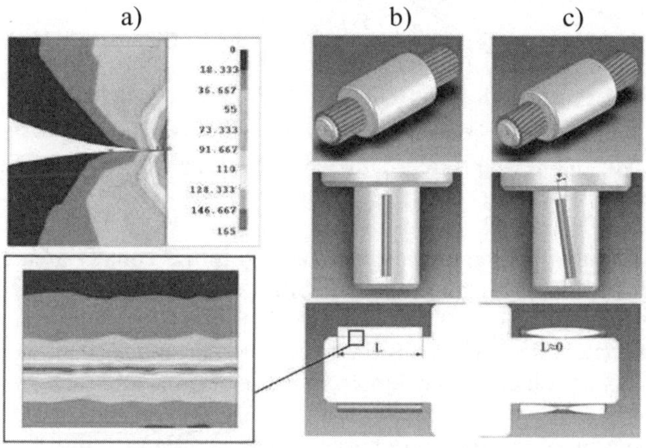

Fig. 1. The pictorial diagram of the roller bearing. a) – the spread of stresses for the connecting rollers with parallel axes, b) – the axes of the parallel elements, c) – the axes of the elements moved by the angle φ.

The area of the friction surface of the elements of bearing is influenced not only by the adequate mutual positioning of the cooperating elements, but also by the changeable relation of the diameters of the needle and the roller, changeable values of Young's modulus, Poisson's ratio of the materials used, and the change of load.

The basic theoretical perspective assumed while calculating contact stresses and the width of the contact area between co-working bodies was based on Hertz's theory, drawing on the following premises: the contacting elements are made from a homogenous, isotopic material; they are limited by the smooth surfaces with a regular curvature; and within the contact area some deformations occur [1].

The analysis of the change in the contact area of the elements of needle bearing presented below was prepared on the basis of all the premises of Hertz's theory concerning the work of two rollers (the shaft pin, the needle roller) with parallel axes, working under static load in dry environment. In the examination of the factors which have an impact on the measure of contact area of the elements cooperating in the form of 'the shaft pin – the needle roller' interaction is particularly important as the contact area influences the change in the moment of motion resistance in the process of bearing.

In the first test eight different needles of 1,4 mm up to 2,5 mm in diameter were juxtaposed with four rollers of 10, 20, 30, 40 mm in diameter. The contact area of the cooperating elements was calculated taking as an assumption the constant load of 200 N/mm. Different combinations of elements (needles and rollers) were tested. The bigger the diameter of the needle, the bigger the contact area becomes – this tendency can be measured by means of power equations (Fig. 2).

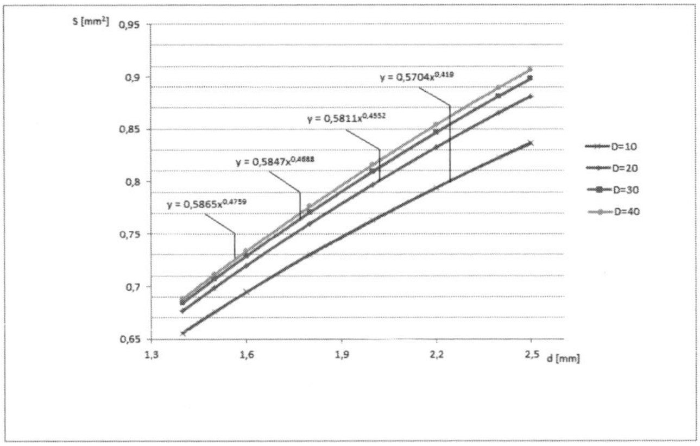

Fig. 2. The influence of the needle roller diameter (d) and the shaft pin diameter (D) on the contact area (S) of the elements of bearing.

In the case demonstrated above, the biggest changes in the contact area can be observed while putting together the needles of bigger diameters with the given set of rollers, i.e. for the needle of 1,4 mm in diameter cooperating with the rollers of 10 and 40 mm in diameter the contact area increased by approximately 5%, while for the needle of 2,5 mm in diameter cooperating with the same rollers the contact area increased by approximately 9%.

The mathematical model of the friction process in the needle bearing

The state of the present studies, the information in scientific publications and the catalogues of leading producers of bearings show that the total moment of resistance in the work of a bearing can be understood as the sum of the elemental moments originating in the bearing friction, the sliding friction, the bearing seal friction and the friction occurring in oil environment.

However, in order to estimate the moment of resistance in the friction sliding we suggest the mathematical model in which the total moment of different types of resistance to motion comprises firstly the moment resulting from the bearing friction, calculated as the product of the load – N and the rolling friction ratio - f, and the moment originating from the sliding friction, taking into account load, the sliding friction - μ and the radius of the roller – r (Eq. 2).

$$M_{Tc} = N \cdot f + N \cdot \mu \cdot r \tag{2}$$

The recommended examination on a stand (the friction machine SMT-1) makes it possible to obtain the values of the moment of resistance created in the process of sliding friction from the total moment of resistance to motion in the bearing. The needle bearing with the outer ring is mounted directly on the shaft pin which is an interior bearing track for the needles. The outer ring is stationary, and the bearing is loaded with radius force. During the work of the roller turning with the rotational speed of 50-100 rotations per minute there exists only the resistance connected with bearing friction. The observations provided by a torque meter help estimate the moment of bearing friction and the value of the ratio *f*. Then, if the number of rotations of the roller is increased, in the total moment there will appear the already mentioned sliding friction, the value of which can be calculated as the difference between the total moment of resistance obtained in the trail work of the bearing and the value of bearing resistance established earlier.

Such an examination helps to estimate the extent to which the bearing friction and sliding friction matter in the overall balance of motion resistance. The next step in the examination may be to check the influence of load with constant rotational speeds, and the influence of the change in the lubricant environment.

The Examination Stand

Figure 3 shows the friction assembly and Figure 4 shows its diagram with the measuring apparatus.

Fig. 3. The examination stand – the friction assembly: a – the strap on the shaft pin, b – the bearing, c – the lubricant measuring needle, d – the load transmitting arm.

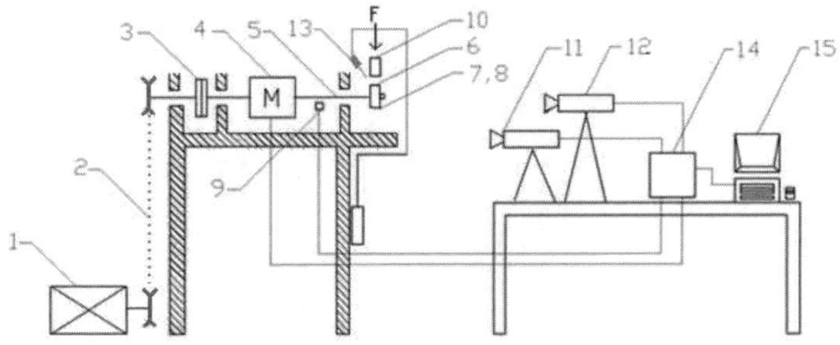

Fig. 4. The diagram of the examination stand: 1 – the electric motor, 2 – the belt transmission, 3 – the clutch, 4 – the torque meter, 5 - the shaft pin with the diameter D, 6 - the needles of the bearing with the diameter *d*, 7 – the sensor of rotations n_2 of the needle around its symmetry axis, 8 – the sensor of rotations n_3 of the needle around the axis of the roller, 9 the sensor of rotations n_1 of the roller, 10 – the arm of radial load, 11 – the thermal camcorder, 12 – the optical camcorder, 13 – the lubricant needle, 14 - the converter, 15 – the computer recording data.

The examination stand (Fig. 4) comprises the electric motor (1) with variable adjustment of rotational speed within the range of 0 – 1500 rotations per minute connected with the roller by means of the belt transmission (2), the clutch (3), the torque meter (4) responsible for constant measurement of the moment of friction, the sensor of rotation of the roller (9), the needle around the axis of the roller (8) and its axis of symmetry (7), the arm (10) responsible for providing load *F* on the bearing, the needle providing lubricant (13), the thermal camcorder (11), the optical camcorder (12), the converter responsible for reading and processing the recorded parameters (14), and the computer recording the data (15). The needle bearing with a fixed outer ring is placed on the strap attached to the shaft pin and loaded with the radial power *F*. The electric motor makes the shaft pin achieve a certain speed, then the sensors of rotation, the optical camcorder, the torque meter and the thermal camcorder register the rotational speed of the roller and the needle, the friction moment and the heat emitted. The examination of bearings with varied internal diameters is possible thanks to the use of changeable straps (3) of different diameters which are placed on the shaft pin (Fig. 5). The straps can also be used to adjust the initial clearance in the friction assembly. The geometry and kinematics of movement of the bearing elements is shown in Figure 6.

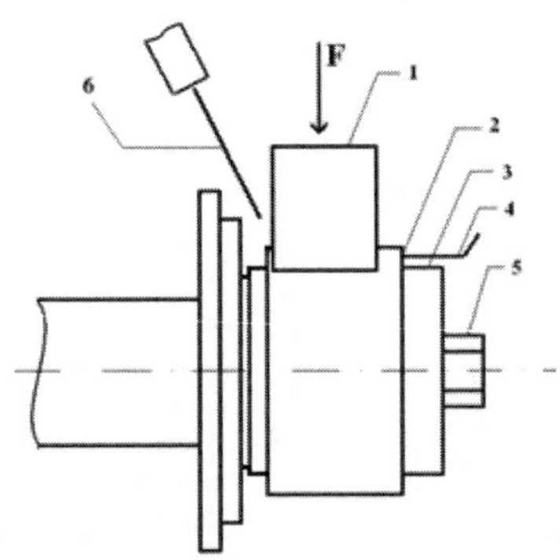

Fig. 5. The friction assembly: 1 - the arm providing load on the sample, 2 – the bearing, 3 – the strap on the shaft pin, 4 – the sensor of the needle rotation, 5 – the screw cup fastening the strap to the shaft pin, 6 – the metering needle providing lubricant.

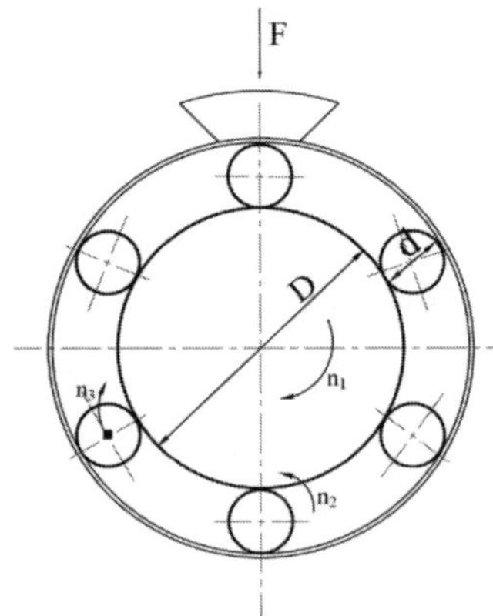

Fig. 6. The geometry of the system: n_1 – the rotation of the roller, n_2 – the rotation of the needle around the symmetry axis, n_3 – the rotation of the needle around the roller axis, *D* – the diameter of the roller, *d* – the diameter of the needle, *F*- the power loading the bearing.

The measurement of the roller wear was done by means of the passameter PMK 25-50, with the accuracy up to 1μm. After the examination in the friction assembly, each sample was measured in four planes, for each of the three cross-sections on the length of the shaft pin (Fig. 7).

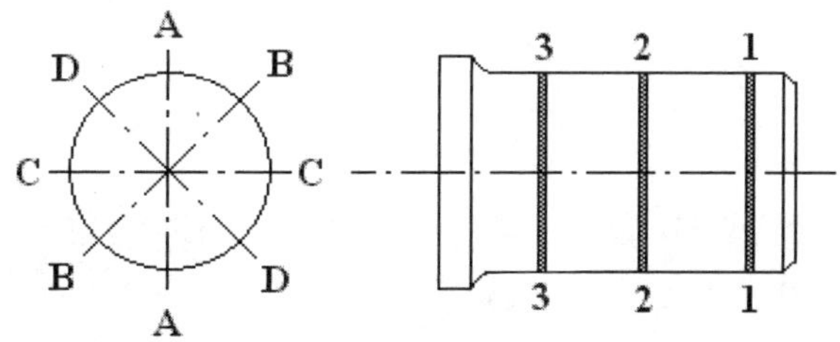

Fig. 7. The diagram of the geometry of the shaft pin wear.

The analysis of the friction process as the function of the initial clearance

The bearing clearance is a very important geometrical parameter which determines the intensity and the type of wear. In order to estimate the influence of clearance on the process of wear the examination was carried out for the nominal clearance of 0.065 mm and other clearance achieved as a sum of the nominal one and the value of reduction of the shaft pin diameter. The associations received in this way made it possible to examine the process of friction with clearances of 0.1, 0.2, 0.3, 0.4, 0.5mm respectively. A hundred-hour examination was carried out (on the friction machine SMT-1) for each value of clearance in bearing; next, the micro-structure of the friction surface was examined under an optic microscope with magnification X300 (Fig. 8.b). On the basis of the results received it can established that the enlargement of clearance between the elements in the friction pair leads to a considerable enlargement of the intensity in the shaft pin wear, which, in turn, results in the worse durability of the connection. Fig. 9. shows the kinetic relations between the wear of the shaft pin and different clearances in the function of work time. The lines 1 and 2, corresponding to the clearances 0.5mm and 0.4mm respectively, indicate the group with a relatively high rate of wear. A significantly smaller rate of wear is represented by the lines 4-7, corresponding to the clearances 0.065mm – 0.3mm respectively (Fig. 9). The dotted line 3, typified as the critical clearance $h = 0.325$mm characterizes the shift from the less intensive, mechano-chemical one, to the more intensive one. The increase of the clearance from the critical $h = 0.325$mm to 0.4mm results in the intensity which is approximately twice bigger. With the clearance 0.5mm the durability of the needle bearing decreases by 50%, if compared with the durability with the critical clearance. Fig. 8.a shows the wear of the shaft pin understood as function of the initial clearance. Point A (Fig. 8.a) as the critical clearance with which the friction pair works in the conditions of normal wear was achieved as a result of crossing the tangents led to the curve of wear in areas I and III [3].

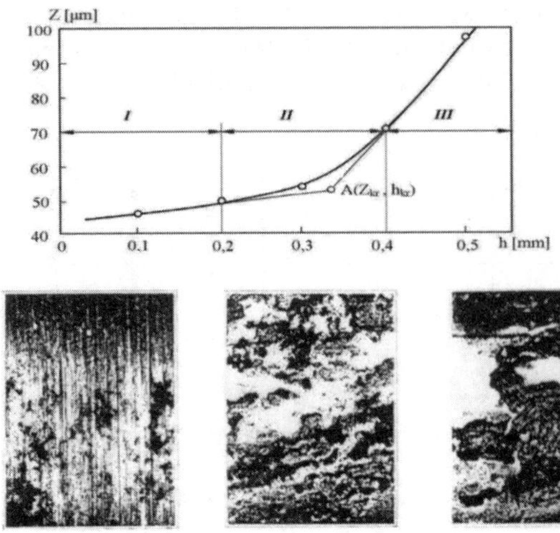

Fig. 8. Demonstration of the shaft wear, a): the shaft wear in the function of initial clearance where: I is normal wear, II is mechano-chemical wear, III is destructive wear, b): the structure of the worn surface of the shaft pin X300.

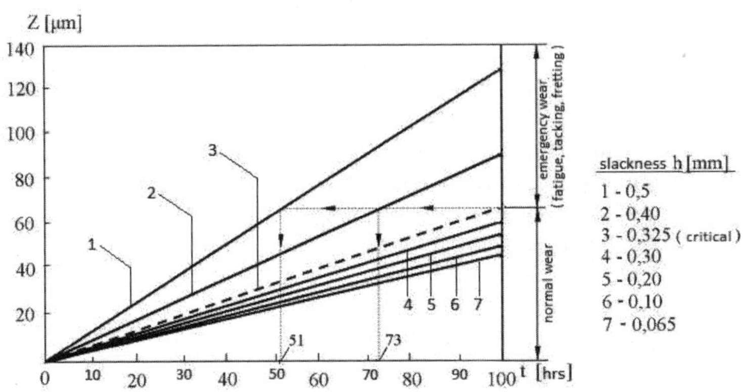

Fig. 9. The kinetic relations in the wear of the needle bearing elements (the operating conditions, different clearances in association)

The intensity of wear for the consecutive values of clearance was examined repeating each trial three times. The results of the examination were subject to mathematical analysis. The data processed in the experiment led to the regression equation of clearance on the degree of wear (Eq.3) for the regression coefficient r = 0,998727.

$$y = \exp(3{,}72157 \ x^3 - 2{,}29608) - 0{,}1 \tag{3}$$

Tab. 1. The results of the static analysis of the experiment

Clearance in the bearing [mm]	0,06	0,1	0,2	0,3	0,4	0,5
Arithmetic mean	0,1	0,10196	0,10478	0,11006	0,12698	0,16102
Standard deviation	0,0	0,00121	0,0018	0,02969	0,00652	0,01
Coefficient of variation with the slackness of 0,06mm [%]	0,0	1,19	1,72	2,70	5,14	6,22
Coefficient of variation in relation to the arithmetic mean [%]	0,0	2,96	4,11	6,04	9,87	10,01

Summing up the results of the mathematical analysis of the experiment it can be said that for the same value of clearance no significant differences between the theoretical and experimental results were noticed. Analyzing the coefficient of variation it can be stated that with the clearance up to 0.3 mm it changes slowly and gradually, but with the clearance of 0.4 mm and 0.5 mm it increases dynamically. This means that with the clearance over 0.3 mm the process of wear assumes a destructive character.

In real conditions, when the surfaces of the bodies in direct contact are uneven, and anisotropy of top layer occurs, the problem of body contact becomes a more complicated one. Further examination will focus on the analysis of friction surface of the needle bearing elements in real conditions, in the lubricated environment and under kinematic load.

Summary

The conducted examination and simulations have demonstrated the changes in the contact area of friction pair elements when two rollers with parallel axes (i.e. the needle roller and the shaft pin) press against each other. Examination of the factors that can have an influence on the contact area of cooperating elements, i.e.: changes in the needle roller diameter and the change in the shaft diameter, are of particular importance because the contact area alters the resistance moment of motion in the bearing. The suggested mathematical model and the examination on a stand help to estimate the extent to which the bearing friction and the sliding friction matter in the overall balance of motion resistance. While examining the shaft neck wear in the function of initial clearance, the conditions of movement from the normal wear to the destructive one were established by means of setting up the critical parameters: the clearance, the intensity of wear in the function of load. It has been observed that the movement from the normal wear to adhesion is characterized by a gradual increase of the intensity of the elements' wear.

References

[1] Dietrich M. (red.): The Basics of Machine Construction, Wydawnictwa Naukowo - Techniczne Wyd. 3. T.2, Warszawa, 2003, s. 391 – 400.

[2] Hebda M ., Wachal A.: Tribology, Wydawnictwa Naukowo-Techniczne, Warszawa, 1980, s. 145-147, 443-446

[3] Nachimowicz J., Baranowski T., Jabłoński D.: Tribological Aspects of Needle Bearing Work, XXIIIth Symposium on the Basics of Machine Construction, Rzeszów – Przemyśl 2007, s. 109 – 115.

[4] Nachimowicz J., Jałbrzykowski M.: Methodological Aspects of Evaluating Tribological Properties of Ball Bearings, Solid State Phenomena Volumes, Mechatronic System and Materials III, Trans Tech Publications, Switzerland, 147-149.

[5] Totten G., Liang H.: Mechanical Tribology, Marcel Dekker, New York 2004, s.64-69, 199-205,

Key Engineering Materials Vol. 490 (2012) pp 296-304

doi:10.4028/www.scientific.net/KEM.490.296

The Integrity Analysis of the Structural Element Made of Ultra High Strength Steel

Marcin Graba[1, a], Timo Nykänen[2,b] and Andrzej Neimitz[3,c]

[1]Kielce University of Technology, Al. 1000 lecia P.P.7, 25-353 Kielce, Poland

[2]Lappeenranta University of Technology, P.O. Box 20, FI-53851, Lappeenranta 85, Finland

[3] Kielce University of Technology, Al. 1000 lecia P.P.7, 25-353 Kielce, Poland

[a]mgraba@tu.kielce.pl, [b]timo.nykanen@lut.fi, [c]neimitz@tu.kielce.pl

Keywords: structural integrity assessment, ultra high strength steel, FITNET

Abstract. In this paper, a FITNET analysis of the structural element made of ultra high strength steel was performed. The critical loadings at failure were estimated at three levels of analysis and two temperatures, -40°C and +20°C. The results of the analysis were verified by full scale experiments at -40°C, which proved that the predictions of the critical moment provide a sufficient safety margin.

Nomenclature

E	Young's modulus
J	J-integral
K_r	The ratio of the applied linear elastic stress intensity factor K_I, to the materials fracture toughness K_{mat}
L_r	The ratio of the applied stress (force) to the stress (force) causeing plastic yielding of the cracked structure.
R_e	yield stress
R_m	ultimate stress
n	Ramberg-Osgood power exponent
σ_r	True stress
ε_r	True strain

Introduction

There is an increasing demand for ultra high strength steels for manufacturing chassis and superstructures for commercial vehicles, booms for forest machinery, crane arms and other lifting equipment, load handling, support and fastening equipment, feeding and unloading hoppers, hook-lift containers and other reliable structures. The S960 QC steel, hot rolled and quenched on line, has improved surface quality and the required yield strength levels (usually above 1000 MPa) in combination with good impact toughness, weldability and formability. However, the behavior of this steel does not always follow the behavior of other ferritic steels. The fracture toughness of the S960 QC steel does not fit into the current Master Curve [2]. Also the experimental correlation between transition temperatures T_0 and T_{28J} leads to very unconservative K_{JC} estimates, indicating that K_{JC} needs to be measured directly [1]. These uncertainties triggered the research program aiming at the estimation of the integrity of structural elements containing cracks, at low and room temperatures. The experimental tests were followed by structural integrity assessments using the FITNET [3] procedure.

Structural element

The structural element was a hollow rectangular structural element 180x135x935 mm of the shape shown in Fig. 1. The wall thickness was 5 mm.

Fig. 1. Cross-section of the element tested

The element was four point bent (Fig. 2) to receive the same stress distribution over the length where cracks were introduced.

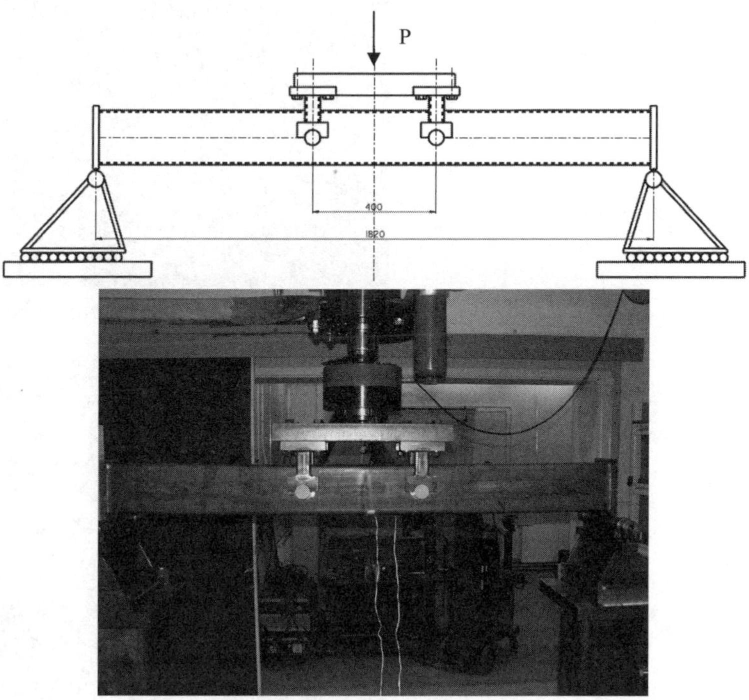

Fig. 2. Experimental set-up

Three lengths of cracks within the profile were tested: a crack through the lower flank of the profile (Fig. 3) – Specimen No. 1, a crack with its front situated in the middle of the side walls (Fig. 4) – Specimen No. 2, and profile with cracks located at the rounded bottom corners of the structural element (Fig. 5) – Specimen No. 3.

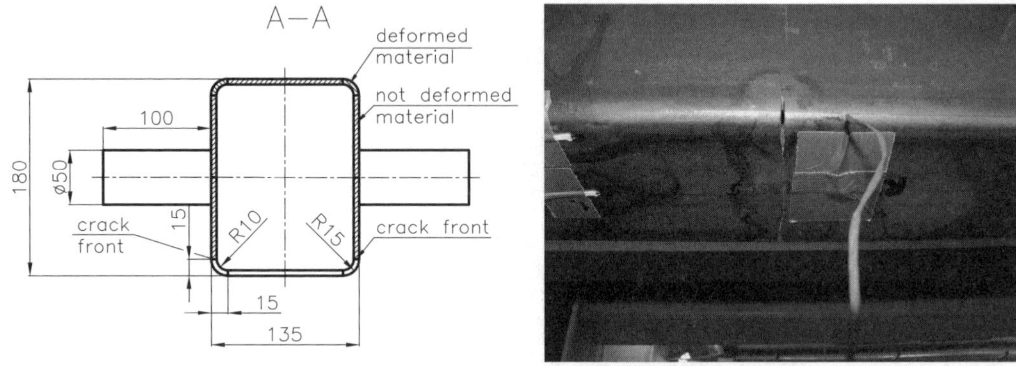

Fig. 3. Crack through the lower flank of the profile, Specimen No. 1

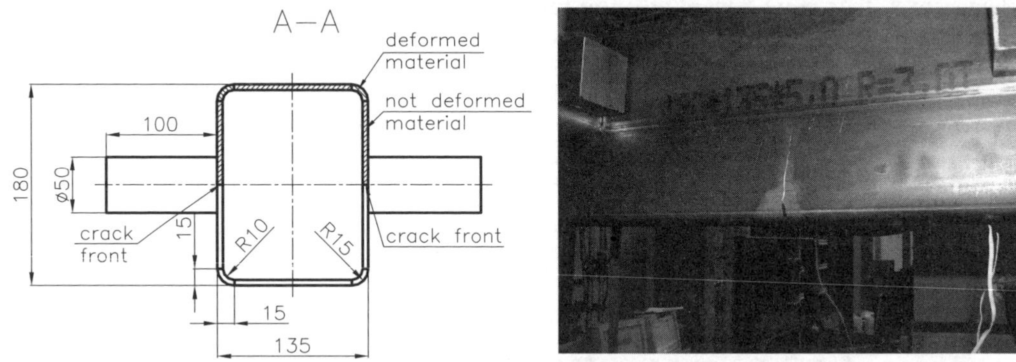

Fig. 4. Crack with its front situated in the middle of the side walls, Specimen No. 2

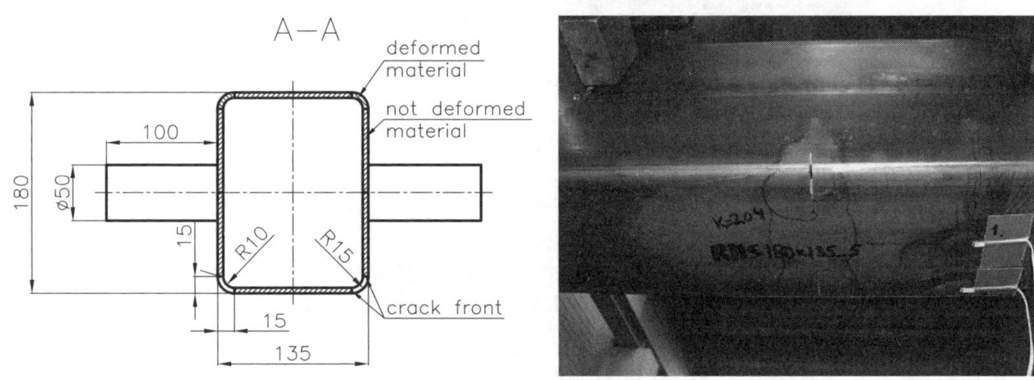

Fig. 5. Profile with cracks located at the rounded bottom corners of the structural element, Specimen No. 3.

The cracks were introduced by sawing the bottom rounded corners of the profile and through the fatigue process. Tests were carried out at -40°C. Subzero temperatures were achieved by circulating a coolant inside the test specimens. The specimens were insulated with plastic foam sheets. The temperature of the test specimens was measured by thermo elements installed on the external surface of the profile close to the crack tip area. The specimens were quasi-static loaded, and during the test, the load and deflections were recorded.

Material

The chemical composition of the S960 QC steel is given in Table 1.

Table 1. Basic chemical composition limits (wt%) and carbon equivalent values of the S960 QC steel

C_{max}	Si_{max}	Mn_{max}	P_{max}	S_{max}	Ti_{max}	CEV typical	CEV max
0.11	0.25	1.20	0.02	0.01	0.07	0.47	0.52
CEV = C+Mn/6+(Cr+Mo+V)/5+(Cu+Ni)/15							

The tensile and fracture properties of the sheet used to produce the structural element are listed in Table 2.

Table 2. Tensile properties and fracture properties of tested materials.

	Material properties	
Material at deformed corners of structure -40°C	E [MPa]	200000
	R_e [MPa]	1162
	R_m [MPa]	1312.50
	n	9.92
Material at undeformed parts of structure -40°C	E [MPa]	200000
	R_e [MPa]	1113
	R_m [MPa]	1210
	n	13.78
	J_c [kN/m],	391
	K_{mat} [MPa·m$^{0.5}$]	279

Analysis

The philosophy of the FITNET approach is that the quality of all input data is reflected in the sophistication and accuracy of the resulting analysis. A series of levels is available, each of increasing complexity and each being less conservative than the previous, lower level. Each level is dependent on the quality and details of the material's property data available. The user should be aware that the higher the level of analysis is, the higher the quality required of the input data and the more complex the analysis is. In turn, the lower the level of analysis, the more conservative the result, but the lowest level which gives an acceptable result implies satisfactory results at higher levels. The level of analysis is characterized mainly by the tensile data of the material.

Since the material was uniform, the computations were performed for three levels of analysis proposed in FITNET: default (only yield strength and Charpy energy are required), 1 (only yield strength and fracture toughness data are required), and 3 (the whole stress-strain curve and fracture toughness data are required). The default level was analyzed, but using the fracture toughness measured according to standard suggestions. However, the thickness 5mm does not satisfy the plane strain requirements. In Fig. 6, the FAD curves are plotted and the formulae to draw these curves are listed.

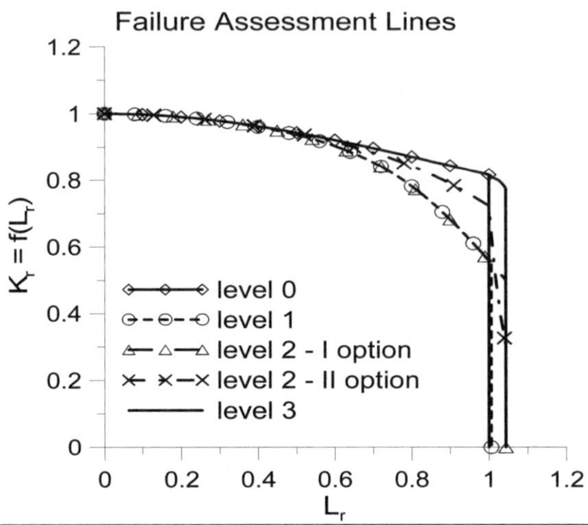

Levels 0 and 1	$f(L_r)=\left(1+0.5L_r^2\right)^{-1/2}\begin{bmatrix}0,\\ 0.3+0.7\exp\left(-\mu L_r^6\right)\end{bmatrix}$	for $L_r \le 1$
Level 1	$f(L_r)=f(1)L_r^{(N-1)/2N}$	for $1 \le L_r \le L_r^{\max}$
Level 3	$f(L_r)=\left[E\varepsilon_r/\sigma_r+0{,}5L_r^2/(E\varepsilon_r/\sigma_r)\right]^{-1/2}$	for $0 \le L_r \le L_r^{\max}$
Level 1	$f(1)=\left(\lambda+\frac{1}{2\lambda}\right)^{-1/2}$	for $L_r = 1$
Level 0	$L_r^{\max}=1+\left(\frac{150}{R_e}\right)^{2,5}$	
Levels 1 and 3	$L_r^{\max}=0{,}5(1+R_m/R_e)$,	
$\mu=\min\left[0{,}001\left(\frac{E}{R_e}\right);0{,}6\right]$, $N=0{,}3\left[1-\frac{R_e}{R_m}\right]$, $\lambda=\left(1+\frac{E\Delta\varepsilon}{R_{eH}}\right)$, $\Delta\varepsilon=0.0375\left(1-\frac{R_{eH}}{1000}\right)$, R_{eH} – upper yield strength		
Fig. 6. The FA lines and appropriate equations for tested materials		

Figure 6 also contains the Failure Assessment Lines – level 2, which are not used in computation since the material is uniform, to show that for non-uniform material the results of the analysis would be more conservative.

In the next step after the Failure Assessment Line is determined, which is dependent on the tensile properties of the material only (except level 4), it is necessary to plot an assessment point or a set of assessment points of co-ordinates $\{L_r, K_r\}$. These coordinates are calculated under the loading conditions applicable (given by the loads, crack size, and material properties), and these points are then compared with the Failure Assessment Line. The Assessment Line defines the envelope for the achievement of a limiting condition for the loading of the cracked structure, and assessment points lying on or within this envelope indicate that the structure, as assessed, is acceptable against this limiting condition. A point which lies outside this envelope indicates that the structure as assessed has failed to meet this limiting condition.

To compute the coordinates of the assessment points, one needs the values of the stress intensity factors (SIF) at several loading levels K_I, the fracture toughness K_{mat}, and the limit load values computed for elastic-perfectly plastic material. Unfortunately, there are not closed form solutions for either the SIF or the limit load for the geometries under consideration. Thus, they will be sought using the finite element (FE) method analysis. The SIF was determined from values of the J-integral computed for several loadings and converted to the SIF units through the formula $K_I = \sqrt{\frac{EJ_I}{(1-\nu^2)}}$.
The virtual crack extension method was applied and the weighted average of the J-integral was computed. The results of the limit load are shown in Table 3. The linear-elastic model of material was used. The FE model of one of the geometries analyzed is shown in Fig. 7.

Computations were carried out using the ADINA SYSTEM 8.5. A contact analysis was performed. Because of the symmetry, only a quarter of the profile was modelled. The finite element mesh was filled with 20-node and 8-node 3D brick elements. The mesh consisted of three layers of finite elements throughout the thickness of the specimen. The profile was modelled using 12834 finite elements and 20000÷30000 nodes.

The limit load was defined at the moment when the whole unbroken cross-section was yielding. An example of the computations is shown in Fig. 8.

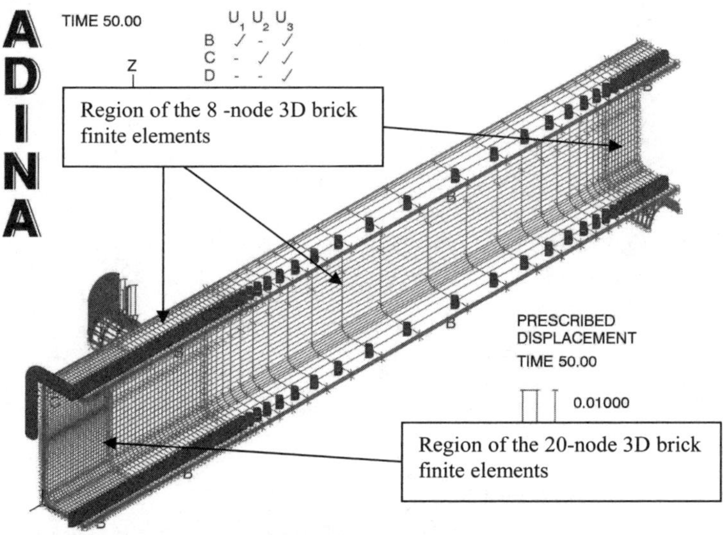

Fig. 7. The finite element model of Specimen No.2

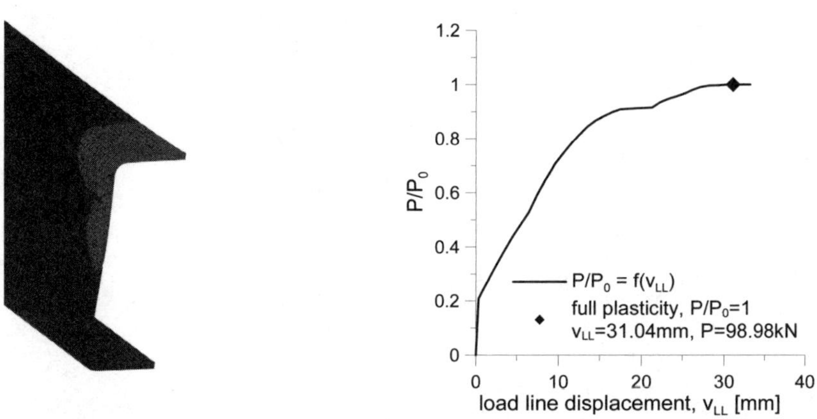

Fig. 8. The moment when the whole ligament starts yielding, Specimen No. 2

Table 3. Limit loads for three tested specimens

	Limit load [kN]	Load point displacement [mm]
Specimen No. 1	209,33	52,41
Specimen No. 2	98,98	31,4
Specimen No. 3	299,32	77,19

Now, all necessary data are available and the results of the analysis are shown in Figures 9-11.

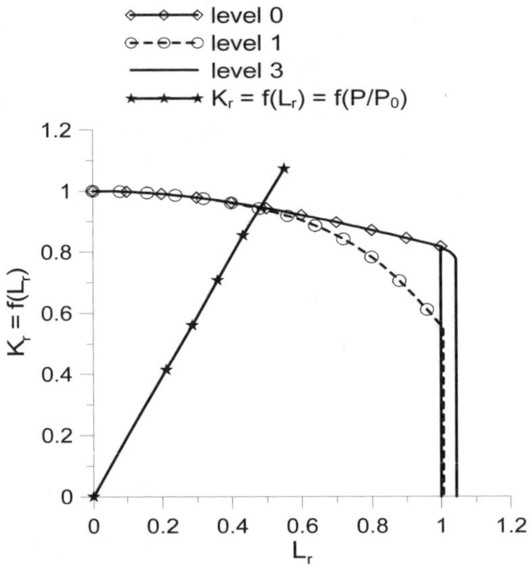

Fig. 9. The FAD analysis of Specimen No. 1

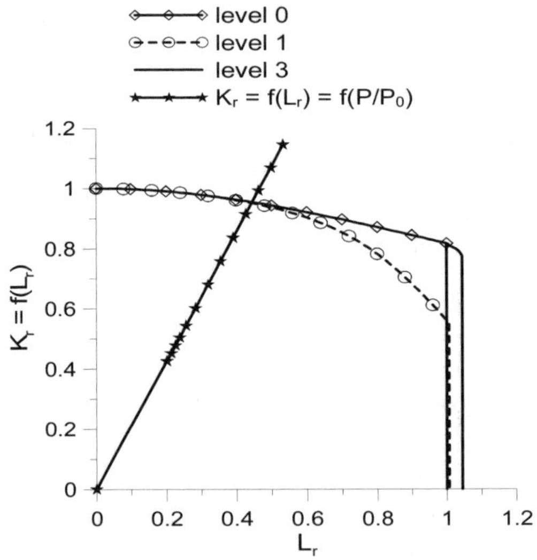

Fig. 10 The FAD analysis of Specimen No. 2

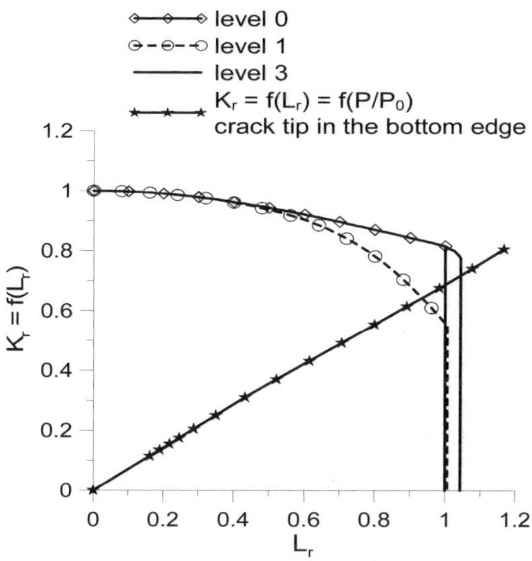

Fig. 11. The FAD analysis of Specimen No. 3

The FAD analysis of Specimen No. 1 (Fig. 9) shows that it can fail at an external load equal to 100.50 kN, which is approximately 48% of the limit load P_0 (P_0=209.33 kN). The SIF is equal to 275.63MPa·m$^{0.5}$ (approximately 94% of 293.22 MPa·m$^{0.5}$). The critical load measured in the experiment was roughly 160 KN. Thus, the conservatism is less than a factor of 2 – it is equal to 1.6.

For comparison, critical values are given for the temperature +20°C. The limit load is 196.2 kN, the critical load 104 kN (53% of P_0), K_{mat}=305.67 MPa·m$^{0.5}$, the critical value of K_I=0.935 K_{mat}

The FAD analysis of Specimen No. 2 (Fig. 10) suggests that it may fail at a load equal to 44.05 kN, which is 44.50% of the limit load (P_0=98.98 kN). The SIF is equal to 280.03 MPa·m$^{0.5}$ (it is approximately 95.5% of the K_{mat}=293.22 MPa·m$^{0.5}$). The critical load measured at Lappeenranta University of Technology was roughly 100 kN. Thus, the conservatism is equal to the factor 2.25.

At the temperature of +20°C, the limit load is P_0=93.87 kN, the critical load is 45.53 kN, K_r=0.945

In both specimens (1 and 2), the selection of the analysis level has almost no influence on the final result. The specimens fractured according to the ductile mechanism by voids nucleation and growth along the slip planes which are slant (45°) with respect to the wall with a rectangular profile. The selection of the analysis level may become important when Specimen No. 3 is considered (Fig. 11). Here, two cases were analyzed, since a crack can propagate in two directions: along the bottom flange and along the side walls. The limit load computed for the bottom flange – in between the two crack tips – is equal to 238 kN. The limit load for the whole cross-section is equal to 299 kN. The critical force necessary to break the lower flange is equal to 221.8 kN (93% of P_o), K_r=0.65. These values are according to level 1 of the FITNET analysis. When the lower flank is broken, we encounter the situation which was already discussed – the crack at the bottom edge. However, now the external loading is greater than the limit load for Specimen No. 1, and we observe a sudden failure of the whole cross-section, which was preceded by the yielding of the material (Fig. 12). According to the FAD analysis at levels 0 and 3, the situation does not change. However, the results are less conservative. $P_{crit.}$=238 kN and 248 kN for the default level and level 3, respectively. Unfortunately, these results were not fully verified experimentally. The specimen had to be unloaded at the external loading 203 kN. It did not fail. FITNET confirms that this loading is still safe. The profile was broken later by fatigue loading.

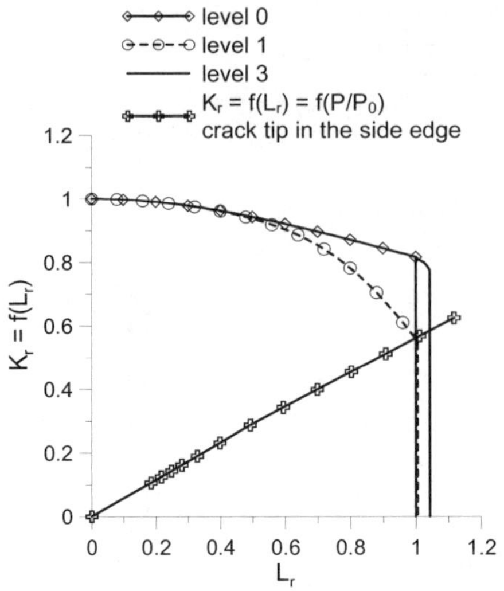

Fig. 12. Fracture of the whole cross-section of Specimen No. 3.

The results of the analysis at +20°C are as follows. Specimen No. 1 should not be loaded by a force grater than 211.4 kN, 215.7 kN and 229.9 kN according to levels of analysis 1, 0 and 3, respectively.

Conclusion. The FITNET analysis has proved to be useful when applied to the ultra high strength steel S960 QC. The experimental tests show that the predictions of the critical moment provide a sufficient safety margin, taking into account the possible scatter of the fracture toughness in the transition temperature range.

Acknowledgements. The financial support of Rautaruukki Oyj and Fimecc Oy are gratefully acknowledged.

Literature

[1] P. Nevasmaa1, P. Karjalainen-Roikonen1, A. Laukkanen, T. Nykänen, A. Ameri T. Björk, T. Limnell, J. Kuoppala. "Fracture characteristics of new ultra-high-strength steel with yield strengths 900 – 960 MPa", Rautaruukki research Report, www.ruukki.com/product /and solutions/steel products/hot rolled steels/optim QC.

[2] ASTM E1921-02. Standard test method for determination of reference temperature, T_0, for ferritic steels in the transition range. Annual book of ASTM standards, vol. 03.01, 2003, pp. 1128-1147

[3] FITNET Fitnet for Service, Vol. I, Eds. M. Kocak, S. Webster, J. J. Janosh, R. A. Ainsworth, R. Koers, ISBN P78-3-940923-00-4, 2008.

Key Engineering Materials Vol. 490 (2012) pp 305-311
doi:10.4028/www.scientific.net/KEM.490.305

The Method of Finite Differences in Solutions Concerning Circular Plate

Henryk G. Sabiniak

Technical University of Lodz, Al. Politechniki 6, 90-924 Łódź, Poland
henryk.sabiniak@p.lodz.pl

Keywords: finite differences, circular plate, worm teeth.

Abstract: Finite difference method in solving classic problems in theory of plates is considered a standard one [1], [2], [3], [4]. The above refers mainly to solutions in right-angle coordinates. For circular plates, for which the use of polar coordinates is the best option, the question of classic plate deflection gets complicated. In accordance with mathematical rules the passage from partial differentials to final differences seems firm. Still final formulas both for the equation (1), as well as for border conditions of circular plate obtained in this study and in the study [3] differ considerably. The paper describes in detail necessary mathematical calculations. The final results are presented in identical form as in the study [3]. Difference of results as well as the length of arm in passage from partial differentials to finite differences for mixed derivatives are discussed. Generalizations resulting from these discussions are presented. This preliminary proceeding has the purpose of searching for solutions to technical problems in machine building and construction, in particular finding a solution to the question of distribution of load along contact line in worm gearing.

Introduction

The proper way of replacing partial differentials with finite differences in the case of formulae expressed in polar coordinates is presented. The derived formulae are widely applied in the theory of the circular plates. Attention is drawn to the mistakes that occur in the literature of the subject. The use of the finite differences method in the solution of classical plate problems is regarded as standard [1], [2], [3], [4]. It mainly refers to the analysis performed in Cartesian coordinates. For circular plates, where the use of polar coordinates is natural, the problem of the classical plate deflection becomes more complex. In this case, the literature presents a formula transformed from Cartesian coordinates into polar coordinates, namely:

$$\left(\frac{\partial^2}{\partial r^2}+\frac{1}{r}\frac{\partial}{\partial r}+\frac{1}{r^2}\frac{\partial^2}{\partial \varphi^2}\right)\left(\frac{\partial^2 \omega}{\partial r^2}+\frac{1}{r}\frac{\partial \omega}{\partial r}+\frac{1}{r^2}\frac{\partial^2 \omega}{\partial \varphi^2}\right)=\frac{q}{D}. \tag{1}$$

Where: ω - plate deflection,
q – load,
D – plate stiffness; D=Eh^2/12(1-v^2),
E – Young's modulus,
h – plate thickness,
v – Poisson's ratio.

The deflection of the plate described by formula (1) was derived taking into account the following assumptions:

- the load is perpendicular to the plate surface,
- the edges can be displaced freely in the plate plane,
- the deflections are small in relation to the plate thickness.

The transition from partial differentials to finite differences, according to the rules of mathematics, seem univocal. However, the formulae resulting form equation (1) and the boundary conditions of the circular plate, obtained in this paper and the ones that were obtained in paper [3] differ substantially.

Generally equations

The first step in the transition from partial differentials to finite differences is to perform differentiation in formula (1). The following is obtained [5]:

$$\frac{\partial^4\omega}{\partial r^4}+\frac{2}{r}\frac{\partial^3\omega}{\partial r^3}-\frac{1}{r^2}\left(\frac{\partial^2\omega}{\partial r^2}-2\frac{\partial^4\omega}{\partial r^2\partial\varphi^2}\right)+\frac{1}{r^3}\left(\frac{\partial\omega}{\partial r}-2\frac{\partial^3\omega}{\partial r\partial\varphi^2}\right)+\frac{1}{r^4}\left(4\frac{\partial^2\omega}{\partial\varphi^2}+\frac{\partial^4\omega}{\partial\varphi^4}\right)=\frac{q}{D}. \tag{2}$$

Now numerical methods can be utilized, replacing partial differentials with finite differences. To do this, the following coordinate system and division of the plate were assumed (Fig. 1).

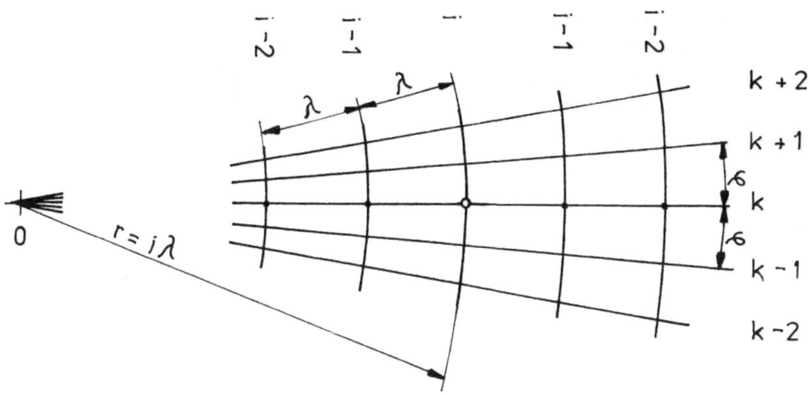

Fig. 1: Division of the plate. Coordinate system

Node (i,k) is defined by coordinates $i\lambda$ and $k\varphi$. The distance along axis r between the adjacent points is λ, and along axis φ is $i\lambda\varphi$. Before replacing partial derivatives in formula (2) with finite differences, it is helpful to transform each component of this equation.

$$\begin{aligned}
&\frac{\partial\omega}{\partial r}\approx\frac{\omega_{i+1,k}-\omega_{i-1,k}}{2\lambda},\\
&\frac{\partial^2\omega}{\partial r^2}\approx\frac{\omega_{i+1,k}-2\omega_{i,k}+\omega_{i-1,k}}{\lambda^2},\\
&\frac{\partial^3\omega}{\partial r^3}\approx\frac{\omega_{i+2,k}-2\omega_{i+1,k}+2\omega_{i-1,k}-\omega_{i-2,k}}{2\lambda^3},\\
&\frac{\partial^4\omega}{\partial r^4}\approx\frac{\omega_{i+2,k}-4\omega_{i+1,k}+6\omega_{i,k}-4\omega_{i-1,k}+\omega_{i-2,k}}{\lambda^4},\\
&\frac{\partial^3\omega}{\partial r\partial\varphi^2}\approx\frac{\omega_{i+1,k+1}-2\omega_{i+1,k}+\omega_{i+1,k-1}-\omega_{i-1k+1}+2\omega_{i-1k}-\omega_{i-1,k-1}}{2\lambda\varphi^2},\\
&\frac{\partial^4\omega}{\partial r^2\partial\varphi^2}\approx\frac{\omega_{i+1,k+1}-2\omega_{i+1,k}+\omega_{i+1,k-1}-2\omega_{i,k+1}+4\omega_{i,k}-2\omega_{i,k-1}+\omega_{i-1,k+1}-2\omega_{i-1k}+\omega_{i-1k-1}}{\lambda^2\varphi^2},\\
&\frac{\partial^2\omega}{\partial\varphi^2}\approx\frac{\omega_{i,k+1}-2\omega_{i,k}+\omega_{i,k-1}}{\varphi^2},\\
&\frac{\partial^4\omega}{\partial\varphi^4}\approx\frac{\omega_{i,k+2}-4\omega_{i,k+1}+6\omega_{i,k}-4\omega_{i,k-1}+\omega_{i,k-2}}{\varphi^4}.
\end{aligned} \tag{3}$$

It seems that in [3] the transition to finite differences was undertaken when the equation of the plate deflection was expressed by formula (1). This approach is erroneous, since there is no symmetry along coordinate φ. This could be done [6] if there was symmetry with respect to two axes perpendicular to each other (Cartesian coordinates). After transformation from Cartesian coordinates to polar coordinates this symmetry is lost [1]. There are no identical terms with respect to r and φ as there were for x and y. Therefore, first we should perform differentiation in formula (1) transforming it into (2), then we should obtain the finite differences.

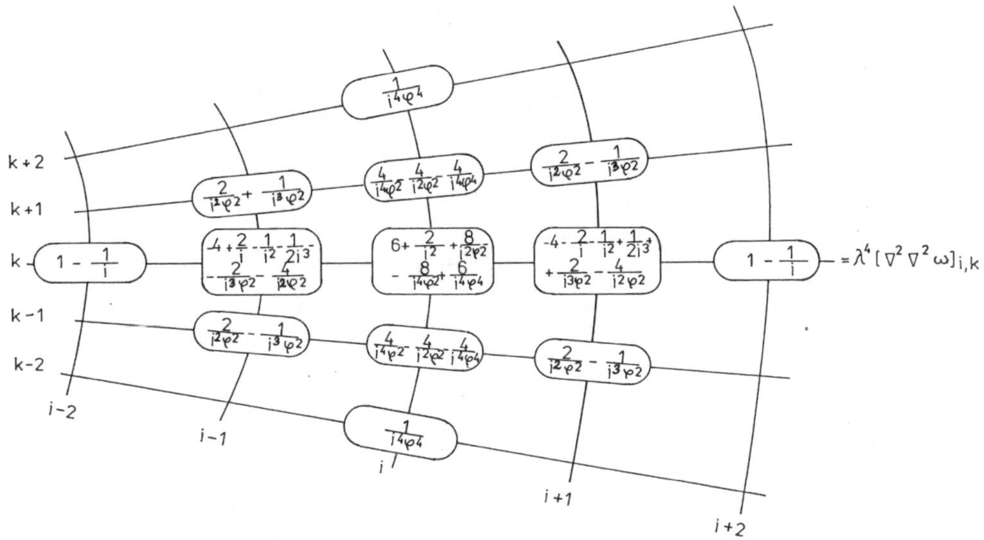

Fig. 2: The equation of the plate deflections

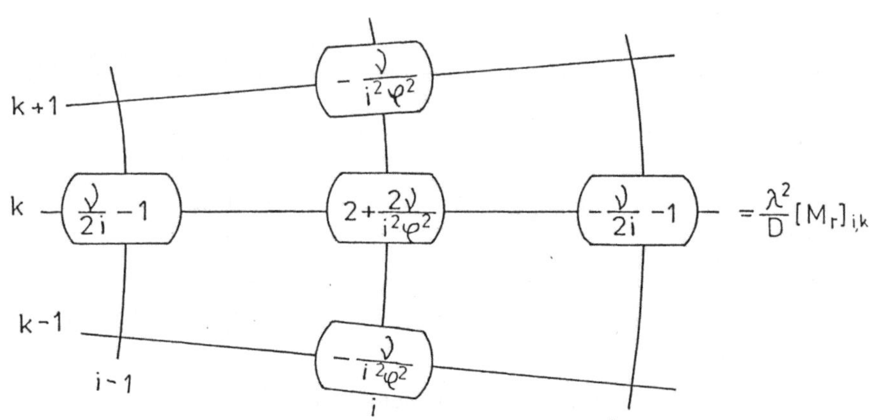

Fig. 3: Bending moment M_r

If the partial differentials are of the second order or lower, there is no discrepancy in the forms of final formulae obtained during the transition from partial differentials to finite differences, since the replacement of differentials with differences is the same for derivatives of the second order or lower. The equation of the plate deflection, however, is of the fourth order and divergencies between final formulae in the form of finite differences are substantial. Fig. 2 presents schematically a formula for the calculation of the plate deflections after partial derivatives have been replaced with finite differences, and the same scheme is applied as in [3]. The above remarks also refer to boundary conditions

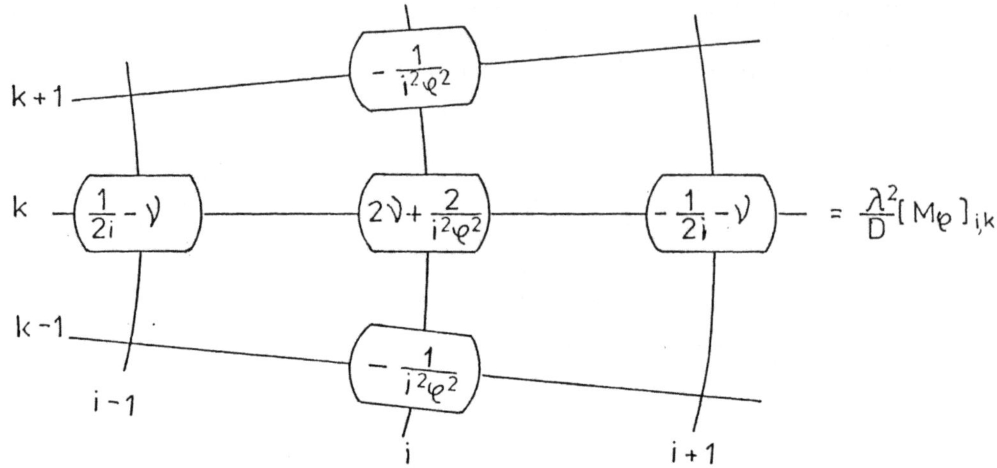

Fig. 4: Bending moment M_φ

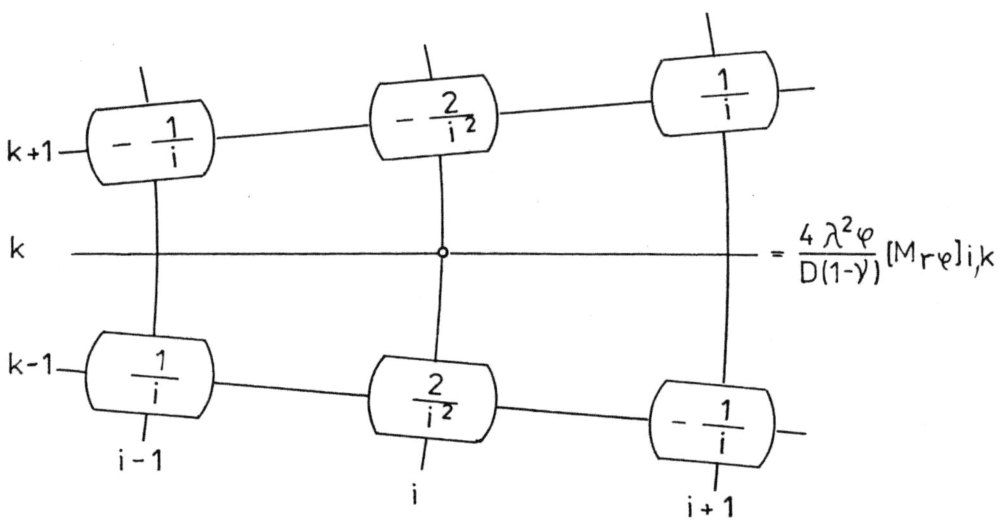

Fig. 5: Twisting moment $M_{r\varphi}$

of the plate. The bending moments [1] at the plate edges are of the second order and there are no divergencies here in final formulae in relation to [3], (Figs. 3 and 4). However, there is a difference in the final formulae for the twisting moment [1], although it is of the second order (Fig. 5). This difference seems to be a technical error. After transformation into finite differences, one component of the formula was missed, namely:

$$-\frac{1}{r^2}\frac{\partial \omega}{\partial \varphi}$$

For shearing forces [1] the discrepancies in the formulae are clear (Figs. 6 and 7). These formulae are of the third order. The presented formulae used in the calculations of deflections of circular plates and boundary conditions by the method of finite differences have different final forms in relation to [3] and they will certainly yield different results in the calculations of specific cases of plates. Therefore we should be aware of the errors they introduce. In the presented formulae the errors result from simplifying assumptions (transition from differentials to differences, lattice density), and in the case of formulae [3] mathematical errors are introduced on top of that.

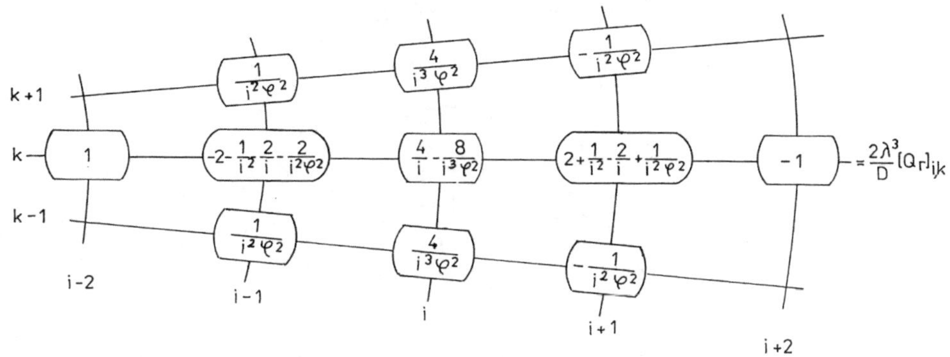

Fig. 6: Shearing force Q_r

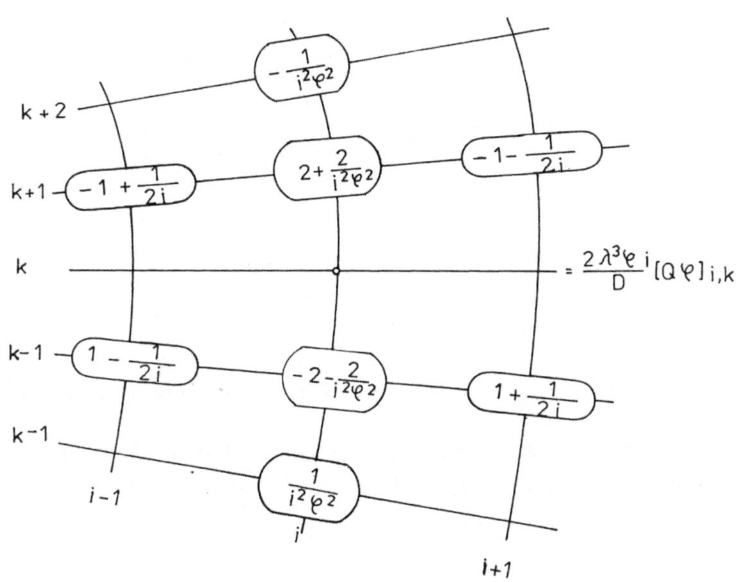

Fig. 7: Shearing force Q_φ

During the transition from partial differentials to finite differences there emerges the problem of determining the length of the radius or, to be more precise, of finding out which circle (i+1, i or i-1) is involved. This problem emerges with higher mixed derivatives. Let us consider this problem using an example of finite differential:

$$\frac{\partial^2 \omega}{\partial r \partial \varphi}$$

If we first replace the differential with a finite difference along the radius, then the finite difference along angle φ shall be computed for radius $r_1=(i+1)\lambda$ and $r_2=(i-1)\lambda$, (Fig. 8). If however we first replace the differential with a finite difference along the angle φ, and we proceed to determine the finite differences along the radius, then the radius for both terms will be equal to $r_1=r_2=i\lambda$ (Fig. 9).

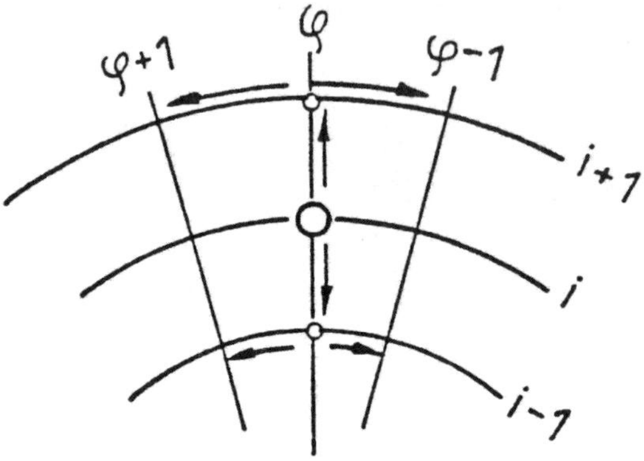

Fig. 8: The way of passage from partial differentials to finite differences

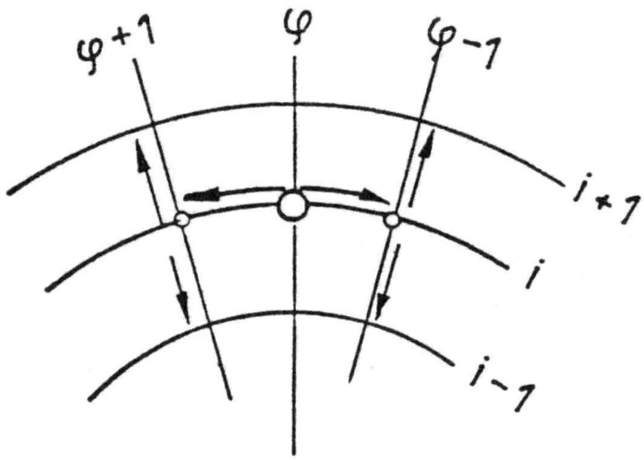

Fig. 9: The way of passage from partial differentials to finite differences

For numerical cases when the plate deflection is sought for long radii with dense discreditsation of the plate, the presented problem becomes unimportant. For some cases, such as, for example, shown in Fig. 10, where:

$$\frac{b}{a} \geq \frac{2}{3}$$

and $i \geq 15$ (np: in worm teeth), the problem can be neglected and all coefficients $i+1 \approx i-1 = i$ can be accepted. In such cases, the formulae presented in Figs. 2, 5, 6 and 7 can be reduced. In engineering calculations such approximation is sufficient and, in some cases, it can also yield satisfactory results in scientific studies.

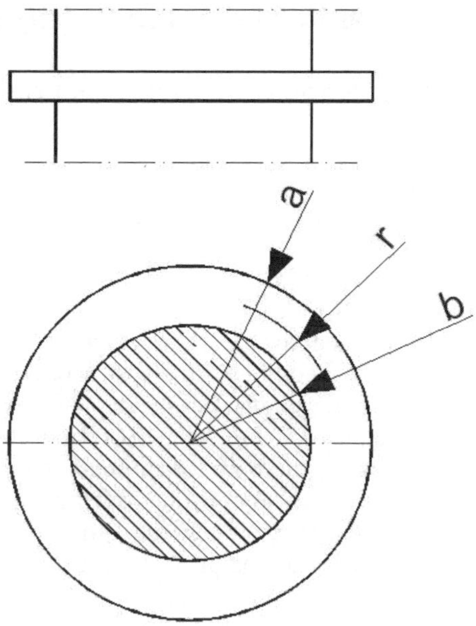

Fig. 10: The range of generalization the length radius

References

[1] S. Timoshenko, S. Woinowsky-Krieger - Theory of plates and shells (Polish edition), Arkady, Warsaw 1962.

[2] Łukaszewicz S.: Concentrated load of plates, shields and shells. (Polish edition), PWN. Warsaw 1976.

[3] Kączkowski Z.: Static calculations. (Polish edition), Arkady. Warsaw 1980.

[4] Kovalenko A.D.: Wheel plates with variable thickness. (Russian language). GIFML. Moscow 1959.

[5] Benda J.: Calculations of wheel plates of finite differences method. (Czech language). Notebook of Technical High School in Brno, nr 4, pp. 175-189, 1964.

[6] Sabiniak H.G.: Load distribution along the contact line between two bracket plates, treated as models for teeth of cylindrical and globoidal toothed gears. (German language). PhD dissertation Technical University in Chemnitz, 1992.

Key Engineering Materials Vol. 490 (2012) pp 312-322
© (2012) Trans Tech Publications, Switzerland
doi:10.4028/www.scientific.net/KEM.490.312

The Numerical Methods in Geometric Construction Mapping

Henryk G. Sabiniak, Robert Cichowicz

Technical University of Lodz, Al. Politechniki 6. 90-924 Łódź. Poland
Technical University of Lodz, Al. Politechniki 6. 90-924 Łódź. Poland
henryk.sabiniak@p.lodz.pl; robert.cichowicz@p.lodz.pl

Keywords: worm teeth, finite element method.

Abstract: A dynamic development of numerical methods enables even so complicated geometrical modelling as modelling of hypoidal teeth, and in particular of worm teeth. Building on the basis of the finite element method of this type of geometrical mathematical model enables observing and analysing physical phenomena taking place in the teeth, e.g. changes of pressure distribution along the contact lines depending on the phase of meshing for particular cooperating pairs of teeth. Due to limited technical possibilities and mathematical apparatus such analyses were carried out only on the basis of the theory of plates with constant thickness. An unquestionable advantage of such numerical modelling is a short time needed to obtain final results, which enables profound analysis of the transmitted load. The knowledge of course of pressure distribution along the contact lines of particular pairs of intermeshing teeth in worm gears already in the design phase enables taking into consideration in adequate correction or even modification of the working surface of teeth, with the aim to equalise the distribution of pressure.

Introduction

The development in calculation possibilities of computers has meant that numerical methods are increasingly more often applied. There are many varieties of numerical methods, some more popular. Among the methods that are currently most often used for this purpose are [1]:

- The Finite element method (FEM),
- The Boundary elements method (BEM),
- The Finite difference method (FDM),

One cannot explicitly determine which of these methods is always the most beneficial. If one considers universality as the criterion, one can assume that the finite element method (FEM) is currently the most comprehensive and most widely used computer method for solving engineering problems.

Currently, there is some research on combining particular methods, which can lead to the creation of hybrid methods that will combine the advantages, and thus, simultaneously reduce the drawbacks of current methods [2], [3].

Meshing worm modeling with FEM

The dynamic development of numerical methods currently allows for such complex geometric modeling of solids as hypoid gear, in particular, the meshing worm [4]. Constructing such a model based on the finite element method enables to track and analyze the phenomena occurring in the meshing, primarily, the contribution of the load carried by particular sets of teeth working together.

To show the possibilities of this method, an example of the evolvent gear teeth boundary was chosen. The evolvent boundary is one of the most popular in worm gears. This is due to its technological characteristics (ease of turning, grinding and checking) and its wear and tear after being adjusted properly [5].

An example of a meshing gear calculation model was built with standard solid elements contained in the ANSYS program - SOLID92. To build a geometric model of a meshing gear, the following parameters [4], [6] were applied:

- The boundary type of a gear tooth: evolvent;
- Axial module: $m_o = 7$ mm;
- Axis distance: a = 125 mm;
- Number of gear teeth: $z_1 = 1$;
- Number of teeth of the worm wheel: = 30;
- Axial angle of contact: $\alpha = 20°$;
- Diameter of the core gear: $d_{f1} = 29.45$ mm;
- External diameter of the screw: $d_{a1} = 60.25$ mm;
- Diameter of the worm wheel teeth: $d_{f2} = 186.95$ mm;
- External diameter of the worm wheel: $d_{a2} = 217.75$ mm.

The ANSYS program, based on the finite element method, allows modeling of any working gear boundaries and their corresponding gear teeth (Fig. 1). This is done by entering the exact dimensions of the worm gear into a previously prepared computation module. As a consequence, it is necessary to design a tooth gear in advance or to obtain information about its geometrical parameters.

To analyze the phenomena that occur in an operating gear, it is necessary to implement such a model, which would be the closest possible to the actual geometry of the structure. Therefore, the most important stage of modeling is the transition from design to the actual physical structure, which accurately reflects the meshing geometry. This is achieved by simulating the engineering process of reducing the teeth gear on computer screen [7], as if it was done in reality.

a) b)

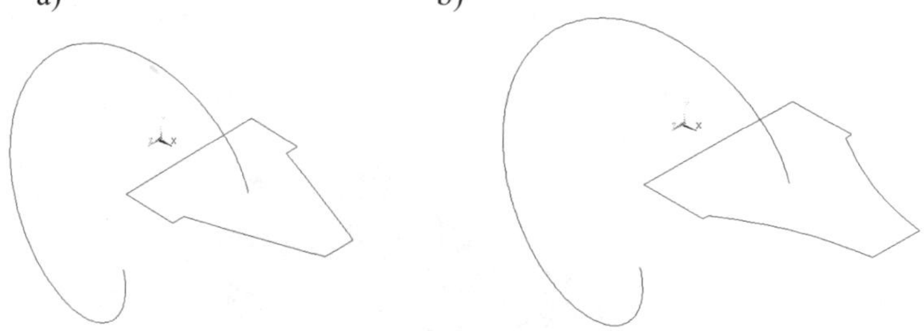

Fig. 4. Tooth worm boundary: a) evolvent, b) circular-concave.

The implementation of the worm gear can be complex and difficult. In the process of modeling, as in reality, the proper execution of the worm cutter, which will be used to cut the gear teeth is one of the most important steps (Fig. 2).

a) b)

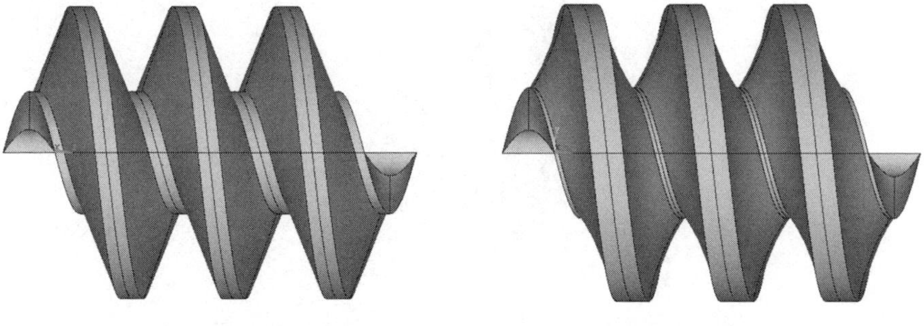

Fig. 2. The gear boundary: a) evolvent, b) circular-concave.

When preparing the engineering process of cutting the worm wheel teeth, the cutter should be divided into layers (Fig. 3). In the cutting process, the following steps are then necessary to program [7]:

- The implementation of the worm cutter model with dividing it into layers;

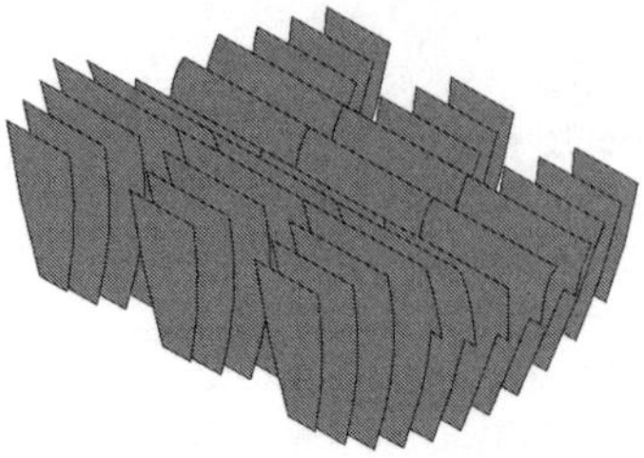

Fig. 3. Model of cutter cutting worm gear teeth divided into layers

- The preparation of the worm wheel rim (Fig. 4);

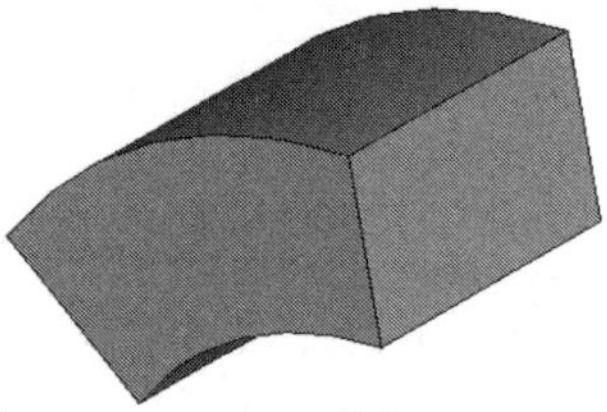

Fig. 4. The component of the worm wheel rim

- Dividing the worm wheel rim into layers (Fig. 5);

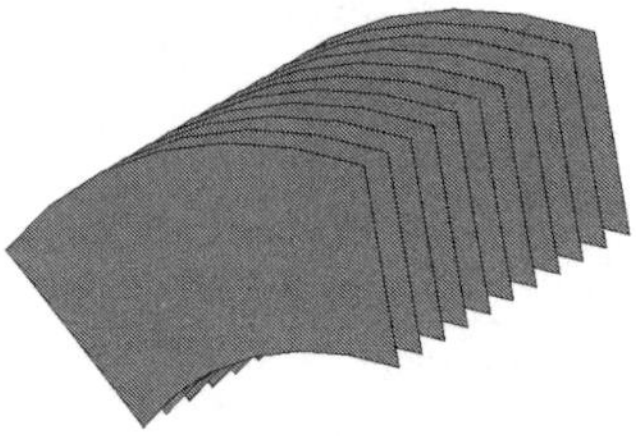

Fig. 5. Piece of the worm wheel rim divided into layers

- Cutting the worm wheel teeth in particular layers with the corresponding layers of the cutter (Fig. 6);

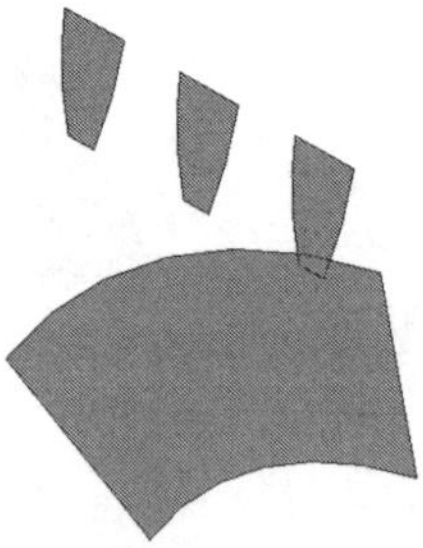

Fig. 6. Milling one layer of the worm wheel rim

- Combining the pieces of the cut worm wheel teeth (Fig. 7);

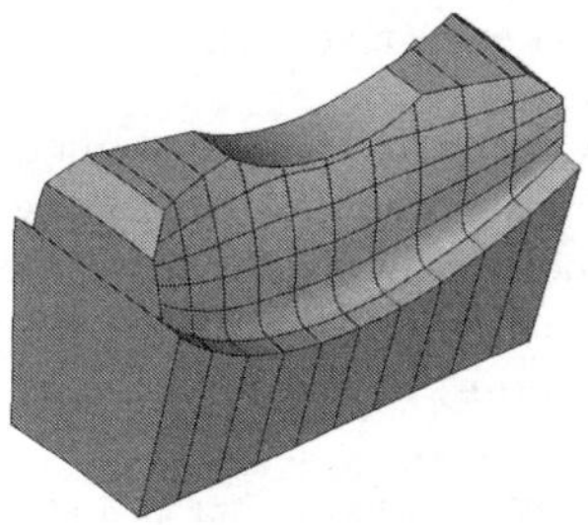

Fig. 7. Worm wheel tooth

- Combining worm wheel teeth and dividing them into finite elements (Fig. 8);

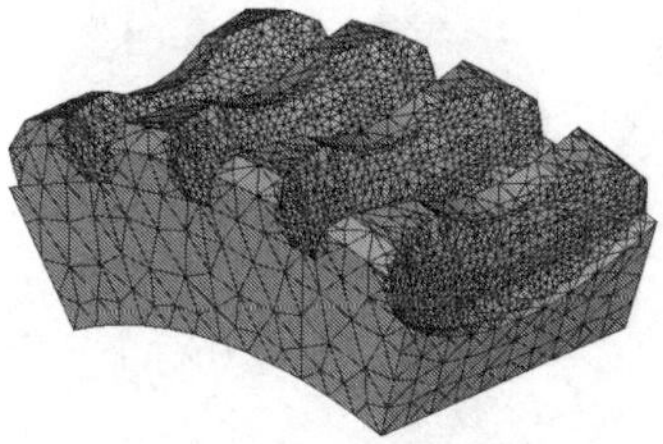

Fig. 8. A cut piece of the worm wheel with marked elements of the division

- Combining the worm and the worm wheel (Fig. 9);

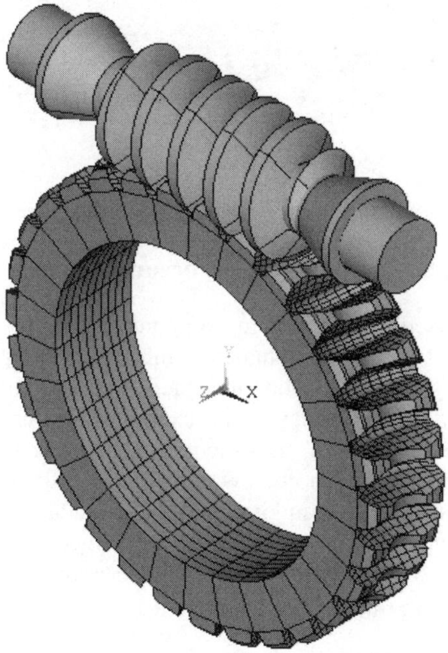

Fig. 9. Worm gear drive assembly

The worm/worm wheel assembly and the simulation of its functioning aims to test the design of the elements (Fig. 9).

Boundary conditions and the worm gear drive load

In numerical methods, and so in the finite element method (ANSYS), the boundary conditions, the area of action and the application of the load need to be defined for each model to obtain a solution. The boundary conditions must clearly indicate the location of a structure in space [15], in this case, the worm gear.
By fulfilling these conditions, the worm becomes loose against the Y and Z axes (Fig. 9). The worm wheel becomes loose against all axes).
The load application point can be implemented either in the form of nodal forces or by means of pressure distribution on the surfaces of elements (either way, the pressure will be reduced to nodal forces by the ANSYS program). In the analyzed worm gear drive, it was assumed that the load is applied as a pressure to the core of the warm (Fig. 10).

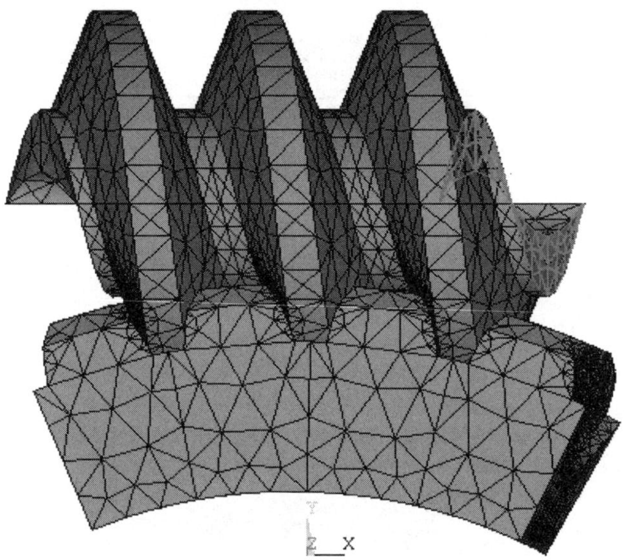

Fig. 10. The load application point in the gear drive – seen in red

The geometry of the worm gear meshing and the occurring forces

Worm gears are characterized by complex meshing geometry. Despite this, the construction of a geometric model, which would reflect the actual structure design to a large extent, is not a problem anymore as it was ten years ago [6]. The presented and developed method of building any geometric model of the worm meshing may be an example. The procedure for designing a worm mill to be used in cutting the gear teeth, as shown in chapter 1, makes it possible to build any geometric model of a worm meshing. The below algorithm for the construction of worm meshing models allows for the analysis of any random working boundaries of worm teeth in the design process [8], as well as, mapping of the worm wheel gear teeth cooperating with the wheel.
Figure 11 presents a piece of worm tooth with the wheel worm in the position known as "zero." This position was applied to determine the action field load with ease and the following meshing stages.

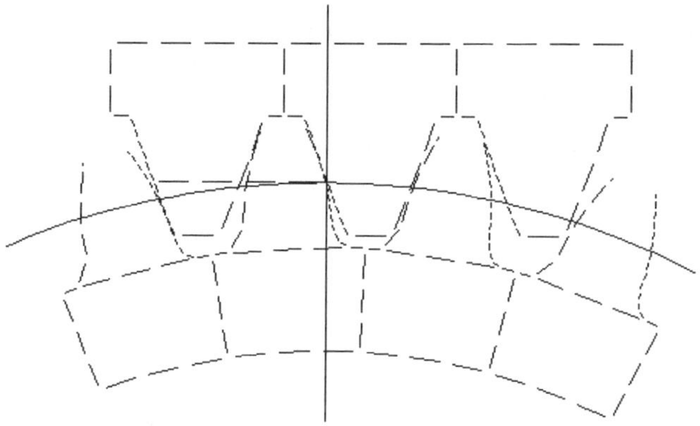

Fig. 11. Worm and worm wheel in the neutral position

The calculation of teeth force distribution for each cooperating set and load transmission for each set along the contact line have been carried out in both directions of rotation of the worm wheel from the "zero" position. In the analyzed meshing, there may be a maximum of three sets of teeth at the same time. However, there may be at least two pairs of teeth working together on the whole action field. In Figure 12 the worm gear teeth are numbered, which will be useful in further analysis of the tested meshing.

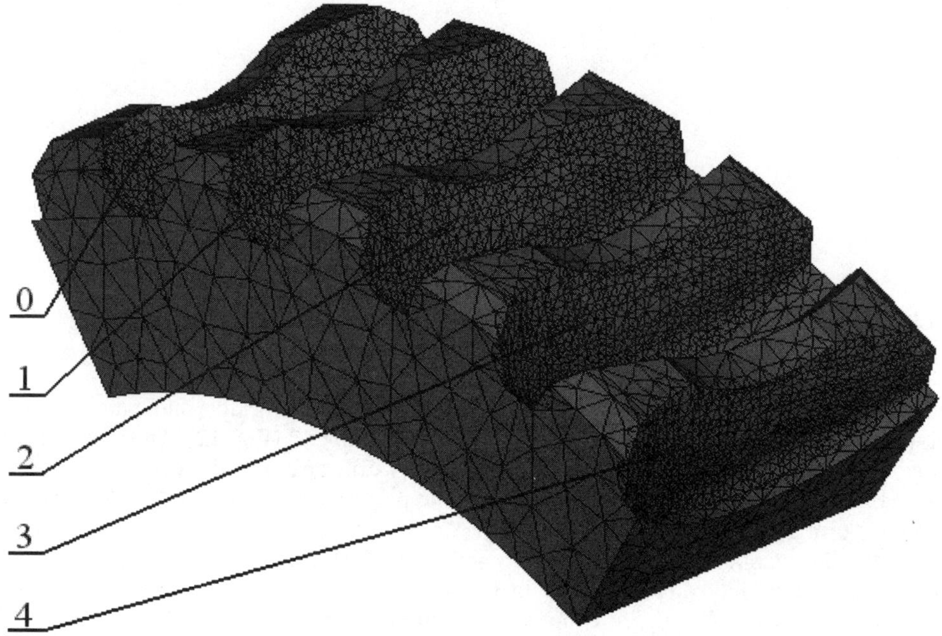

Fig. 12. The applied numbering of the worm gear teeth

Figure 13 shows the percentage of transmission of teeth force by individual pairs of teeth in cooperation depending on the meshing phase (the worm gear rotation angle). On this basis, one can also determine the degree of coverage [6].

The evolvent boundary: tooth 0, tooth 1, tooth 2, tooth 3, tooth 4.

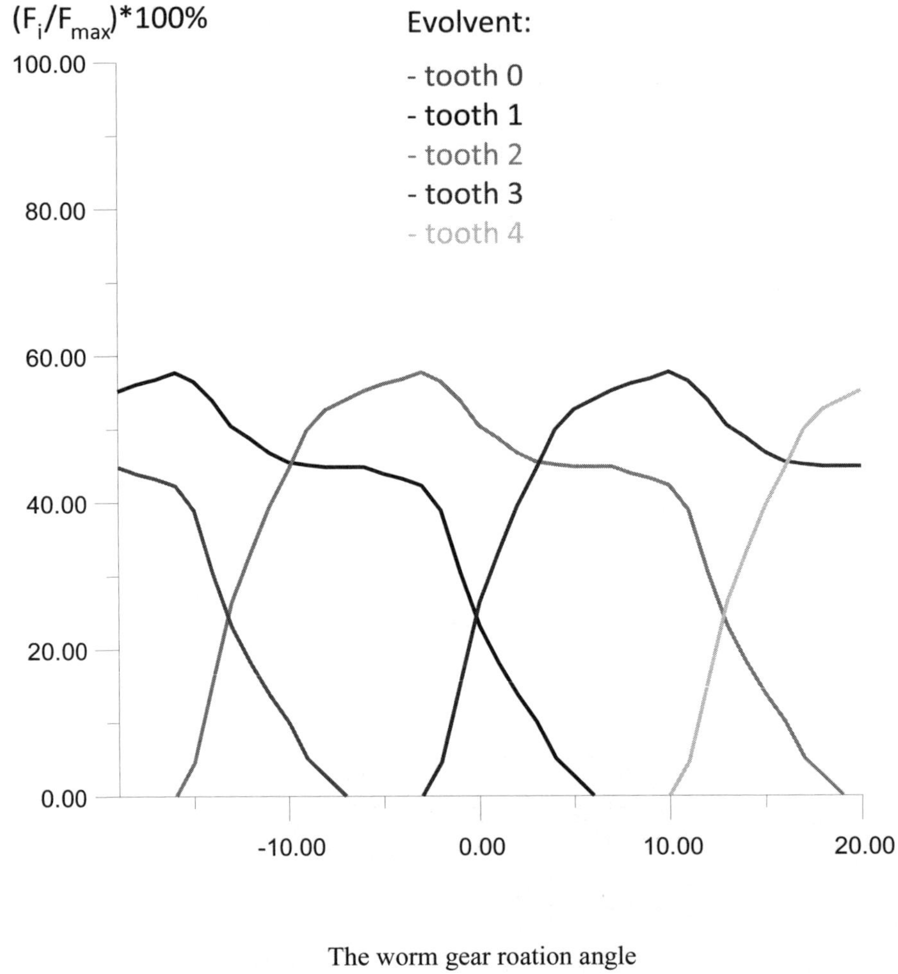

Fig. 13. The percentage of teeth force transmission by individual pairs of teeth in cooperation (for phase 1).

Figure 13 shows, essentially, only one full cycle of meshing for one pair of teeth. The cooperating worm wheel tooth of the pair is labeled with number 2 (Fig. 12). The moment when the set starts to cooperate is found at -16 degrees angle of worm gear and the cooperation finishes an angle of +19 degrees. The area of cooperation for one pair of teeth involves the 36-degree rotation angle of the worm gear.

Figure 14 shows contact lines of the worm gear teeth for +1 degree angle of rotation (Fig. 11). The darker blue color indicates the working tooth surfaces and the surfaces not involved in the cooperation are indicated in gray. At this stage of the meshing, tooth number "1" takes 18.14%, tooth no. "2" 48.78% of the tooth "3 " 33.08% of the teeth force (Fig. 14) [3].

The application of the finite element method in this way, i.e. to build geometric models of the meshing worm, and then to track their functioning, for the first time made it possible to determine the course of the changes and the distribution of the teeth force on particular teeth pairs in cooperation with each other along with its analysis for any meshing phase and for the working boundary of the surfaces of worm teeth. Therefore, for the first time, it is possible to speak of introducing a quasi-dynamic analysis to worm gear construction [17].

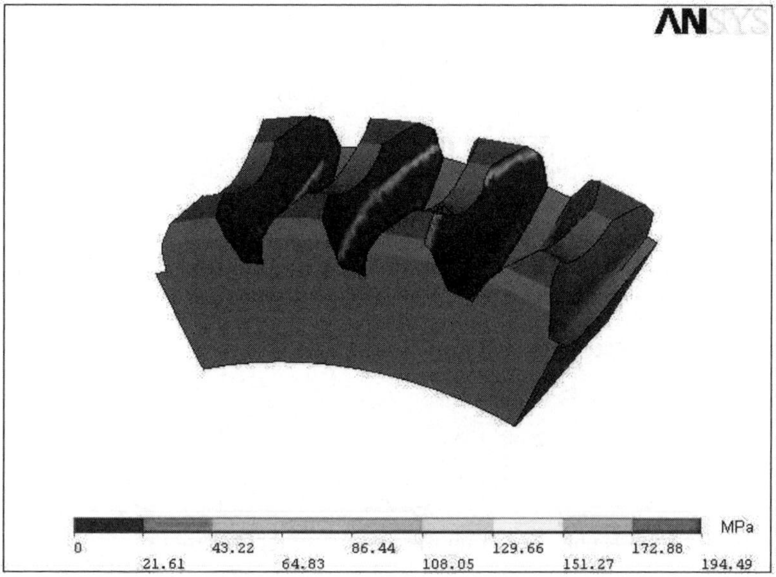

Fig. 14. The path of the contact line for +1 rotation angle of the worm wheel

However, figures 18, 19 and 20 show the load changes along the line of contact for teeth: "1 ", "2" and "3" for the position of worm wheel +1 degree (in red). The blue line is the force distributed equally for this pair of teeth and for this meshing phase. The areas located between the red line and the axis of abscissa, as well as the blue line and the axis of abscissa, are equal. The size of these fields corresponds exactly to the transmitted teeth force by each pair of teeth at this stage of meshing.

Tooth 1

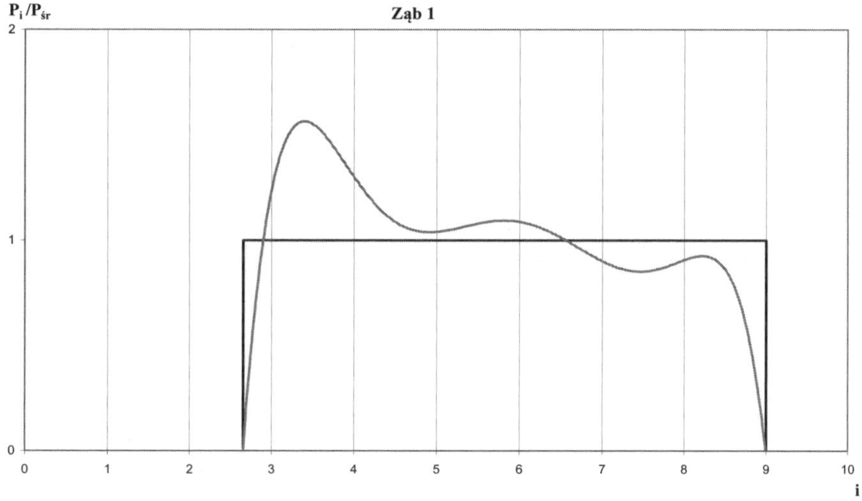

Fig. 18. The loads changes along the line of contact for tooth '1' (+ 1 degree)

Tooth 2

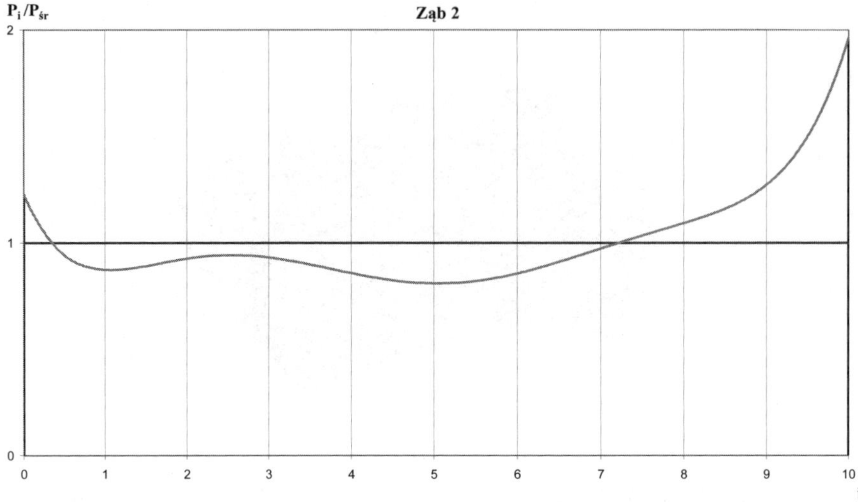

Fig. 19. The loads changes along the line of contact for tooth '2' (+ 1 degree)

Tooth 3

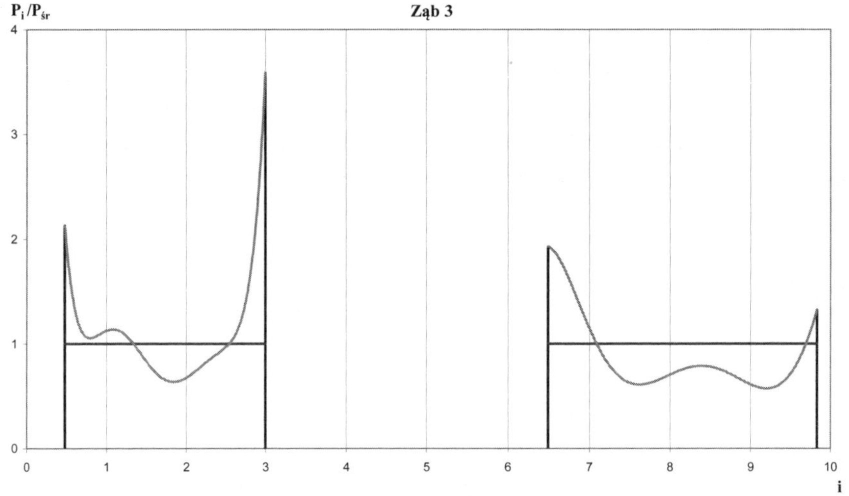

Fig. 20. The loads changes along the line of contact for tooth '3' (+ 1 degree)

Summary

The theoretical and experimental research on worm gears carried out in this chapter and the results analysis of load distributions obtained in this way, both for individual pairs of teeth and along the lines of contact, are grounds for these comments and conclusions:

- The presented procedure allows to make worm teeth boundary modules and worm wheel for theoretical conclusions as it is done in reality (mapping the turning and grinding of worm threads and worm wheel teeth hobbing).

Due to the fact that worm gear modeling was done along with the technological requirements of its manufacture using the machine tools, it can be concluded that:

  - The lines of contact between the cooperating worm teeth and worm wheel are the same as real ones not the approximate, imaginary or supposed lines;
  - The received loads are distributed on the surface, as it is in reality. However, one dimension in relation to another one is minimal (the contact width against the length) - but it is still a surface. It reflects the actual functioning of the working surfaces of the teeth and their actual contact, and not as it was previously assumed that the load distribution is along a line - it was a one-dimensional matter.

- The presented numerical method for worm gear analysis enables to make any sort of modification of the worm teeth's working boundaries and to reflect it through milling the boundaries of the cooperating worm wheel teeth.
- It seems that, currently, the theoretical method of analysis of the worm teeth gear presented above is the closest to technological reality [9].
- The existing theoretical solutions presented in literature and in experimental studies [9] confirm the phenomenon occurring in a worm gear, such as: a high degree of coverage, an uneven load distribution along the line of contact with the occurrence of accumulation and the related pitting fatigue.

The presented methods of worm gear analysis is very helpful not only in science, in terms of identifying and clarifying of this issue, but primarily in the engineering design of the entire worm gear, and not only the meshing itself. The developed program permits checking whether the models were designed along with their technological processes. It allows for an in-depth analysis through a simulation of cooperation of the designed worm and worm wheel, which is so helpful in the work of designers, technologists and operators of these kind of worm gears.

References

[1] Rusiński E., Czmochowski J., Smolnicki T., Advanced methods of finite elements in load-bearing structures. (Polish language). Publishing house of Technical University in Wroclaw. 2000.

[2] Cichowicz R., Load distribution in worm gearing, PhD dissertation. (Polish language). Technical University in Lodz, September 2005.

[3] Burczyński T., Boundary element method in mechanics. Computer aided CAD/CAM. WN-T. Warsaw 1995.

[4] Sabiniak H.G., Cichowicz R., Mathematical modelling of worm gearing. 21st Symposium of Fundamentals of Machinery Design. Volume 2. WN-T 2003, p. 203-208.

[5] Marciniak T. The effect of pressures distribution on teeth fatigue in warm gears with involute and concave-circular contours. PhD dissertation. (Polish language). Technical University in Lodz, May 1984.

[6] Sabiniak H.G., Cichowicz R., Load transmitted by particular pairs of teeth in worm teeth. Proceedings of the International Scientific Conference. MECHANICS 2004, p. 341-344.

[7] Sabiniak H.G., Cichowicz R., The influence of modelling in worm gear meshing numerical methods on the stress distribution accuracy. (Polish language). Technical and Sicentific Conference. Simulation 2004. Rydzyna. October 2004.

[8] Sabiniak H.G., Limits of worm gears load capacity. (Polish language). The Mechanical Engineering Review. No 14. 2000, p. 17-18.

[9] Sabiniak H.G., Load distribution along the contact line between two bracket plates, treated as models for teeth of cylindrical and globoidal toothed gears. (German language). PhD dissertation. Technical University in Chemnitz, 1992.

Key Engineering Materials Vol. 490 (2012) pp 323-333

doi:10.4028/www.scientific.net/KEM.490.323

Wear of Ultra-thin DLC or Tungsten-modified DLC Coatings under Reciprocating Sliding

Krzysztof Druet[1, a], Jacek Igor Lubinski[2, b]

[1, 2]Gdansk University of Technology, Faculty of Mechanical Engineering
11/12 Gabriela Narutowicza St., 80-233 Gdansk, Poland

[a]kdruet@pg.gda.pl, [b]chinczyk@pg.gda.pl

Keywords: wear, diamond-like carbon, reciprocating sliding, carbon coatings

Introduction

The application of various coatings on traditional material substrata (e.g. metal alloys) is a well established technology used to modify surface characteristics/performances of various components in technical systems exposed to adverse conditions, e.g. corrosive environment, high temperature, erosion and/or wear, etc. [1… 10]. Coatings can also provide a desired (mostly low) friction coefficient between solids forming various tribological systems. In recent years there has been a particular increase in interest in carbon-based coatings, either crystaline (diamond or graphite) or amorphous. This applies especially to amorphous carbon with high content of sp^3 bonds characteristic of the diamond - known as DLC (diamond-like carbon). This coating can be "pure" form of DLC, if it contains only sp^3 bonded carbon, but may also contain other elements and crystalline structures; Hydrogen, carbon with the "graphite" sp^2 bonds, metals and/or other modifiers, which can provide the required specific properties of the coating and reduce the cost of production.

Unlike most hard coatings (for example ceramic, nitrides), which are extremely resistant to wear, but at the same time cause high friction, or the soft layers (for example MoS_2), which allow to obtain small friction coefficient (due to easy shear of the material shell), but their wear is fast, the amorphous carbon coatings allow to obtain both low (or even extremely low) friction and high resistance to wear (Fig. 1) [2].

The Department of Machine Design and Vehicles at the Faculty of Mechanical Engineering of Gdansk University of Technology for several years have conducted research on hard coatings in various friction systems [11, 12]. A series of experiments, including two types of amorphous carbon coatings deposited on stainless steel, was carried out as part of this research.

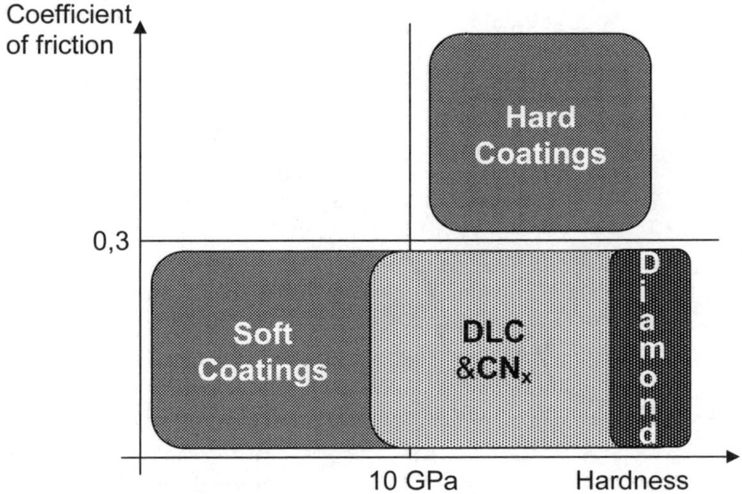

Fig. 1. Classification of coatings with respect to hardness and friction coefficient [2]

Tested materials

Two types of amorphous carbon coatings, deposited on X30Cr13 (consistent with EN 10088-1:2005) or AISI 316 stainless steel specimens were tested:

- diamond-like carbon (DLC) coating deposited in vacuum with a hybrid ARC-MAG-RF PACVD method, i.e. cathode arc evaporation (ARC), pulse reactive magnetron sputtering (MAG) and radio frequency (RF) excited plasma assisted chemical vapor deposition (PACVD) [13],
- Tungsten-containing hydrogenated carbon (W-C: H, W content 5% wt.) coating deposited in vacuum with pulse reactive magnetron sputtering (MAG) [14].

The specimens made from X30Cr13 steel were quenched and tempered before deposition process; a hardness of approximately HRC 52.

The obtained coating thicknesses were in the range of between 1 and 3 μm.

Deposition processes were performed at the Department of Materials Science and Engineering of Koszalin University of Technology.

Coating adhesion tests

The coatings were pre-tested to assess their adhesion to the steel substratum. These studies were performed by scratch-test method, consisting of scratching the surface of the coating with the diamond cone of Rockwell (type C) geometry and define the so-called, critical load, which is a quantitative measure of adhesion. Tests were performed at the Department of Materials Science and Engineering University of Koszalin University of Technology using a CSEM Revetest® .

Specific damage of coating is determined by means of the critical load L_c. In described case, L_{c1} is value of load under which first cracks of coatings are observed and L_{c2} is value of load resulting penetration of coating in the scratch groove. Determined critical loads based on the results of optical observation and measurement of distinct changes of friction forces which occur in specific locations of the test scar of coating.

The test conditions were as follow:

range of normal load	0 – 100 N
speed of increase of normal load	100 N/min
velocity of feed of specimen support	10 mm/min
length of scratch	10 mm

The results of adhesion of both coatings are presented in graphs – Fig. 2 and Fig. 3. The graphs show that for W-C: H coatings, slightly - by about 6% higher values of critical loads appear.

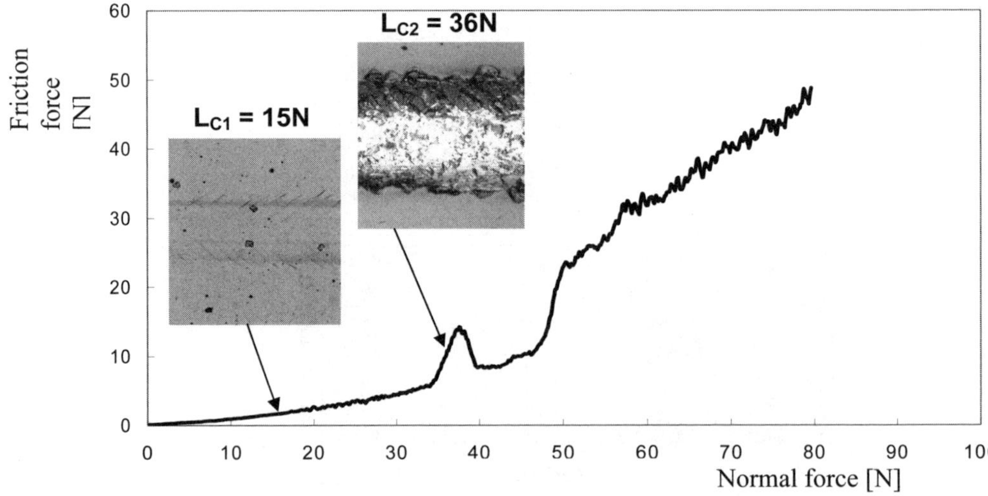

Fig. 2. The values of critical load for DLC coating

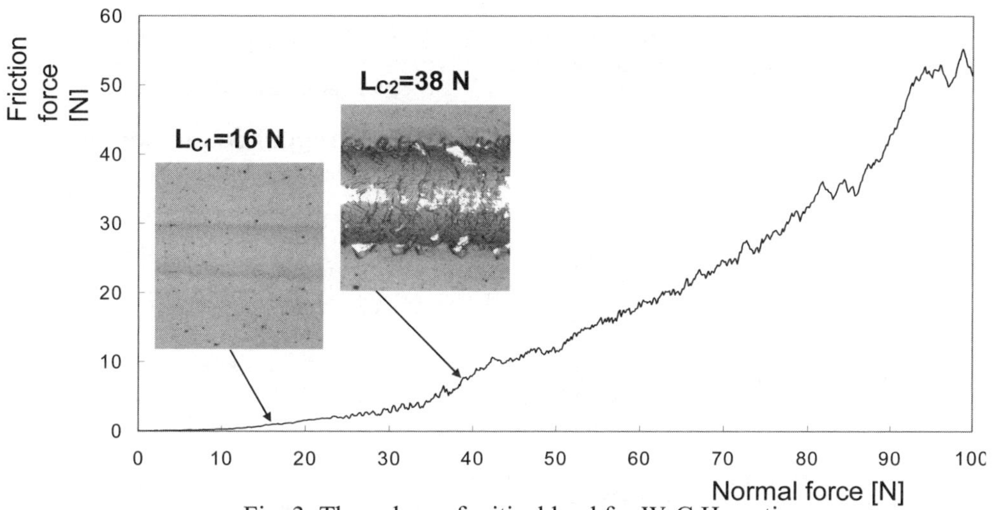

Fig. 3. The values of critical load for W-C:H coating

Wear testing method

A concentrated contact (ball-on-flat block) reciprocating sliding system was used for investigation of wear of coatings. In these researches the test rig TPZ-1 [15] was used. A general view of the TPZ-1 tribometer can be seen in Fig. 4. A schematic illustrating of the principle operation is shown in Fig. 5.

The upper, ball specimen (2) is attached to a plunger (1), which is supported in a shell (5) restrained elastically against the chassis (6). Load is exerted on test contact by weight acting on a plunger (1). The lower, flat block specimen (3) is attached to a reciprocating sledge (4) driven by a motor through crank mechanism. The motor takes power from a frequency inverter, which allows for steeples velocity change. The sledge (4) houses the lower specimen holder, the lubricant container and a heating/cooling semiconducting element. The computer system allows for programmed temperature and velocity changes in test.

Fig. 4. General view of the TPZ-1 tribometer

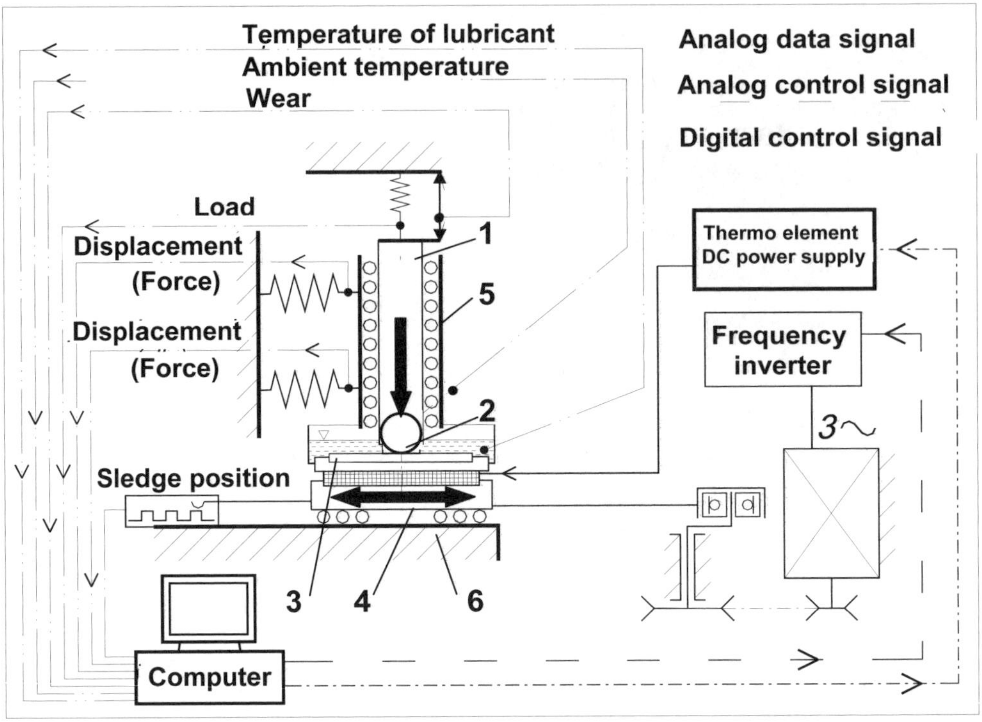

Fig. 5. Principle of operation of the TPZ-1 tribometer

In TPZ – 1 tribometer the measured parameters are friction and wear. Friction is measured as a deflection of the frames supporting the plunger housing (5); the horizontal displacement of the upper specimen is also measured in the process. As linear wear of specimens is significant, it may be measured as the vertical displacement of the plunger (1) (Fig. 5). In described case, changes of thickness of worn coatings were too small and quantitative assessments of wear extent were based on the microscope measurement of the area of friction traces on the surfaces of ball specimens. There are also measured: sledge travel, temperature of specimens and lubricants. All data are digitally recorded and post processed.

Experimental procedure

The coatings were deposited both on surfaces of ball specimens (∅5 mm, AISI 316 steel) and on flat surfaces of block specimens (X30Cr13 steel).

Before the start of the study the specimens and other components of the test system were thoroughly degreased with warm water and strong detergent. Remains of cleaning agents were removed by rinsing in distilled water. Dried specimens were then washed with isopropanol in an ultrasonic washer. The washing process described also was also used when preparing specimens for microscope measurement of wear extent and reuse in the next test run.

Sliding contacts of specimens were lubricated either with distilled water or with a medical saline solution (0,9% NaCl in water).

The wear test conditions were as follow:

- Hertzian stress in the contact zone 1,55 GPa, constant
- sliding velocity $0{,}08 \pm 0{,}02$ m/s
- stroke 40 mm, constant
- lubricant temperature 30 ± 3°C, constant

The sliding velocity was constant - almost throughout the stroke of reciprocating motion, except in the vicinity of turning points.

Each test was conducted as a series of short consecutive runs with wear measurements carried out on specimens in between runs. Each test was broken up into three stages:

- initial stage of 60 s with 10 s intervals,
- intermittent stage continued till accumulated time reached 10 min, with 1 min intervals,
- extensive stage continued to accumulated time of 30 min and later with consecutive 30 min intervals.

The test schematic was developed in earlier experiments with various ceramic materials and is based on experience. Least resistant materials rarely pass the initial stage; with the increase of wear resistance tests can last through the intermittent stage and past that stage it is not unusual, that the extensive stage exceeds 30 hours of accumulated run time.

The minimum number of replications was arbitrarily set to 3 (applied only in case of materials most resistant to wear), but in some cases as much as 15 test replications were executed for a single material.

The testing sequence material wise was random in order to avoid potential systematic errors.

The course of wear of the ceramic coating was evaluated quantitatively and qualitatively by microscope observation. The amount of wear measured as surface area of wear mark on the ball. Wear mark morphology was studied with the use of SEM and optical microscopy. Primary objects for wear testing were the coated surfaces of balls. The reciprocating block specimen faces were examined with same techniques as an auxiliary data source.

Results and discussion

The course of wear was typically a very rapid increase of the wear scar on the ball surface in the initial phase of test. In the latter course of the test the wear mark growth was slow or immeasurable with a clearly distinguishable threshold, beyond which wear process accelerated to a radical, catastrophic damage developing on the coated sliding surface.

Initial, rapid wear scar growth followed by a rapid decrease of wear gradient is understandable, taking into account the type of contact of specimens: point contact of a sphere against a flat surface. As a result of wear on the ball the concentrated contact develops into a distributed, conformal contact, resulting in a decrease of the surface stress in the contact zone between the specimens.

Wear marks on balls were elliptical in shape with the longer axis perpendicular to the direction of sliding. In rare cases circular wear marks were observed.

A variety of wear forms were observed: scratches, cracks, delamination, pits and steel substratum exposure. Some of wear forms observed can be contributed to imperfections in coatings. Certain wear marks were probably caused by wear particles drawn into the contact zone causing scratching and denting in the surface.

In some cases the damage developed from earlier defects on coating surface, whiles the minor breaches or cracks, not penetrating through the coating, could disappear as a result of material removal by mild wear, before the steel substratum was exposed.

The presence of defects, which reduce significantly carrying portion of the surface of contact, usually does not accelerate growth of wear traces in the mild wear (slow) phase of the test . The rapid growth occurs only when the destruction of the coating is very extensive and results from exposure of the substratum material (stainless steel).

The reciprocating motion resulted in fatigue cracks development in the coating. All the wear forms developing would usually not cause any noticeable increase in wear rate measured as the surface area of the elliptic wear mark on the ball surface while in mild wear phase of the test.

It can be proposed, that numerous forms of wear encountered should be classified into wear types in a systematic approach based on such factors (in the case of crack propagation involving cases) as e.g.: cause of crack initiation, crack type (instantaneous or fatigue), characteristics of the breakage surface, location of the initiation point, crack propagation direction etc.

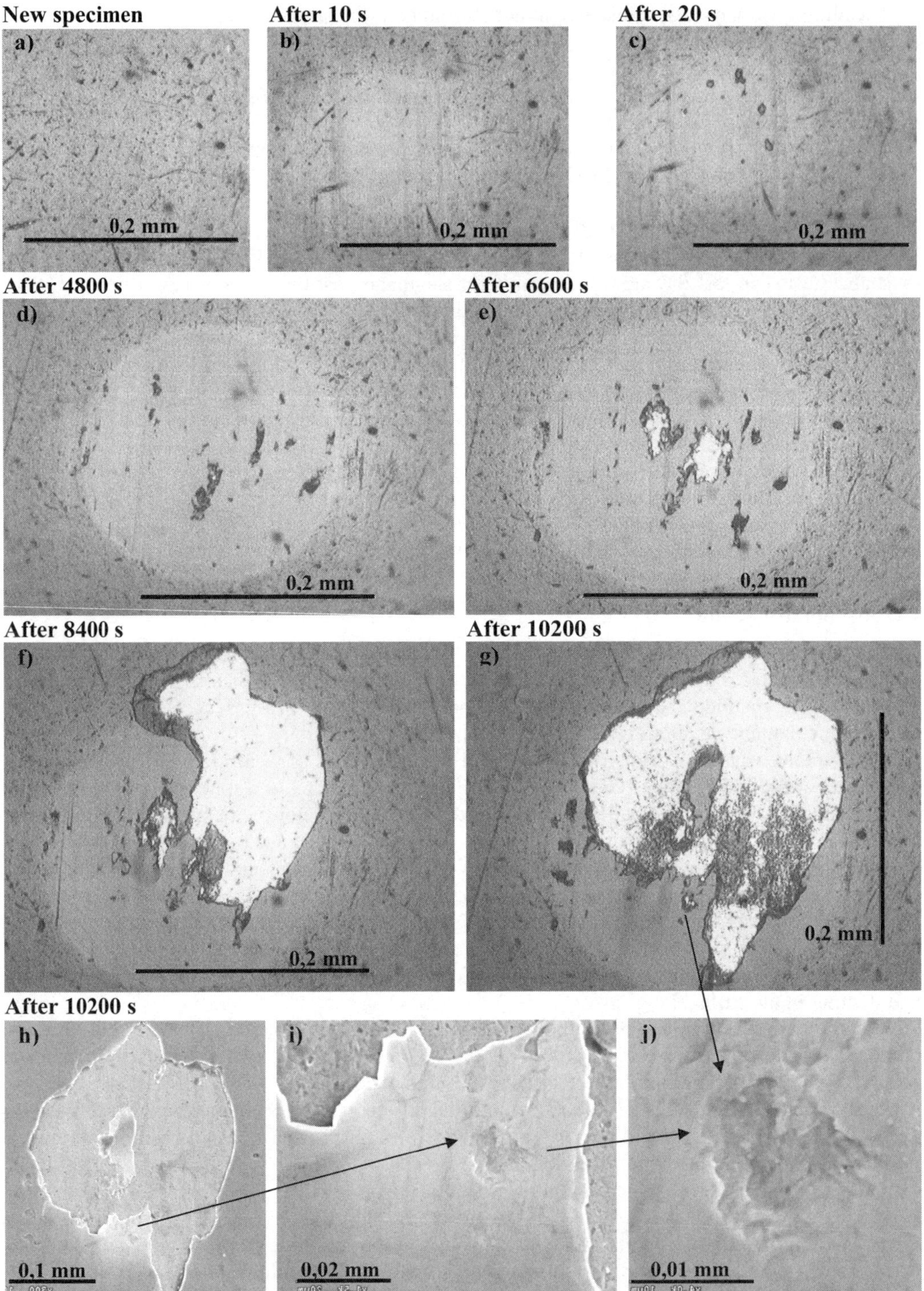

Fig. 6. Development of failure on W-C:H coating on ∅5 mm ball. Test carried out with aqueous saline solution (NaCl 0,9%). Pictures *a* throughg are optical microscope images. Pictures h, i and j are SEM images of the final damage

Fig. 6 shows an example of damage to the W-C: H coating (deposited on ∅5 mm ball) in a saline solution lubricated test; The experiment was continued until the exposure of a large area of substratum. Details of the final damage of the coating are shown in scanning electron microscope photographs - Fig. 6 h, i, j. Moreover it is interesting that even significant damage to the coating, combined with the exposure of a large area of steel never resulted in a sudden termination of the experimental run. It can be assumed that the portion of coating remaining in the contact area provides sufficient load carrying capacity for uninterrupted low friction slidining conditions.

The courses of wear, in other words – the sizes of the trace areas on the balls – are characterized by a relatively great scattering, but it is not uncommon in the study of friction. The studies of the probability distribution of the random variable representing the wear of the coating agreemee well with the normal distribution. Kolmogorov-Smirnov test, Lilliefors test and Shapiro-Wilk test, suggest that there is no reason to reject the hypothesis of normal distribution of wear amount detected with the TPZ-1 tribometer.

In Fig. 7 and 8 compound result graphs of the development of wear scar area on tested coatings as a function of test time are presented for a time range up to 50 400 s (14 h). Point clouds obtained by wear measurements were subjected to curve fitting with simple functions, so the relationships between working time of sliding coupling and wear trace area on the ball were established. The regression functions were fitted by the least squares. A particullarily good correlation with test results can be achieved by logarithmic and polynomial curve fitting, including a linear relationship. This means that acceptable values of multiple correlation coefficients were obtained for these functional dependencies. The resultant correlation coefficient, which should be as close to 1 as possible, exceeded 0.8.

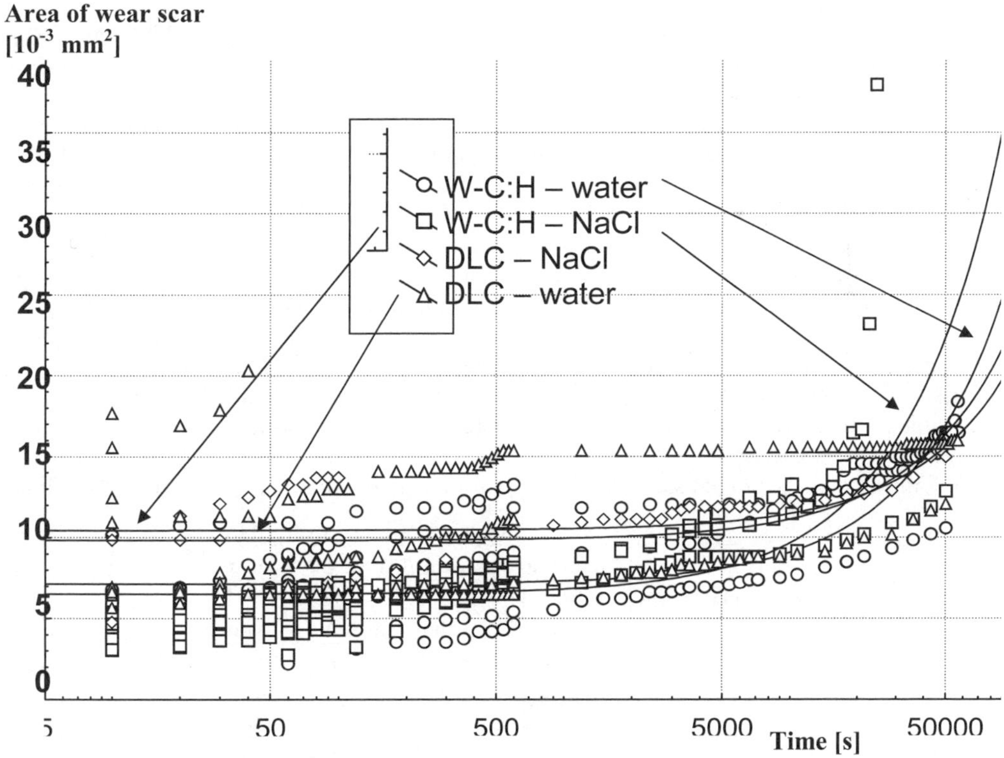

Fig. 7. Linear function describing the development of area of elliptic wear mark on DLC or W-C:H coated 5∅ mm balls. Logarithmic scale for the axis of abscissa (time axis) results in curving up of graph lines.

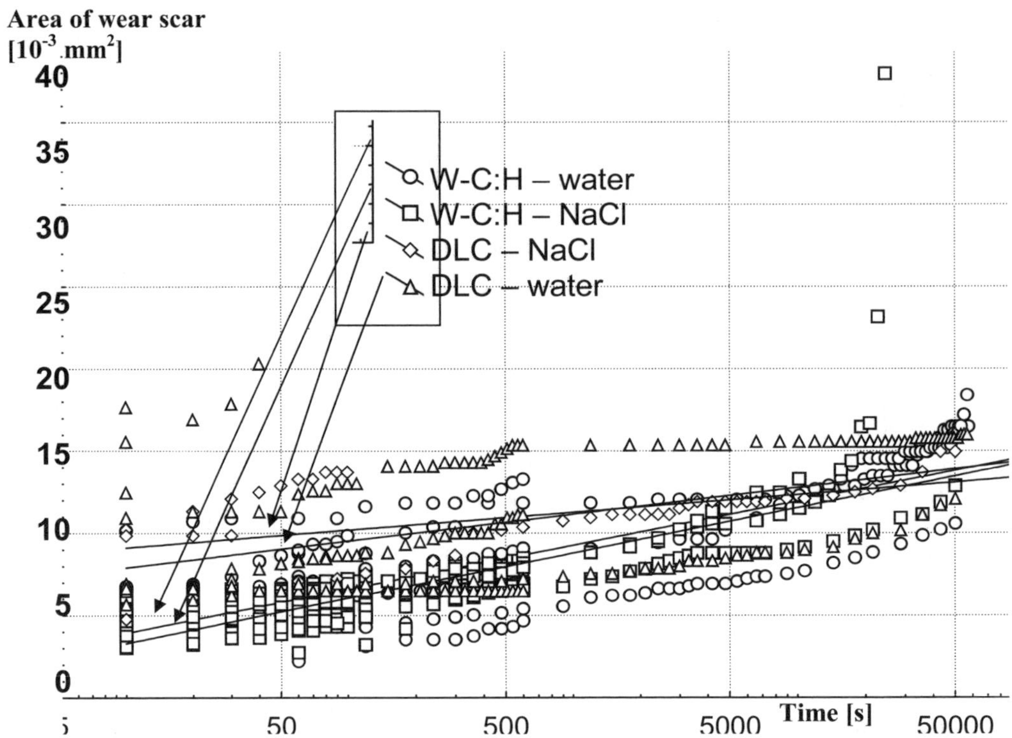

Fig. 8. Logarithmic function describing the development of area of elliptic wear mark on DLC or W-C:H coated ∅5 mm balls

Choosing the right form of the regression function is difficult to rely only on the multiple correlation coefficients, because a large dispersion of results and uneven distribution of points (the congestion for short working times of examined friction couplings) occur. Choice of a regression function should be primarily associated with the substantial hypothesis resulting from the phenomenological wear process model, which requires further research.

Approximating lines are fitted assuming a linear relationship – in Fig. 7, whereas logarithmic relationship is used – in Fig. 8. Both plots are drawn in a semi-logarithmic co-ordinate system. The graphs in Fig. 7 clearly show runaway wear growth in the final phase of test, especially in longest test runs, e.g. for W-C: H. The obtained least square approximated fitting curves intersect, which is an issue important to the formulating conclusions. A more detailed analysis allows for a presumption, that such a situation can be attributed to the strong influence of single test run series upon the resultant curve fit.

A comparison of averaged surface area of wear marks on ball surface is presented in Table 1.

Table 1 Averaged surface area of wear marks

Test time [s]	Mean surface area of wear mark [μm^2]			
	DLC saline	DLC dist.water	W-C: H saline	W-C: H dist.water
60	7087,4	6936,6	4146,9	4068,4
600	10367,3	7235,1	6754,4	5403,5
3600	11881,5	8331,5	8796,5	6927,2
7200	11881,5	8824,7	8796,5	7370,2
14400	12264,8	9157,7	9490,8	8168,1
21600	12666,9	10009,1	10194,5	8821,6
36000	13681,6	11102,4	11102,4	9848,9
50400	14954,0	12073,1	12839,7	10565,2

From the comparison it can the following conclusions can be drawn:

1) the wear rate of unsuplemented DLC is greater than of tungsten supplemented material for both lubricants: pure water and medical saline solution,
2) NaCl content in water accelerates wear of both tested materials,
3) the wear rate of tungsten supplemented DLC lubricated with saline can be greater than that of pure DLC lubricated with distilled water.

The described tendencies are sufficiently strong not to attempt a detailed confidence level analysis for conclusion formulation on differences in wear rates of the tested sliding pairs. Such approach was possible thanks to a concurrent tendency observed in a different experimental system –unidirectional, rotational sliding in a flat annular contact (hollow cylinder faces in rotational sliding) [11, 12].

In general, it is found that the W-C:H coating is more wear resistant than DLC coating and NaCl content in water (only concentration of 0,9% tested) may accelerate wear process. When lubricated with a medical saline solution, catastrophic destruction of the coating followed quickly, and areas of wear trace were more extensive. The deleterious effect of NaCl content was not confirmed in the case of W-C:H coatings.

The values of the friction coefficient for DLC coatings ranged from close to zero values to about 0.23 while sliding contacts were lubricated with distilled water or from 0.05 to 0.13 while lubricated with an aqueous solution of NaCl;In both cases with a tendency to decrease within the test.

The values of the friction coefficient for W-C:H coatings ranged from 0,03 to about 0.12 while sliding contacts were lubricated with distilled water, with a tendency to decrease during the tests or from 0.11 to 0.14 while lubricated with an aqueous solution of NaCl, with a tendency to slight increase during the course of a test.

In the case of some more general conclusions it is worth noting, that tribological properties of bodies (e.g. resistance to wear of solids, or anti wear properties of lubricants) are not only material related properties, but properties manifested by tribological systems. It is thus crucial to maintain careful approach when attempting generalised statements, since direct correlation between the experimental system and real life 'technical' system remains problematic still.

The use of "ball-on-block" contact configuration in reciprocating sliding is common in tribological research. It is, probably, only „ball-on-disk" systems, that are more commonplace in available literature. The ball-on-flat general arrangement is favoured for the lack of problems resulting from 'tilting' of one of the specimens and resulting misalignment of the sliding contact, which is commonplace in pin-on-flat contacts, where edge contact instead of flat-on-flat can occur; This effect is decreased or eradicated along the test as a result of running in, but at the cost of decreased accuracy. In tests on thin coatings there is no room for running in, because the coating would be destroyed before it could be worn in to achieve the desired contact configuration. In ball-on-flat contact the problem is minimised, despite the high contact stress early in the test. Authors' own research includes flat-on-flat testing on thin coatings, but in a ring-on-flat configuration (hollow cylinder faces in unidirectional rotational sliding) with the use of PT series tribometer, which, thanks to self-aligning, hydrostatic support of the experimental set-up eliminates the running in problem [11, 12]. As mentioned, the DLC and W-C:H coated specimens were also tested with the use of that method and the results confirm findings from tests in ball-on-flat configuration and reciprocating sliding.

Among questions arising from the obtained experimental results one of the more prominent is the one about the possible relationship between the wear resistance and concentration of tungsten. So far only the concentration of 5% was considered, but subsequent research on coatings with different tungsten concentrations are under way and results will soon be published.

Conclusions

The experimental data is indicative of better wear resistance of tungsten-containing hydrogenated carbon (W-C:H) coating than non-supplemented diamond-like carbon (DLC) coating.

The NaCl content in water may accelerate wear process, especially on DLC coating.

A variety of wear forms were observed: scratches, cracks, delaminating, pits and steel substrate exposure. Some of observed wear forms could be put down to imperfections in coating. Wear marks on balls were elliptical in shape with the longer axis perpendicular to the direction of sliding. In rare cases circular wear marks were observed.

Best curve fitting for experimental data (the wear mark surface area versus time i.e. sliding distance) was obtained with linear or logarithmic fit. The exact phenomenological model of the wear process remains yet to be assessed.

The measurement of wear intensity on coatings with the use of methods described above proved very effective. Simple observation by optical microscopy aided by digital photography provide sufficient data for qualitative and quantitative analysis.

Summary

A ball-on-plate system was used for investigation of wear of ultra-thin hard coatings. DLC or tungsten doped DLC (W-C:H) coated stainless steel balls (Ø 5 mm) were sliding under constant load at nominal surface pressure 1,155 GPa on a this same material coated face of stainless bar moving in a rectilinear reciprocating motion (sliding velocity 0,08 ±0,02 m/s, stroke 40 mm).

Sliding contacts were lubricated either with distilled water or with aqueous solution of NaCl.

The course of wear of the coating deposited on the ball-specimen was evaluated by means of optical and scanning electron microscope observations. Wear marks on balls were elliptical in shape with the longer axis perpendicular to the direction of sliding. In rare cases circular wear marks were observed. A variety of wear forms were observed: scratches, cracks, delamination, pits and steel substrate exposure. Some of observed wear forms could be put down to imperfections in coating.

The amount of wear was assessed as surface area of wear mark on the spherical specimen. The kinetics of wear was analysed as the wear mark surface area versus time (i.e. sliding distance). The best curve fitting for experimental data was obtained with linear or logarithmic fit.

The better wear resistance was ascertained of tungsten-containing hydrogenated carbon (W-C:H) coating than diamond-like carbon (DLC) coating.

The NaCl content in water may accelerate of wear process, especially on DLC coating.

Acknowledgments

The financial support from the Polish Ministry for Higher Education and Informatics in the form of research grants is hereby gratefully acknowledged.

References

[1] K. Bobzin, N. Bageivan, PVD component coatings for environmentally friendly tribosystems. World Tribology Congress 2009, Kyoto, Japan, Sept. 6–11, 2009, p. 340.

[2] J. Fontaine, A. Erdemir and C. Donner, Fundamental phenomena in tribology of DLC: how to reduce friction and wear, World Tribology Congress 2009, Kyoto, Japan, Sept. 6–11, 2009, p. 407.

[3] R. Hauert, An overview on the tribological behavior of diamond-like carbon in technical and medical applications, Tribology International 37 (2004), pp. 991–1003.

[4] Jin Woo Yi and others, Tribological performance of hydrophilic DLC surfaces in aqueous and biological environment, World Tribology Congress 2009, Kyoto, Japan, Sept. 6–11, 2009, p. 473.

[5] E. Liu and others, Tribological behaviour of different diamond-like carbon materials, Surface and Coatings Technology 106 (1998), pp. 72-80.

[6] M. Madej, D. Ozimina, E. Zdravecka and W. Piekoszewski, The characterization of diamond-like carbon coatings used for artificial joints, Proc. Int. Conf. Tribochemistry, Lanzhou, China, 2009, p. 28.

[7] P. Platon, S. Fournier and R. Rouxel, Tribological behaviour of DLC coatings compared to different materials used in hip prostheses, Wear 250 (2001), pp. 227-236.

[8] H. Ronkainen, K. Holmberg, Friction and Wear Properties of DLC Films – The Enironmental and Mechanical Aspects, World Tribology Congress 2009, Kyoto, Japan, Sept. 6–11, 2009, p. 669.

[9] R. Tietema, D. Doerwald, R. Jacobs and T. Krug, Diamond-like Carbon Coatings for tribological applications on Automotive Components, World Tribology Congress 2009, Kyoto, Japan, Sept. 6–11, 2009, p. 26.

[10] M. Tokoro and others, Some surface aspects of DLC-coatings as tribomaterial durable for water lubrication, World Tribology Congress 2009, Kyoto, Japan, Sept. 6–11, 2009, p. 456.

[11] J. I. Lubinski, K. Druet, A. Olszewski, A. Neyman and J. Sikora, A Multi Rig Screening Test for Thin Ceramic Coatings in Bio-Tribological Applications, Journal of Biomechanics, Vol. 43, supp. 1, 2010, Abstracts of the International Conference on Orthopaedic Biomechanics Clinical Applications and Surgery (OBCAS), London, United Kingdom, June 6-9, 2010, p. 52.

[12] A. Neyman et. al., Friction and wear behaviour of self mated DLC and DLC-W coatings in water or salt in water solution lubricated sliding, iCAT 2008, Advanced Tribology 2008, 2nd International Conference, Extended abstracts, Singapore, December 3–5, 2008, pp. 76-78.

[13] O. Wänstrand, M. Larsson and A. Kassman-Rudolphi, Mechanical and tribological properties of vapour deposited low friction coatings on Ni plated substrates, Tribology International, Volume 33, Issue 10, October, 2000, pp. 737-742.

[14] W. Karaszewski, A, Neyman, Cr_2O_3 coating for sliding bearings, Nordtrib`98, Proceedings of the 8th International Conference on Tribology, 1998 vol. 2, pp. 881-888.

[15] K. Druet, T. Lubinski and J. I. Lubinski, Reciprocating motion tribometer, 2nd World Tribology Congress 2001, Vienna, Austria, Sept. 3–7, 2001, p. 663.

Keyword Index

Author Index